Statistical Regression with Measurement Error

KENDALL'S LIBRARY OF STATISTICS

Published titles:

1. Multivariate Analysis
   Part 1: Distributions, Ordination and Inference
   WJ Krzanowski (University of Exeter) and FHC Marriott (University of Oxford)
   1994, ISBN 0 340 59326 1
2. Multivariate Analysis
   Part 2: Classification, Covariance Structures and Repeated Measurements
   WJ Krzanowski (University of Exeter) and FHC Marriott (University of Oxford)
   1995, ISBN 0 340 59325 3
3. Multilevel Statistical Models
   Second edition
   H Goldstein (University of London)
   1995, ISBN 0 340 59529 9
4. The Analysis of Proximity Data
   BS Everitt (Institute of Psychiatry) and S Rabe-Hesketh (Institute of Psychiatry)
   1997, ISBN 0 340 67776 7
5. Robust Nonparametric Statistical Methods
   Thomas P Hettmansperger (Penn State University) and Joseph W McKean (Western Michigan University)
   1998, ISBN 0 340 54937 8
6. Statistical Regression with Measurement Error
   Chi-Lun Cheng (Academia Sinica, Republic of China) and John Van Ness (University of Texas at Dallas)
   1999, ISBN 0 340 61461 7

In preparation:

Latent Variable Models and Factor Analysis
Bartholomew and Knott

Statistical Inference for Diffusion Type Processes
Rao

Statistical Regression with Measurement Error

Chi-Lun Cheng
Academia Sinica, Republic of China

and

John W. Van Ness
University of Texas at Dallas

A member of the Hodder Headline Group
LONDON • SYDNEY • AUCKLAND

Co-published in the USA by
Oxford University Press Inc., New York

First published in Great Britain in 1999 by
Arnold, a member of the Hodder Headline Group,
338 Euston Road, London NW1 3BH

http://www.arnoldpublishers.com

Co-published in the USA by
Oxford University Press Inc.,
198 Madison Avenue, New York, NY 10016

British Library Cataloguing in Publication Data
A catalogue record for this book is available from the British Library

Library of Congress Cataloging-in-Publication Data
A catalog record for this book is available from the Library of Congress

ISBN 0 340 61461 7

1 2 3 4 5 6 7 8 9 10

Commissioning Editor: Nicki Dennis
Production Editor: James Rabson
Production Controller: Sarah Kett
Cover Design: Terry Griffiths

Typeset in 10/12 pt Times by Focal Image Ltd, Torquay
Printed and bound in Great Britain by St Edmundsbury Press, Bury St Edmunds, Suffolk
and MPG Books, Bodmin, Cornwall

To the memory of my father, Tsing-Ching Cheng, and to my mother, Kuei-Fang Li Cheng.

– Chi-Lun Cheng

To the memory of my daughter, Julia Williams Van Ness, and my father, Winslow J. Van Ness; and to my mother, Ada Van Ness, and my family, Nancy, Karen and David.

– John Van Ness

Contents

Preface

Arnold invited us to contribute to a radical expansion of the three-volume series, *Kendall's Advanced Theory of Statistics* written in various editions by Sir Maurice Kendall and Alan Stuart or Alan Stuart and Keith Ord. This new expansion includes two general central volumes in the tradition of the past editions plus a series of monographs, each covering some specialized statistical topic. We were asked to write a monograph expanding on the topic heretofore covered in Chapter 29 of Volume 2 (see Kendall and Stuart, 1979, pp. 399–443). Chapter 29, entitled 'Functional and Structural Relationship' in the fourth edition of Volume 2, is not included in the fifth edition of the central volumes, but instead, is expanded to this monograph, which is designed to provide a comprehensive coverage of measurement error models.

This book covers measurement error models, also called errors-in-variables models, and a closely related model called the Berkson model. Three basic types of measurement error models are discussed: the structural model, the functional model and the ultrastructural model. Coverage includes models with and without equation error, vector explanatory variable models, and linear and polynomial models. Topics include model identifiability, parameter estimation, confidence intervals, asymptotic theory, finite sample properties of estimates, modified least squares techniques, non-normal models, robust estimation methods, prediction, and statistical calibration. Orthogonal regression is discussed as a special case of the measurement error model.

This volume is intended to be written at the level of and in the style of the past editions of *Kendall's Advanced Theory of Statistics*. Our goal was to follow Kendall's original aim of providing a comprehensive coverage of the subject that emphasizes the ideas and the practical implementation of the theory without too great an emphasis on the theorem–proof format. This book puts particular emphasis on comparing the various measurement error models side-by-side and on contrasting and unifying the various estimation methods used in measurement error models. Additional mathematical details on the general measurement error model can be found in Schneeweiss and Mittag (1986) and Fuller (1987). Carroll *et al.* (1995) concentrates on nonlinear measurement error models.

This book is intended to be accessible to readers who have a background equivalent to a year's course in probability and theoretical statistics. The book does not require knowledge of measure theory. Drafts of this book have been used for a one-semester graduate course; however, there is more than enough material for a two-semester course. Problems are included at the end of chapters.

In classical regression, the notation $\mathbf{x}_i$ for the independent variable and y_i for the dependent variable seems to be almost universal. But the corresponding notation for the ME model, unfortunately, is not so universal. In this book we try to keep the notation as parallel as possible to that used in classical regression. Greek letters are used to denote (unknown) true parameters, true values, and errors whenever possible. There are a few exceptions, such as q_i representing the equation error, $e_i = \varepsilon_i + q_i$ representing the sum of measurement error and equation error and v_i representing the 'residual'.

The authors would like to thank University of Texas at Dallas graduate students Marko Beljak and Ryan Gill for their considerable assistance in performing many of the computations for the examples in the book and for reading through the manuscript. We would also like to thank Nancy Van Ness for reading several drafts of the book. We received many helpful comments from Prof. Hans Schneeweiss.

Chi-Lun Cheng gratefully acknowledges financial support from the Institute of Statistical Science, Academia Sinica and the National Science Council, Republic of China (Taiwan).

1

Introduction to Linear Measurement Error Models

1.1 Preliminaries

Measurement error (ME) models (also called errors-in-variables models) are a generalization of standard regression models. For the simplest linear ME model, the goal is to estimate from bivariate data a straight line fit between the two variables, both of which are measured with error. Adcock (1877; 1878) is usually regarded as the first person specifically to consider such models. ME models occur very commonly in data analysis, far more commonly than allowed for by many data analysts who tend to use the more familiar standard regression models. ME models have both advantages and disadvantages compared to standard regression models. This chapter introduces ME models and discusses some of these strengths and weaknesses as well as some of the elementary properties and typical applications of these models.

An important special case of measurement error models, called the Berkson model, will be introduced in Section 1.6. We will refer to that model as the Berkson model and reserve the term 'ME model' for the standard functional, structural, and ultrastructural models introduced in this section.

The standard regression model with one explanatory variable is given by

$$y = \beta_0 + \beta_1 \xi + \varepsilon, \tag{1.1}$$

where the independent variable, ξ, is either fixed or random and the error, ε, has mean zero and is uncorrelated with ξ. Given a set of independent observations,

$$\begin{pmatrix} \xi_1 \\ y_1 \end{pmatrix}, \ldots, \begin{pmatrix} \xi_n \\ y_n \end{pmatrix},$$

the unknown intercept, β_0, and slope, β_1, are usually estimated using least-squares techniques or some more robust procedure.

The corresponding standard linear ME model assumes that the variables ξ and η are related by

$$\eta = \beta_0 + \beta_1 \xi, \tag{1.2}$$

but that the variables ξ and η are unobservable and can only be observed with additive errors. Thus, instead of observing ξ and η directly, one observes the variables

$$x = \xi + \delta \quad \text{and} \quad y = \eta + \varepsilon, \tag{1.3}$$

where ξ and the errors, δ and ε, are uncorrelated. A sample of size n for this model would be

$$\begin{pmatrix} x_1 \\ y_1 \end{pmatrix}, \ldots, \begin{pmatrix} x_n \\ y_n \end{pmatrix}. \tag{1.4}$$

Because both ξ and η are measured with error, this model is called the measurement error model. We call this the 'standard' ME model because it does not have equation error, which is introduced in Section 1.5.

The variables ξ and η are sometimes called **latent variables** in some fields of application. Applications in which the ξ-variable is measured with error are perhaps more common than those in which ξ is measured precisely. Most medical variables, such as blood pressure, pulse rate, temperature, and blood chemistries, are measured with nonnegligible error. Agricultural variables such as rainfall, soil nitrogen content, degree of pest infestation, farm crop acreage allocation, and the like cannot be measured precisely. In management sciences, social sciences, and nearly every other field many variables can only be measured with error.

For a sample of size n, the linear univariate ME model (1.2)–(1.3) can be formulated as follows. The unobservable 'true' variables, (ξ_i, η_i), satisfy

$$\eta_i = \beta_0 + \beta_1 \xi_i, \qquad i = 1, 2, \ldots, n; \tag{1.5}$$

however, one observes (x_i, y_i), $i = 1, 2, \ldots, n$, which are the true variables plus the additive errors, $(\delta_i, \varepsilon_i)$:

$$x_i = \xi_i + \delta_i, \quad \text{and} \quad y_i = \eta_i + \varepsilon_i, \qquad i = 1, 2, \ldots, n. \tag{1.6}$$

For now, assume that $\delta_1, \ldots, \delta_n, \varepsilon_1, \ldots, \varepsilon_n$ all have finite variances are uncorrelated, and (without loss of generality) have mean zero:

$$\begin{gathered} E(\delta_i) = E(\varepsilon_i) = 0, \quad \text{var}(\delta_i) = \sigma_\delta^2, \quad \text{var}(\varepsilon_i) = \sigma_\varepsilon^2, \qquad \text{for all } i, \\ \text{cov}(\delta_i, \delta_j) = \text{cov}(\varepsilon_i, \varepsilon_j) = 0, \qquad \text{for all } i \neq j, \text{ and} \\ \text{cov}(\delta_i, \varepsilon_j) = 0, \qquad \text{for all } i, j. \end{gathered} \tag{1.7}$$

Most theoretical results require that the variables be independent and not just uncorrelated. Frequently the errors, $(\delta_i, \varepsilon_i)$, are assumed to be normally distributed and the uncorrelated assumption automatically becomes an independence assumption. The normality assumption turns out to have important ramifications with respect to the identifiability of the model, as discussed in the next section and in Chapter 4. Nonnormal models are also treated in Chapter 4. Correlated errors are discussed in Section 1.4.

The restrictive assumption on the errors, δ_i and ε_i, are that they each have common means and variances. If these common means are not zero (i.e. there are systematic errors), they can be absorbed into β_0, so the assumption that the errors have mean zero is not restrictive. Such an absorption might, however, convert a no-intercept model into an intercept model.

There are three separate models of the form (1.2)–(1.3) depending on the assumptions about ξ. If the ξ_i are unknown constants, then the model is known as a **functionalmodel**; whereas, if the ξ_i are independent identically distributed random variables and independent of (not just uncorrelated with) the errors, the model is known as a **structural model**. For the structural model, we denote

$$E\xi_i = \mu \quad \text{and} \quad \text{var}(\xi_i) = \sigma^2.$$

A third model, the **ultrastructural model** (Dolby, 1976), assumes that the ξ_i are independent random variables as in the structural model but not identically distributed, instead having possibly different means, μ_i, and common variance, σ^2. The ultrastructural model is a generalization of the functional and structural models: if $\mu_1 = \ldots = \mu_n$, then the ultrastructural model reduces to the structural model; whereas if $\sigma^2 = 0$, then the ultrastructural model reduces to the functional model.

In the functional and ultrastructural cases, the ξ_i or μ_i are treated as nuisance parameters and their number increases with sample size. This can be handled asymptotically in several ways; for example, by a condition such as (1.48). If the μ_i (or ξ_i) are unequal, then it also makes sense to speak of replications. For example, one might have r replications at each 'level' i giving the model:

$$\eta_{ij} = \beta_0 + \beta_1\xi_{ij}, \qquad i = 1, \ldots, n; \; j = 1, \ldots, r, \tag{1.8}$$

with observations

$$x_{ij} = \xi_{ij} + \delta_{ij} \quad \text{and} \quad y_{ij} = \eta_{ij} + \varepsilon_{ij}, \tag{1.9}$$

where $E(\xi_{ij}) = \mu_i$ for all j, the ξ_{ij} are independent with variance σ^2, and the assumptions analogous to (1.7) hold. The unreplicated model and the replicated model are usually studied separately (see, for example,). This book will concentrate on the unreplicated model.

At first sight, the ME model looks like a regression model. Regression, however, is a special case of the ME model when the error, δ, is identically zero. If one tries to write (1.2) and (1.3) as an ordinary regression model one obtains

$$y = \beta_0 + \beta_1 x + (\varepsilon - \beta_1\delta) \equiv \beta_0 + \beta_1 x + \zeta. \tag{1.10}$$

This is not the usual regression model, x is random (for any ME model) and it is correlated with the error term, ζ: $\text{cov}(x, \zeta) = -\beta_1\sigma_\delta^2$. This covariance is only zero when $\sigma_\delta^2 = 0$, which is the regression model, or when $\beta_1 = 0$, which is the trivial case. If one attempts to use ordinary regression estimates (least squares) on ME model data, then one obtains inconsistent estimates (see (1.21)). Consistent estimates are defined as follows.

Definition 1.1 *Denote by $\hat{\theta}_n$ an estimator of a parameter, θ, based on a sample of size n, that is, $\hat{\theta}_n = \hat{\theta}_n(X_1, \ldots, X_n)$. As the sample size increases, the sequence of estimators, $\{\hat{\theta}_n\}$, is said to be consistent if for any constant, $\epsilon > 0$*

$$P_\theta(|\hat{\theta}_n - \theta| > \epsilon) \to 0 \qquad \text{as } n \to \infty.$$

If θ is a vector, then $|\hat{\theta}_n - \theta|$ is replaced by $\|\hat{\theta}_n - \theta\|$.

In other words, an estimate is consistent if, as the sample size increases to ∞, the estimate converges (in probability) to the true parameter (see, for example, Bickel and Doksum, 1977). Consistency is usually considered a minimal requirement (necessary condition) for a 'reasonable' estimator. It is not a sufficient condition for a reasonable estimator because it is an asymptotic (as $n \to \infty$) requirement and does not guarantee good finite-sample properties of the estimator. Because the least-squares estimates are not consistent, they are not considered reasonable estimates for the parameters of the ME model.

Example 1.1 (*Kendall and Stuart, 1979, p. 399*)
Boyle's law relating the pressure (P) to the volume (V) for the adiabatic expansion of a gas has the form $PV^\gamma = c$, where γ and c are parameters to be estimated. Taking logarithms of both sides and letting $\eta = \ln(P)$, $\xi = \ln(V)$, $\beta_0 = \ln(c)$ and $\beta_1 = -\gamma$, gives the form (1.2). There are certainly experiments conducted, involving Boyle's law, in which both variables can only be measured with error. Thus, a linear ME model in the logarithms of the variables would be appropriate. Note that an additive error for the ME model for the logarithm variables converts to a multiplicative error in the original variables.

Example 1.2
Suppose one wished to model the crime rate, η, in neighbourhoods as a function of the average family income, ξ, in those neighbourhoods. If one gathers data from randomly selected neighbourhoods, then both variables, crime rate and average family income, are random and can only be measured with error. Because ξ is randomly selected, the structural model would be appropriate.

Example 1.3
The replicated model could be used in an agricultural experimental design in which there are n plots, each treated with a different level of fertilizer, and r plants are randomly selected from each plot. If the level of fertilizer, ξ, and some characteristic of the plant, η, are both measured with error, and if the levels of fertilizer are set by the experimenter, then a replicated functional model is indicated.

Example 1.4
The structural model can arise in calibration experiments. A packaged food manufacturer wants to measure the moisture in a product with a newly available on-line moisture meter. The manufacturer wishes to calibrate this new instrument with the more commonly accepted forced air oven method. Samples are obtained by randomly sampling the product coming off the production line over a period of time.

The moisture in each sample is measured by both the forced air oven method (ξ) and the on-line meter (η). Both methods of measuring moisture are subject to error. A structural model seems well suited to relate the two measurements (see Sections 2.5.2 and 7.4.3; and Cheng and Van Ness, 1997b). It might be reasonable to assume a no-intercept, $\beta_0 = 0$, model here.

1.2 Elementary Properties of Measurement Error Models

The seemingly minor change between model (1.1) and model (1.2)–(1.3) has some important practical and theoretical ramifications. These will be discussed throughout this book, but it is useful to point out here some properties that distinguish ME models from regression models.

1.2.1 Identifiability of the model

One of the most important differences between ME models and ordinary regression models concerns model identifiability. It is common to assume that all the random variables in the ME model are jointly normal. In this case, the structural model (1.2)–(1.3) is not identifiable. This means that different sets of parameters can lead to the same joint distribution of x and y. In this situation, it is impossible to estimate consistently the parameters from the data, (1.4), because the limit of a consistent estimator has to be unique. In other words, it is impossible for $\hat{\theta}_n \to \theta_1$ and $\hat{\theta}_n \to \theta_2$ with $\theta_1 \neq \theta_2$.

Formally, if $\mathbf{Z}$ is a random vector whose distribution is from some family $\mathcal{F} = \{F_\theta; \boldsymbol{\theta} \in \Theta\}$, then the parameter, θ_i, the ith component of the vector, $\boldsymbol{\theta}$, is identifiable if and only if no two values of $\boldsymbol{\theta} \in \Theta$, whose ith components differ, lead to the same distribution of $\mathbf{Z}$. The vector of parameters, $\boldsymbol{\theta}$, is said to be identifiable if and only if all its components are identifiable. The model is said to be identifiable if $\boldsymbol{\theta}$ is identifiable. Some additional details about identifiability can be found in Appendix A.

Example 1.5

Suppose that $E\xi = 0$, then, because $Ex = 0$ and $Ey = \beta_0$, the simple linear normal structural model is completely determined by β_0, β_1, and the covariance matrix

$$\mathbf{C} = \operatorname{cov}\left(\begin{pmatrix} x \\ y \end{pmatrix}\right) = \begin{bmatrix} \sigma^2 + \sigma_\delta^2 & \beta_1\sigma^2 \\ \beta_1\sigma^2 & \beta_1^2\sigma^2 + \sigma_\varepsilon^2 \end{bmatrix}. \tag{1.11}$$

The model with $\beta_0 = 0$ and $\beta_1 = \sigma^2 = \sigma_\delta^2 = \sigma_\varepsilon^2 = 1$ has the covariance matrix

$$\begin{bmatrix} 2 & 1 \\ 1 & 2 \end{bmatrix}$$

but so does, for example, the model with $\beta_0 = 0$, $\beta_1 = 2/3$, $\sigma^2 = 3/2$, $\sigma_\delta^2 = 1/2$, and $\sigma_\varepsilon^2 = 4/3$. The model is not identifiable because more than one set of parameters lead to

$$\begin{pmatrix} x \\ y \end{pmatrix} \sim N\left(\begin{pmatrix} 0 \\ 0 \end{pmatrix}, \mathbf{C}\right).$$

There is a vast amount of material on identifiability in the econometrics literature. See, for example, Reiersol (1950), Bowden (1973), Deistler and Seifert (1978), and Aigner *et al.* (1984). Identifiability of a parameter does not guarantee the existence of a consistent estimator of that parameter (see the discussion of the functional model below). For this reason, some authors (such as Malinvaud, 1970, p. 401) define identifiability differently. They define a parameter to be identifiable if it has a consistent estimator.

There are six side assumptions found in the literature, any one of which will make the normal structural model identifiable:

$$\begin{array}{ll} \text{(a)} & \text{the ratio of the error variances, } \lambda \equiv \sigma_\varepsilon^2/\sigma_\delta^2, \text{ is known;} \\ \text{(b)} & \text{the } \xi \text{ reliability ratio, } \kappa_\xi, \text{ is known;} \\ \text{(c)} & \sigma_\delta^2 \text{ is known;} \\ \text{(d)} & \sigma_\varepsilon^2 \text{ is known;} \\ \text{(e)} & \text{both of the error variances, } \sigma_\varepsilon^2 \text{ and } \sigma_\delta^2, \text{ are known; or} \\ \text{(f)} & \text{the intercept, } \beta_0, \text{ is known and } Ex \neq 0. \end{array} \tag{1.12}$$

Here, the ratio,

$$\kappa_\xi \equiv \frac{\sigma^2}{\sigma^2 + \sigma_\delta^2} = \frac{\sigma^2}{\mathrm{var}(x)}, \tag{1.13}$$

is called the **reliability ratio** (Fuller, 1987, Section 1.1) and is related to the bias of the standard regression estimate of β_1 (see (1.21)).

In selecting side conditions, one does not wish to assume knowledge of β_1 whose estimation is the primary objective. The parameter, μ, is already identifiable because $\bar{x}$ is a consistent estimate of μ by the law of large numbers, so one cannot improve matters there. In fact, in the normal structural model with no side conditions, μ is the only parameter that is identifiable. The only assumption about β_0 that is commonly used is that $\beta_0 = 0$, which is the no-intercept model. This leaves assumptions about the error variances. Assumption (1.12a) is the most popular of these additional assumptions and is the one with the most published theoretical results. It has a long history dating back to Adcock (1877; 1878). It also leads to one of the more interesting ME models – see Section 1.2.3. Assumption (1.12b) is commonly found in the social science and psychology literatures. It is referred to as **heritability** in genetics. Assumption (1.12c) has gained attention recently and is a popular assumption when working with nonlinear models (see Chapter 6). It is also likely to be realistic in many applications because it is not uncommon to have replications of x, which can be used to estimate σ_δ^2. This assumption is closely related to assumption (1.12b). It is commonly used with the equation error model introduced in Section 1.5. Assumption (1.12d) is less useful and cannot be used to make the equation error model or the ME model with more than one explanatory (x) variable identifiable. Assumption (1.12e) frequently leads to the same estimates as those for (1.12a) and also leads to an overidentified model. The assumption (1.12f) does not make the normal model, with more than one explanatory variable, identifiable. This book will cover, in some detail, the use of each of these assumptions.

The assumption that σ^2 is known also makes the normal structural model identifiable, but it is not used because it is hard to imagine a situation in which one would know the variance of the unobservable ξ.

Example 1.6

Suppose that β_0 is known to be zero (the no-intercept model) and that $Ex = \mu \neq 0$; then $\beta_1 = Ey/Ex$ and β_1 is identifiable because it can be consistently estimated by $\bar{y}/\bar{x}$. The other parameters are also identifiable because one can solve for them given β_1 and the mean vector and covariance matrix of (x, y). Thus the normal structural model is identifiable in this case.

It might seem odd, to people familiar with ordinary regression, that one needs to put additional restrictions on the normal structural model because there is no need for this in ordinary regression. The normal ME model is unidentifiable because it has an additional unknown parameter that the regression model does not have. This additional 'degree of freedom' allows more than one model to yield the same joint distribution of (x, y) (you have the same number of equations but one more unknown). Without identifiability of a parameter, very little can be done. The identifiability of a parameter in the structural model is equivalent to the existence of a consistent estimator of that parameter (Deistler and Seifert, 1978). As shall be seen in the paragraphs below, this is not true for the other models. There are alternative forms of additional information besides (1.12) that can be used; for example, one can use instrumental variables as discussed in Chapter 4. One possible source of information for providing an identifiability assumption is from replications in the experiment (see Madansky, 1959).

If the linear structural model is not normal then side conditions are unnecessary for identifiability. Reiersol (1950) proved (see Appendix A) that if (δ, ε) are jointly normal, then the nonnormality of ξ is necessary and sufficient for the identifiability of β_0 and β_1 (see also Madansky, 1959; and Bunke and Bunke, 1989, p. 233). Moreover, if ξ is normal and δ and ε are independent, β_0 and β_1 are identifiable if and only if neither δ nor ε has a distribution that is divisible by a normal distribution. A distribution of a random variable, W, is said to be divisible by the distributions of U and V if their corresponding characteristic functions satisfy $\varphi_W = \varphi_U \cdot \varphi_V$. Reiersol (1950) also showed that ξ being nonnormally distributed is sufficient for β_0 and β_1 to be identifiable in the structural model when δ and ε are independent.

The functional model has $n+4$ unknown parameters. Even without any side conditions, β_0 and β_1 are identifiable (see Bowden and Turkington, 1984, p. 3), even for (δ, ε) jointly normal. For the functional model, unfortunately, this does not guarantee the existence of consistent estimates of β_1. Wald (1949) showed that the identifiability of β_1 in the structural model with $(\xi, \delta, \varepsilon)$ normal is equivalent to the existence of a consistent estimate of β_1 in the functional model with (δ, ε) normal. Reiersol's (1950) result, combined with this, shows that if (δ, ε) are normal, the functional model cannot have a consistent estimate without some side condition or other additional information. For the normal functional model, side conditions (1.12a) or (1.12e) can be used to obtain consistent maximum likelihood estimates (see Section 1.3.2). The consistent estimates resulting from the other normal structural model side conditions do not come from maximum likelihood. Condition (1.12b) is not well defined for the functional model.

Example 1.7

Consider the normal functional model, (1.2) and (1.3), where ξ is fixed. Given $n > 1$ independent (but not identically distributed) observations, (1.4), then

$$E\begin{pmatrix} x_i \\ y_i \end{pmatrix} = \begin{pmatrix} \xi_i \\ \beta_0 + \beta_1 \xi_i \end{pmatrix}, \qquad \text{cov}\begin{pmatrix} x_i \\ y_i \end{pmatrix} = \begin{pmatrix} \sigma_\delta^2 & 0 \\ 0 & \sigma_\varepsilon^2 \end{pmatrix}. \tag{1.14}$$

Given two parameter vectors,

$$\boldsymbol{\theta} = \begin{pmatrix} \beta_0 \\ \beta_1 \\ \sigma_\delta^2 \\ \sigma_\varepsilon^2 \\ \xi_1 \\ \vdots \\ \xi_n \end{pmatrix} \quad \text{and} \quad \tilde{\boldsymbol{\theta}} = \begin{pmatrix} \tilde\beta_0 \\ \tilde\beta_1 \\ \tilde\sigma_\delta^2 \\ \tilde\sigma_\varepsilon^2 \\ \tilde\xi_1 \\ \vdots \\ \tilde\xi_n \end{pmatrix},$$

the distribution of (1.4) is the same under $\boldsymbol{\theta}$ and $\tilde{\boldsymbol{\theta}}$, and (1.14) implies that

$$\xi_i = \tilde\xi_i, \qquad i = 1, \ldots, n;$$

$$\sigma_\delta^2 = \tilde\sigma_\delta^2; \qquad \sigma_\varepsilon^2 = \tilde\sigma_\varepsilon^2;$$

and

$$\beta_0 + \beta_1 \xi_i = \tilde\beta_0 + \tilde\beta_1 \tilde\xi_i = \tilde\beta_0 + \tilde\beta_1 \xi_i, \qquad i = 1, \ldots, n. \tag{1.15}$$

Clearly ξ_i, σ_δ^2, and σ_ε^2 are identifiable. Thus the model will be identifiable when β_0 and β_1 are.

Consider two cases. First, suppose that all the ξ_i are equal, $\xi_i = c, i = 1, \ldots, n$; then (1.15) gives one equation in two unknowns and β_0 and β_1 are not identifiable. In this case, the model is

$$x_i = c + \delta_i \quad \text{and} \quad y_i = \beta_0 + \beta_1 c + \varepsilon_i.$$

This can be considered a special case of both the structural and the functional models, but in this book we will consider this a structural model. Then the previous statement about the identifiability of the functional model is not violated by this example.

Second, if not all the ξ_i are equal, say $\xi_1 \neq \xi_2$, then

$$\beta_0 + \beta_1 \xi_1 = \tilde\beta_0 + \tilde\beta_1 \xi_1$$
$$\beta_0 + \beta_1 \xi_2 = \tilde\beta_0 + \tilde\beta_1 \xi_2$$

and subtracting these equations yields $\beta_0 = \tilde\beta_0$ and $\beta_1 = \tilde\beta_1$ and the model is identifiable.

The normal ultrastructural model behaves much like the functional model because the number of unknown parameters, $n+5$, also grows with the sample size. Another side condition found in the ultrastructural model literature is that both $\lambda \equiv \sigma_\varepsilon^2/\sigma_\delta^2$ and $\nu = \sigma^2/\sigma_\delta^2$ are known (see Section 1.3.3).

1.2.2 The role of variables in the λ known ME model

Compare, for example, the ME model with λ known with the standard regression model. The ME model treats the x and the y variables symmetrically. There is no explanatory variable nor dependent variable. This can be seen most easily by a change of parameters, $\beta_\xi = -\beta_1/\beta_0$ and $\beta_\eta = 1/\beta_0$ (for $\beta_0 \neq 0$); then (1.2) can be written

$$\beta_\xi \xi + \beta_\eta \eta = 1 \tag{1.16}$$

and the ME model becomes completely symmetric in x and y. Ordinary regression does not do this; the regression line obtained by regressing y on ξ is not the same as the regression line obtained by regressing ξ on y (see Figure 1.1). Section 1.2.3 shows that, under suitable assumptions, the estimation problem for the ME model with λ known (and the variables rescaled so that $\lambda = 1$) leads to orthogonal regression. Orthogonal regression gives the same line no matter which variable is considered the 'explanatory variable'.

Side conditions (b), (c) and (d) in (1.12) obviously do not treat the variables symmetrically. The symmetry also does not hold for equation error models (see Section 1.5).

1.2.3 ME regression with λ known and orthogonal regression

One of the interesting things about ME models is that under the assumption that λ is known and the data are scaled so that $\lambda = 1$, the maximum likelihood solution of the normal ME regression problem is orthogonal regression. Orthogonal regression minimizes the sum of squares of the orthogonal distances from the data points to the regression line instead of the sum of squares of the vertical distances, as in standard regression (see Figure 1.1).

If λ is known, one can rescale so that $\lambda = 1$:

$$\dot{x} = \sqrt{\lambda} x.$$

For the remainder of this subsection, we drop the dot notation and assume that the scale has been chosen so that $\lambda = 1$. For the functional model, the likelihood function is

$$\begin{aligned} &L(\beta_0, \beta_1, \sigma_\delta^2, \xi_1, \ldots, \xi_n) \\ &\quad \propto \; \sigma_\delta^{-2n} \exp\left[\frac{-1}{2\sigma_\delta^2}\left\{\sum (x_i - \xi_i)^2 + \sum (y_i - \beta_0 - \beta_1 \xi_i)^2\right\}\right] \end{aligned} \tag{1.17}$$

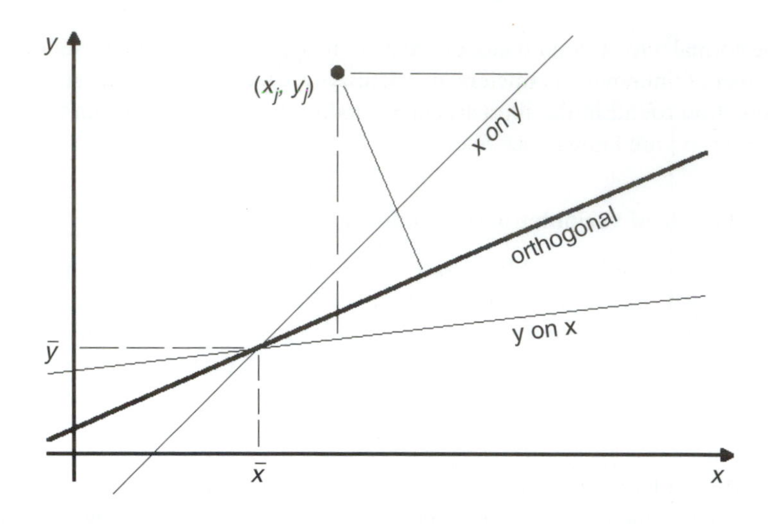

Figure 1.1: Comparison of the three regression distances.

where $\propto$ is read 'is proportional to' and means that the normalizing constant is omitted. Note that $\delta_i = x_i - \xi_i$ and $\varepsilon_i = y_i - \beta_0 - \beta_1 \xi_i$ so that maximizing (1.17) requires minimizing $\sum(\delta_i^2 + \varepsilon_i^2)$, which by Pythagoras' theorem means that the sum of squares of the **orthogonal** distances from the data points to the line is minimized (see Figure 1.2).

For the structural model, a little algebra shows that the maximum likelihood estimates of β_0 and β_1 are obtained by minimizing the following sum of weighted squares:

$$(\hat{\beta}_0, \hat{\beta}_1) = \arg \min_{(\beta_0, \beta_1)} \sum_{i=1}^{n} \left(\frac{y_i - \beta_0 - \beta_1 x_i}{K(\beta_1)} \right)^2 \tag{1.18}$$

where $K(\beta_1) = \sqrt{1 + \beta_1^2}$. Standard regression leads to

$$(\hat{\beta}_{0R}, \hat{\beta}_{1R}) = \arg \min_{(\beta_0, \beta_1)} \sum_{i=1}^{n} (y_i - \beta_0 - \beta_1 x_i)^2. \tag{1.19}$$

(The notation '$\arg\min_x f(x)$' is read 'the argument, x, which minimizes $f(x)$'.) The $1/K(\beta_1)$ term in (1.18) is a regression weight, which causes the minimization to select the sum of squares of the orthogonal distances from the data points to the line (see Figure 1.2). That is, from analytical geometry, the squared distance from a point, (x_i, y_i), to a straight line with intercept β_0 and slope β_1 is just $(y_i - \beta_0 - \beta_1 x_i)^2/(1 + \beta_1^2)$.

Figure 1.1 shows three regression lines: the standard regression of y on x, the standard regression of x on y (inverse regression), and orthogonal regression. Figure 1.1 also shows the distances measured from a typical point, A, to the three regression lines. Note that all three regression lines pass through the point $(\bar{x}, \bar{y})$.

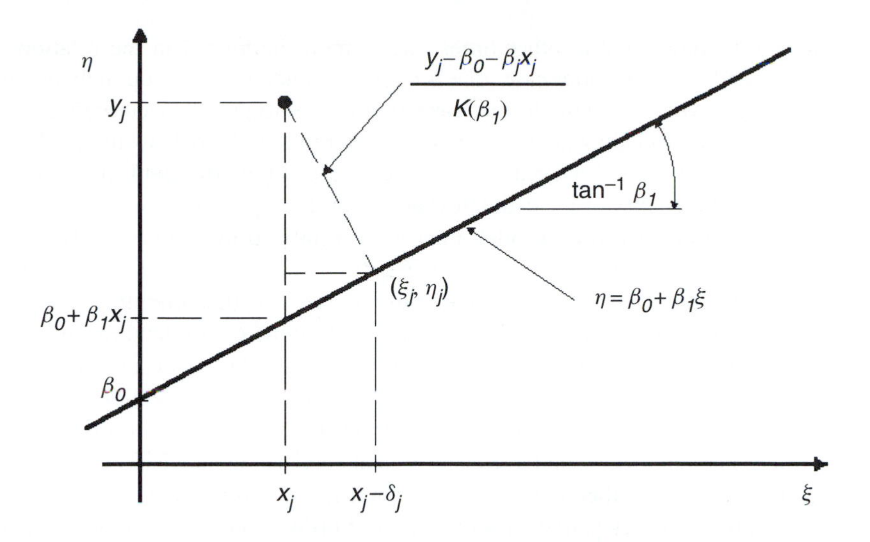

Figure 1.2: Orthogonal regression distance.

Orthogonal regression, like the general ME model (except for the β_0 known case), has the property that the ME regression line always lies between the standard regression line of y on x and the standard regression line of x on y. In fact, it will be shown (see (1.28)) that if $\hat{\beta}_{1R}$ is the estimated slope parameter from a standard regression of y on x, $\hat{\beta}_{1I}$ is the estimated slope from the standard regression line of x on y (i.e. the estimated regression line is $x = \hat{\beta}_{0I} + \hat{\beta}_{1I}y$, which implies $y = -\hat{\beta}_{0I}/\hat{\beta}_{1I} + x/\hat{\beta}_{1I}$), and $\hat{\beta}_1$ is any ML estimate for the normal ME model including the orthogonal regression estimate, then always

$$|\hat{\beta}_{1R}| \le |\hat{\beta}_1| \le |1/\hat{\beta}_{1I}|. \tag{1.20}$$

If one uses standard regression (of y on x) on data from a normal structural model, the estimate of β_1 is

$$\hat{\beta}_{1R} = \frac{\sum(x_i - \bar{x})(y_i - \bar{y})}{\sum(x_i - \bar{x})^2},$$

which has expectation (from (1.11)),

$$E(\hat{\beta}_{1R}) = \frac{\text{cov}(x, y)}{\text{var}(x)} = \beta_1 \frac{\sigma^2}{\sigma^2 + \sigma_\delta^2} = \beta_1 \kappa_\xi < \beta_1 \tag{1.21}$$

and $\hat{\beta}_{1R}$ will always be biased towards zero. The measurement error attenuates the regression slope estimate. As σ_δ^2 increases, the bias increases.

1.2.4 Prediction under ME models

In Section 2.5, we will take up another curious property of ME models, which concerns prediction. Sometimes one constructs a regression model for the purpose of predicting y from x and at other times one is more interested in the relationship between y and x. In the latter case, one is primarily interested in accurately estimating β_0 and β_1, whereas, in the former case, one wishes to predict accurately a new y knowing the corresponding new x. It would appear that both goals would lead to the same estimates of β_0 and β_1, but this is not the case. Quite different loss functions (measures of the quality of the estimates) are involved.

For the **normal structural model**, x and y are jointly normal and the best expected mean squared predictor of y given x is $E(y|x)$, which is given by the standard regression of y on x because the data are normal. That is, the conditional (given x) minimum mean squared error predictor is attained by the conditional mean and the conditional mean is linear for normal random variables. Thus, $E(y|x) = \alpha_0 + \alpha_1 x$, where $\alpha_1 = \sigma_{xy}/\sigma_x^2$ and $\alpha_0 = Ey - \alpha_1 Ex$. One would, therefore, not want to use $\beta_0 + \beta_1 x$ as the predictor if the goal is to minimize the expected mean squared error! Thus, those data analysts who have been 'mistakenly' using standard regression for prediction when in fact their data were from a normal structural model, were actually using the optimal mean squared error procedure. Furthermore, this best mean squared predictor can be found even if the model is not identifiable. On the other hand, if one wishes to predict y (or η) given the true value, ξ, which could happen in some applications, then one would use $E(y|\xi)$, which is the ME predictor, $\beta_0 + \beta_1 \xi$, and the regression solution is no longer optimal. Prediction is discussed further in Section 2.5.

Furthermore, suppose the linear structural model holds and is not normal. Then $E(y|x)$ is no longer even linear and ordinary linear regression would probably not work well. If the errors are independently normally distributed, then $E(y|x)$ is a linear function of x if and only if x is normally distributed (Lindley, 1947; Kendall and Stuart, 1979, p. 438). If the joint distribution is not normal, then the ordinary regression of y on x still gives the best mean squared error predictor among all linear predictors.

Under the **functional model**, $E(y|x)$ no longer has the form of the regression linear predictor under any circumstances (see Aigner *et al.*, 1984; and Section 1.3.5). The observations might be independent and normal, but they are not identically distributed. In this situation, Buonaccorsi (1995) suggested using the usual functional model estimates for β_0 and β_1.

1.2.5 Practical model building

Investigators using regression simply to find a linear relationship between two variables usually cannot accept two regression lines and do not understand why it matters which variable is called x and which is called y. Some have in fact suggested calculating both standard regression lines and then taking some weighted average of the two (see, for example, Rabinowicz, 1970, Chapter 3). Rabinowicz writes:

> It remains for me to add that about one-third of all the least-squares solutions I have seen have been wrong in that they have assumed the x-axis to be error-free, whereas, in fact, the y-axis had less error.

For example, one suggestion has been to base the weighting of lines on the reliability ratios, κ_ξ and $\kappa_\eta = \sigma_\eta^2/(\sigma_\eta^2 + \sigma_\varepsilon^2)$. ME regression provides an analytically sound solution to the problem of finding a single line. ME models are also frequently better matched to the physical situation. Regression does not purport to give a functional relationship between mathematical variables or a structural relation between random variables. In the fixed effects case, it presents a relation between the mean of the dependent variable and the value of the independent variable; and, in the random effects case, it exhibits a property of a bivariate distribution.

In fixed effects standard regression, the values of the independent variable, ξ, may be fixed at will; for example, at equal intervals. For the functional model, the values of ξ can be determined by fixed selected experimental situations, but it is assumed that those values of ξ can only be measured with error.

Example 1.8
Suppose it is desired to relate the logarithm of the light intensity, η, of certain stars to the logarithm of their respective temperatures, ξ (see Figures 7.1 and 7.4). The temperatures are fixed but cannot be measured precisely and the functional model would be appropriate.

Finally, a comment concerning the computational aspects of ME modelling. Probably one of the main reasons why standard regression is so much more common than ME modelling is that it is computably and analytically simpler than ME modelling. The computational difficulties for ME models were particularly true if one wished to use some more complex form of regression such as robust regression. For the λ-known case of the ME model, Ammann and Van Ness (1988) introduced a fast converging iterative algorithm which converts nearly any standard regression routine into a corresponding orthogonal regression routine (see Section 7.4). Thus, essentially any regression procedure available for standard regression is now also available for ME regression with $\lambda = 1$ known. This means that one can easily compute robust orthogonal regressions, L_1 (mean absolute deviation) orthogonal regressions, weighted orthogonal regressions, and so on. So these computational procedures exist at least for the λ-known case of the ME model. The theoretical aspects of the ME model are indeed more difficult, but much has now been developed in the literature, as we shall see.

1.3 Maximum Likelihood Estimation in the Univariate Normal Measurement Error Model

If one is willing to make the further assumption that the observables (x_i, y_i) are jointly normally distributed, the maximum likelihood estimates of the parameters can be derived for the so-called 'univariate model', the model with only one ξ-variable. This joint normality assumption is equivalent to the assumption that the ξ_i, δ_i, and ε_i are jointly normal.

Maximum likelihood (ML) estimates, obtained by maximizing the likelihood function with respect to the parameter vector, $\boldsymbol{\theta}$, will be denoted by $\hat{\boldsymbol{\theta}}$. A common way of finding the ML estimate is to differentiate the likelihood function with respect to the parameters, and set the derivatives equal to zero. The resulting equations will be called the **likelihood equations**. Several possibilities present themselves. The likelihood equations can have a unique solution (critical point) that does indeed maximize the likelihood function; this solution gives the ML estimates. The likelihood equations can have one or more solutions, none of which are a maximum. Such a critical point might be a minimum, a saddle point, a local maximum, or a local minimum of the likelihood function. In this case the likelihood function might have no maximum or it might have a maximum on the boundary of the parameter space. If it has a unique maximum on the boundary, then the ML estimate exists, but it is not obtained by solving the likelihood equations. If the likelihood equations have a solution that is not a maximum but a saddle point or local maximum, then one can still consider the solution as an estimator. Such estimators will be called **likelihood equation (LE) estimates**. LE estimates might or might not be useful and they occur in ME models, as we shall see. Another possibility is that the likelihood equations have more than one solution, one of which is a unique global maximum. In this case, this global maximum is the ML estimate. If there are solutions with no global maximum, then it may be that one of the solutions is useful and can be selected as an LE estimate. The most common situation in statistics in which LE estimates give 'good' estimates is in the normal mixture problem where the likelihood function is unbounded but there are consistent LE estimates (see, for example, Lehmann, 1983, p. 442).

1.3.1 Maximum likelihood estimation for the structural relationship

Equations (1.2)–(1.3), (1.7), and the definition of the structural relationship give:

$$\begin{aligned}
Ex &= E\xi = \mu, \\
Ey &= E\eta = \mu_y = \beta_0 + \beta_1 \mu, \\
\text{var}\,(x) &= \sigma_x^2 = \text{var}(\xi) + \sigma_\delta^2 = \sigma^2 + \sigma_\delta^2, \\
\text{var}(y) &= \sigma_y^2 = \text{var}(\eta) + \sigma_\varepsilon^2 = \beta_1^2\sigma^2 + \sigma_\varepsilon^2, \\
\text{cov}(x, y) &= \sigma_{xy} = \text{cov}(\xi, \eta) = \beta_1\sigma^2.
\end{aligned} \tag{1.22}$$

Because the means and variances are fixed, the data (1.4) are independent and identically distributed. This is not true for the functional model in which the data are independent but not identically distributed.

One could write down the ML function for the structural model and maximize it directly, or one can use the functional invariance property of ML estimates (see, for example, Kendall and Stuart, 1979, Chapter 18) to derive the estimates from the familiar ML estimates for the means, variances, and covariances. The latter approach is adapted here. The ML estimates for the corresponding parameters in the bivariate normal distribution are:

$$\begin{aligned}
\hat{\mu} &= \bar{x} = \frac{1}{n}\sum x_i, \\
\hat{\mu}_y &= \bar{y} = \frac{1}{n}\sum y_i,
\end{aligned}$$

$$
\begin{aligned}
\hat{\sigma}_x^2 &= s_{xx} = \frac{1}{n}\sum(x_i - \bar{x})^2, \\
\hat{\sigma}_y^2 &= s_{yy} = \frac{1}{n}\sum(y_i - \bar{y})^2, \\
\hat{\sigma}_{xy} &= s_{xy} = \frac{1}{n}\sum(x_i - \bar{x})(y_i - \bar{y}).
\end{aligned}
\tag{1.23}
$$

By the invariance properties of ML estimates, if one could solve uniquely

$$
\begin{aligned}
&(a) & \bar{x} &= \hat{\mu}, \\
&(b) & \bar{y} &= \hat{\beta}_0 + \hat{\beta}_1\hat{\mu}, \\
&(c) & s_{xx} &= \hat{\sigma}^2 + \hat{\sigma}_\delta^2, \\
&(d) & s_{yy} &= \hat{\beta}_1^2\hat{\sigma}^2 + \hat{\sigma}_\varepsilon^2, \text{ and} \\
&(e) & s_{xy} &= \hat{\beta}_1\hat{\sigma}^2,
\end{aligned}
\tag{1.24}
$$

for the six estimated parameters, $\hat{\mu}$, $\hat{\beta}_0$, $\hat{\beta}_1$, $\hat{\sigma}^2$, $\hat{\sigma}_\delta^2$, and $\hat{\sigma}_\varepsilon^2$, and obtain valid estimates as answers, then one would have the ML estimates for the normal structural model. By 'valid answers' one means that the solutions for the variances $\hat{\sigma}^2$, $\hat{\sigma}_\delta^2$, and $\hat{\sigma}_\varepsilon^2$ must be nonnegative. For $\hat{\sigma}_\delta^2$ to be nonnegative one must have, from (1.24c) and (1.24e), that

$$
s_{xx} - \frac{s_{xy}}{\hat{\beta}_1} \geq 0. \tag{1.25}
$$

For $\hat{\sigma}_\varepsilon^2 \geq 0$ one must have, from (1.24d) and (1.24e), that

$$
s_{yy} - \hat{\beta}_1 s_{xy} \geq 0. \tag{1.26}
$$

Similarly, adding the requirement that $\hat{\sigma}^2 \geq 0$ and assuming $\sigma^2 > 0$, one obtains the following set of five restrictions:

$$
\begin{aligned}
&(a) & s_{xx} &\geq s_{xy}/\hat{\beta}_1, \\
&(b) & s_{yy} &\geq \hat{\beta}_1 s_{xy}, \\
&(c) & s_{xx} &\geq \hat{\sigma}_\delta^2, \\
&(d) & s_{yy} &\geq \hat{\sigma}_\varepsilon^2, \\
&(e) & \text{sign}(s_{xy}) &= \text{sign}(\hat{\beta}_1).
\end{aligned}
\tag{1.27}
$$

It is obvious that one is unable to solve (1.24) uniquely because there are six unknowns, namely $\hat{\mu}$, $\hat{\beta}_0$, $\hat{\beta}_1$, $\hat{\sigma}^2$, $\hat{\sigma}_\delta^2$, and $\hat{\sigma}_\varepsilon^2$, but only five equations. Model unidentifiability (Section 1.2.1) has struck. With any one of the six additional assumptions in (1.12), however, one can obtain the ML solutions (see Moran, 1971). This will be done in the following subsections.

Example 1.9

Moran (1971) illustrated the lack of uniqueness as follows. Let γ be a small positive quantity less than both $|\hat{\beta}_1|$ and $\hat{\sigma}_\varepsilon^2\hat{\beta}_1^{-1}\hat{\sigma}^{-2}$. Replace the quantities $\hat{\beta}_0$, $\hat{\beta}_1$, $\hat{\sigma}^2$, $\hat{\sigma}_\delta^2$, and $\hat{\sigma}_\varepsilon^2$ in (1.24) by $\hat{\beta}_0 - \gamma\hat{\mu}$, $\hat{\beta}_1 + \gamma$, $\hat{\beta}_1\hat{\sigma}^2(\hat{\beta}_1 + \gamma)^{-1}$, $\hat{\sigma}_\delta^2 + \gamma\hat{\sigma}^2(\hat{\beta}_1 + \gamma)^{-1}$, and $\hat{\sigma}_\varepsilon^2 - \hat{\beta}_1\gamma\hat{\sigma}^2$, respectively. Then the five equations remain unchanged so that if one set of estimates is an ML solution, so is the other.

Before proceeding, it can now be shown that the regression line solving (1.24) lies between the standard regression line of y on x and the standard regression line of x on y, as is intuitively reasonable. Equation (1.24e) implies that $\hat{\sigma}^2 = s_{xy}/\hat{\beta}_1$ if $\hat{\beta}_1 \neq 0$. Substituting into (1.24c) yields

$$\frac{s_{xy}}{s_{xx} - \hat{\sigma}_\delta^2} = \hat{\beta}_1,$$

and substituting into (1.24d) yields (if $s_{xy} \neq 0$)

$$\frac{s_{yy} - \hat{\sigma}_\varepsilon^2}{s_{xy}} = \hat{\beta}_1.$$

Because the estimated variances must be nonnegative, these two equations imply, in the notation of (1.20),

$$|\hat{\beta}_{1R}| = \frac{|s_{xy}|}{s_{xx}} \leq |\hat{\beta}_1| \leq \frac{s_{yy}}{|s_{xy}|} = \left|\frac{1}{\hat{\beta}_{1I}}\right|. \tag{1.28}$$

Case (a): Structural ML estimation when the ratio of the error variances, λ, is known.

Assuming that λ is known, that is, assuming (1.12a), is the classical method of handling the identifiability problem. If $\sigma^2 = 0$, then it is readily seen that β_1 is indeterminate; so assume that $\sigma^2 > 0$. Expression (1.24) yields the quadratic equation,

$$\hat{\beta}_1^2 s_{xy} + \hat{\beta}_1(\lambda s_{xx} - s_{yy}) - \lambda s_{xy} = 0. \tag{1.29}$$

Unless $s_{xy} = 0$ (in which case $\hat{\beta}_1 = 0$ except if $\lambda = s_{yy}/s_{xx}$, which implies $\hat{\beta}_1$ is indeterminate and $\hat{\sigma}^2 = 0$ (see 1.27)) this quadratic has nonzero roots

$$\hat{\beta}_1 = \frac{s_{yy} - \lambda s_{xx} \pm \{(s_{yy} - \lambda s_{xx})^2 + 4\lambda s_{xy}^2\}^{1/2}}{2s_{xy}} \equiv \frac{U}{2s_{xy}}.$$

By (1.24e), $0 \leq \hat{\sigma}^2 = s_{xy}/\hat{\beta}_1 = 2s_{xy}^2/U$ and, therefore, U must be nonnegative, which implies that the positive sign be taken in U:

$$\hat{\beta}_1 = \frac{s_{yy} - \lambda s_{xx} + \{(s_{yy} - \lambda s_{xx})^2 + 4\lambda s_{xy}^2\}^{1/2}}{2s_{xy}}. \tag{1.30}$$

One now needs only to check that (1.27) is satisfied. That $\hat{\sigma}^2 \geq 0$ is already ensured, so all that is needed is that (1.27b) be satisfied, because $\hat{\sigma}_\varepsilon^2 = \lambda\hat{\sigma}_\delta^2$. But this condition is true because

$$\begin{aligned} 2\hat{\beta}_1 s_{xy} &= s_{yy} - \lambda s_{xx} + \sqrt{(\lambda s_{xx} - s_{yy})^2 + 4\lambda s_{xy}^2} \\ &\leq s_{yy} - \lambda s_{xx} + \sqrt{(\lambda s_{xx} - s_{yy})^2 + 4\lambda s_{xx} s_{yy}} = 2s_{yy}. \end{aligned}$$

The ML solutions for the remaining parameters can now be easily derived from (1.24):

$$
\begin{aligned}
&(a) \quad \hat{\beta}_0 = \bar{y} - \hat{\beta}_1 \bar{x}, \\
&(b) \quad \hat{\sigma}_\delta^2 = \frac{s_{yy} - 2\hat{\beta}_1 s_{xy} + \hat{\beta}_1^2 s_{xx}}{\lambda + \hat{\beta}_1^2}, \\
&(c) \quad \hat{\sigma}^2 = s_{xy}/\hat{\beta}_1.
\end{aligned}
\tag{1.31}
$$

There are some equivalent forms of (1.30) appearing in the literature. If $\alpha = 1/\beta_1$ then, from (1.24c)–(1.24e),

$$\hat{\alpha}^2 \lambda s_{xy} + \hat{\alpha}(s_{xy} - \lambda s_{xx}) - s_{xy} = 0.$$

Thus, provided $s_{xy} \neq 0$,

$$\hat{\alpha} = \frac{\lambda s_{xx} - s_{yy} + \{(s_{yy} - \lambda s_{xx})^2 + 4\lambda s_{xy}^2\}^{1/2}}{2\lambda s_{xy}},$$

which gives

$$\hat{\beta}_1 = \frac{2\lambda s_{xy}}{\lambda s_{xx} - s_{yy} + \{(s_{yy} - \lambda s_{xx})^2 + 4\lambda s_{xy}^2\}^{1/2}}.$$

Another form is

$$\hat{\beta}_1 = s(\lambda) + \operatorname{sign}(s_{xy})\{s^2(\lambda) + \lambda\}^{1/2}, \qquad s(\lambda) = \frac{s_{yy} - \lambda s_{xx}}{2 s_{xy}}.$$

Naturally all these forms are equivalent.

Case (b): Structural ML estimation when the reliability ratio, κ_ξ, is known.

(1.12b) and (1.24c) yield

$$
\begin{aligned}
\hat{\sigma}^2 &= s_{xx}\kappa_\xi \\
\hat{\sigma}_\delta^2 &= s_{xx}(1 - \kappa_\xi),
\end{aligned}
\tag{1.32}
$$

which are nonnegative, ensuring (1.27a) and (1.27c)–(1.27e) are satisfied. From (1.24e),

$$\hat{\beta}_1 = \kappa_\xi^{-1} \frac{s_{xy}}{s_{xx}} = \kappa_\xi^{-1} \hat{\beta}_{1R}. \tag{1.33}$$

From (1.24d),

$$\hat{\sigma}_\varepsilon^2 = s_{yy} - \hat{\beta}_1 s_{xy} = s_{yy} - \frac{s_{xy}^2}{\kappa_\xi s_{xx}}. \tag{1.34}$$

This must be assumed to be nonnegative in order to satisfy (1.27b). If it is negative, take $\hat{\sigma}_\varepsilon^2 = 0$ and the ML estimate of the slope becomes

$$\hat{\beta}_1 = \frac{s_{yy}}{s_{xy}} = \operatorname{sign}(s_{xy}) \left[\frac{s_{yy}}{\kappa_\xi s_{xx}} \right]^{1/2}.$$

This is just the reciprocal of the inverse regression estimate (see comments at the end of this subsection). Of course, β_0 is estimated by the equation (1.31a).

Case (c): Structural ML estimation when σ_δ^2 is known.

Additional assumptions (1.12c) and (1.24) immediately give

$$\hat{\beta}_1 = \frac{s_{xy}}{s_{xx} - \sigma_\delta^2} \tag{1.35}$$

if $s_{xx} > \sigma_\delta^2$. This also ensures that (1.27c) and (1.27e) are satisfied. Expression (1.27d) is satisfied because (1.24c) implies $\hat{\sigma}^2 > 0$, but (1.27b) must be required in order for (1.24) to give the ML estimates; that is, $s_{yy} \geq s_{xy}^2/(s_{xx} - \sigma_\delta^2)$ must hold. If these conditions are not satisfied, the ML solution is

$$\hat{\beta}_1 = s_{yy}/s_{xy} \tag{1.36}$$

(see Exercise 1.6). This is, again, just the reciprocal of the inverse regression estimate (see comments at the end of this subsection). The remaining estimates are given by the formulae (1.31a) , (1.31c), and

$$\hat{\sigma}_\varepsilon^2 = s_{yy} - \hat{\beta}_1 s_{xy}. \tag{1.37}$$

Note that as $\sigma_\delta^2 \to 0$, $\hat{\beta}_1 \to \hat{\beta}_{1R}$.

Case (d): Structural ML estimation when σ_ε^2 is known.

Additional assumptions (1.12d) and (1.24) immediately give

$$\hat{\beta}_1 = (s_{yy} - \sigma_\varepsilon^2)/s_{xy} \tag{1.38}$$

if $s_{xy} \neq 0$. If (1.27a) and (1.27b) are true, that is, if $s_{yy} > \sigma_\varepsilon^2$ and $s_{xx} \geq s_{xy}^2/(s_{yy} - \sigma_\varepsilon^2)$, then all of (1.27) is satisfied and (1.24) gives the ML estimates. If these conditions are not satisfied, the ML solution is

$$\hat{\beta}_1 = s_{xy}/s_{xx}. \tag{1.39}$$

This is just the ordinary regression estimate (see comments at the end of this subsection). The remaining estimates are given by the formulae (1.31a), (1.31c), and

$$\hat{\sigma}_\delta^2 = s_{xx} - \frac{s_{xy}}{\hat{\beta}_1}. \tag{1.40}$$

Note that as $\sigma_\varepsilon^2 \to 0$, $\hat{\beta}_1 \to 1/\hat{\beta}_{1I}$, see (1.28), and the absolute value of (1.39) is the lower bound in (1.28).

Case (e): Structural ML estimation when both σ_δ^2 and σ_ε^2 are known.

Additional assumption (1.12e) allows us to derive both (1.35) and (1.38), which are contradictory. Thus (1.24) cannot be used to derive the ML estimates. They must be derived by maximizing the likelihood directly. This was done by Birch (1964). The answer is that $\hat{\beta}_1$ is given by (1.30) and

$$\hat{\sigma}^2 = \frac{s_{yy} + \lambda s_{xx} - 2\sigma_\varepsilon^2 + [(s_{yy} - \lambda s_{xx})^2 + 4\lambda s_{xy}^2]^{1/2}}{2(\lambda + \hat{\beta}_1^2)}$$

provided one or more of the following conditions is satisfied: $s_{xx} > \sigma_\delta^2$, $s_{yy} > \sigma_\varepsilon^2$, or $s_{xx} > (\sigma_\delta^2 - s_{xx})(\sigma_\varepsilon^2 - s_{yy})$. If none of these three conditions holds (which seems unlikely in practice), then $\hat{\sigma}^2 = 0$ and $\hat{\beta}_1$ cannot be estimated via likelihood techniques. This model is overspecified and other reasonable consistent estimates exist; for example, Kendall and Stuart (1979, p. 406) gave

$$\hat{\sigma}^2 = \frac{s_{xx} + \hat{\beta}_1 s_{xy} - \sigma_\delta^2}{1 + \hat{\beta}_1^2}$$

as an estimate of σ^2.

Case (f): Structural ML estimation when β_0 is known.

Finally, apply assumption (1.12f), which assumes that β_0 is known and $Ex \neq 0$. From (1.24b),

$$\hat{\beta}_1 = (\bar{y} - \beta_0)/\bar{x}, \qquad \text{assuming that } \bar{x} \neq 0. \tag{1.41}$$

The remaining parameters are given by the formulae (1.31c), (1.40), and (1.37). Because $\mu \neq 0$, (1.41) is a consistent estimator.

Other side conditions.

If one were to use the identifiability assumption that σ^2 is known, the additional assumption (1.24e) gives

$$\hat{\beta}_1 = s_{xy}/\sigma^2,$$

and (1.27c)–(1.27e) are automatically satisfied because σ^2 is not being estimated. From (1.24c) and (1.24d),

$$\hat{\sigma}_\delta^2 = s_{xx} - \sigma^2,$$

$$\hat{\sigma}_\varepsilon^2 = s_{yy} - \hat{\beta}_1^2 \sigma^2 = s_{yy} - \frac{s_{xy}^2}{\sigma^2}.$$

For these to be the ML solutions, the left-hand sides of these last two equations must be nonnegative.

Comments on the additional restrictions.

The requirement that the variance estimates, $\hat{\sigma}^2$, $\hat{\sigma}_\delta^2$, and $\hat{\sigma}_\varepsilon^2$, must be nonnegative places additional restrictions on the model beyond the identifiability assumptions (1.12) for some cases. This is another requirement not found in ordinary regression that some might find unusual. First, it should be noted that all these restrictions are satisfied if the sample size is large and the model is correct. Second, let us compare, for example, the σ_δ^2 known case with ordinary regression. The restrictions for the structural model are $s_{xx} > \sigma_\delta^2$ and $s_{yy} \geq s_{xy}^2/(s_{xx} - \sigma_\delta^2)$. In regression, $\sigma_\delta^2 = 0$ and the first equation simply says that not all x_i are the same, and the second equation is the result of the Cauchy–Schwarz inequality – so these inequalities are automatically satisfied. Now consider the ME model. From (1.3) it is clear that the assumption $s_{xx} > \sigma_\delta^2$ is quite reasonable. In fact, if it is violated, this would be an indication that the model is not fitting the data. Either the model needs to be changed or more data are required. This also means that, even though the estimate, (1.36), is theoretically correct, before it is used one should first re-examine the modelling situation. It is the ML estimate for the ME model with all $\varepsilon_i = 0$. Similar remarks can be made about the second inequality if one examines (1.22) carefully.

Example 1.10
Suppose one assumes the structural model with the side assumption $\lambda = 1$, and it turns out that $\hat{\sigma}_\delta^2 = 3$. Alternatively, suppose that one makes the side assumption $\sigma_\delta^2 = 3$, and the data just happen to give $\hat{\lambda} = 1$. Will all the estimates be the same, no matter which situation holds? The answer is yes because the same set of equations, (1.24), is being solved. This is true no matter which pair of parameters one works with in (1.12).

Comments on the use of maximum likelihood.

The maximum likelihood approach to estimation is primarily justified by asymptotic (as the sample size goes to ∞) considerations. Maximum likelihood estimates do not necessarily have any finite-sample optimal properties such as unbiasedness, minimum mean squared error, and so on. For finite samples, the ML estimates used here can, therefore, violate the additional restrictions. If the model is true, though, these violations should disappear as the sample size increases.

1.3.2 Maximum likelihood estimation for the functional relationship

The simple normal functional version of model (1.5)–(1.6) assumes that the ξ_i are fixed but unknown constants. Thus, $\sigma^2 = 0$ and there are $n+4$ unknown parameters: the 'structural parameters' are β_0, β_1, σ_δ^2, and σ_ε^2; and the 'incidental parameters' are $\xi_1, \ldots, \xi_n$.

The log-likelihood function is

$$L(\beta_0, \beta_1, \sigma_\delta^2, \sigma_\varepsilon^2, \xi_1, \ldots, \xi_n) \propto -\frac{n}{2}\log\sigma_\delta^2 - \frac{n}{2}\log\sigma_\varepsilon^2 - \frac{1}{2\sigma_\delta^2}\sum(x_i - \xi_i)^2 - \frac{1}{2\sigma_\varepsilon^2}\sum(y_i - \beta_0 - \beta_1\xi_i)^2. \quad (1.42)$$

Taking partial derivatives with respect to $\beta_0, \beta_1, \sigma_\delta^2, \sigma_\varepsilon^2$, and $\xi_1, \ldots, \xi_n$, and equating the results to zero, gives the likelihood equations,

$$\sum(y_i - \hat{\beta}_0 - \hat{\beta}_1\hat{\xi}_i) = 0, \qquad \sum(y_i - \hat{\beta}_0 - \hat{\beta}_1\hat{\xi}_i)\hat{\xi}_i = 0,$$

$$\frac{1}{n}\sum(x_i-\hat{\xi}_i)^2=\hat{\sigma}_\delta^2,\qquad \frac{1}{n}\sum(y_i-\hat{\beta}_0-\hat{\beta}_1\hat{\xi}_i)^2=\hat{\sigma}_\varepsilon^2,\ \text{and} \tag{1.43}$$
$$\frac{(x_i-\hat{\xi}_i)}{\hat{\sigma}_\delta^2}+\frac{\hat{\beta}_1(y_i-\hat{\beta}_0-\hat{\beta}_1\hat{\xi}_i)}{\hat{\sigma}_\varepsilon^2}=0,\qquad i=1,\ldots,n.$$

The solutions are given by

$$\begin{aligned}\hat{\beta}_1^2&=\frac{\hat{\sigma}_\varepsilon^2}{\hat{\sigma}_\delta^2}=\frac{s_{yy}}{s_{xx}},\qquad 2\hat{\sigma}_\delta^2=s_{xx}-\frac{s_{xy}}{\hat{\beta}_1},\qquad 2\hat{\sigma}_\varepsilon^2=s_{yy}-\hat{\beta}_1 s_{xy},\\ \hat{\beta}_0&=\bar{y}-\hat{\beta}_1\bar{x},\qquad 2\hat{\xi}_i=x_i+\frac{y_i-\hat{\beta}_0}{\hat{\beta}_1}=x_i+\bar{x}+\frac{y_i-\bar{y}}{\hat{\beta}_1}.\end{aligned} \tag{1.44}$$

Note that, in these calculations, the sign of $\hat{\beta}_1$ is chosen to be the same as that of s_{xy} because this gives the larger value of the likelihood function.

The solutions (1.44) are LE estimates (see the beginning of Section 1.3), but they do not give a maximum of the likelihood function, and, therefore, they have all sorts of problems. The likelihood function is actually unbounded. This was first shown by Anderson and Rubin (1956). For the no-intercept model ($\beta_0=0$), Solari (1969) pointed out that the solution of these likelihood equations is a saddle point and not a maximum of the likelihood surface. Willassen (1979) obtained the same result for the intercept model.

To see the unboundedness, consider the log-likelihood function, (1.42). Note that

$$-\frac{2}{n}L\propto\log\sigma_\delta^2+\log\sigma_\varepsilon^2+\frac{S_1}{\sigma_\delta^2}+\frac{S_2}{\sigma_\varepsilon^2}, \tag{1.45}$$

where

$$S_1=\frac{1}{n}\sum(x_i-\xi_i)^2,\qquad S_2=\frac{1}{n}\sum(y_i-\beta_0-\beta_1\xi_i)^2.$$

Write (1.45) as

$$-\frac{2}{n}L\propto h(\sigma_\delta^2,S_1)+h(\sigma_\varepsilon^2,S_2), \tag{1.46}$$

where $h(a,b)=\log a+b/a$. Now $h(a,0)=\log a$, so that as $a\to 0$, $h(a,0)\to -\infty$. Thus (1.46), the sum of two such functions, will tend to minus infinity if either of the functions on the right-hand side does so while the other remains bounded. Now consider the values

$$\xi_i=x_i,\quad i=1,\ldots,n;\qquad \sigma_\delta^2\to 0. \tag{1.47}$$

This makes $S_1=0$ and $2L/n\to\infty$, and L itself tends to infinity irrespective of the values assigned to β_1. Thus, ML estimation of β_1 is not possible here.

The estimators (1.44) are also not consistent (for the $\hat{\xi}_i$, the meaning of consistency is not clear). Assume the following limits exist:

$$\xi^*\equiv\lim_{n\to\infty}\bar{\xi}_n,\qquad s_{\xi\xi}^*\equiv\lim_{n\to\infty}\frac{1}{n}\sum(\xi_i-\bar{\xi})^2>0, \tag{1.48}$$

where $\bar{\xi}_n = n^{-1}\sum \xi_i$. It can easily be shown that $\bar{x}$, $\bar{y}$, s_{xx}, s_{xy} and s_{yy} converge to ξ^*, $\beta_0 + \beta_1\xi^*$, $s^*_{\xi\xi} + \sigma^2_\delta$, $\beta_1 s^*_{\xi\xi}$ and $\beta_1^2 s^*_{\xi\xi} + \sigma^2_\varepsilon$, respectively, with probability one as n tends to infinity. Thus, it is clear, for example, that $\hat{\beta}_1$ does not converge to β_1:

$$\hat{\beta}_1^2 = \frac{s_{yy}}{s_{xx}} \overset{w.p.1}{\rightarrow} \frac{\beta_1^2 s^*_{\xi\xi} + \sigma^2_\varepsilon}{s^*_{\xi\xi} + \sigma^2_\delta} \qquad \text{as } n \to \infty.$$

Even the form of the estimate, $\hat{\beta}_1$, appears strange because it is just the ratio of the estimates of the two error variances and does not involve the covariance in any way.

The problem is easily traced back to the presence of the nuisance parameters, $\xi_1, \ldots, \xi_n$, whose number increases with sample size. Neyman and Scott (1948) called such nuisance parameters *incidental parameters*, while the other parameters, which are fixed in number, are referred to as *structural parameters*. It is generally true that ML estimation can break down in the presence of incidental parameters; however, maximization of the likelihood in the presence of incidental parameters is sometimes possible under certain restrictions. The resulting estimators still need not be consistent because the proof of the consistency of ML estimators requires the number of unknown parameters to be fixed as the sample size increases (see, for example, Cramér, 1946, pp. 500 ff.). Thus the consistency of any resulting estimators needs to be checked case by case. The case where the ratio of error variances is known, which follows, exemplifies this situation.

We now examine the behavior of ML estimation under each of the side assumptions (1.12).

Case (a): Functional ML estimation when the ratio of the error variances, λ, is known.

If λ is known, then the log-likelihood function becomes

$$\begin{aligned} L(\beta_0, \beta_1, \sigma^2_\varepsilon, \xi_1, \ldots, \xi_n) \;\propto\; & \frac{n}{2}\log\lambda - n\log\sigma^2_\varepsilon - \frac{\lambda}{2\sigma^2_\varepsilon}\sum(x_i - \xi_i)^2 \\ & - \frac{1}{2\sigma^2_\varepsilon}\sum(y_i - \beta_0 - \beta_1\xi_i)^2, \end{aligned}$$

and (1.45) becomes

$$\frac{2}{n}L \propto \log\lambda - 2\log\sigma^2_\varepsilon - \frac{\lambda S_1 + S_2}{\sigma^2_\varepsilon}, \tag{1.49}$$

and (1.47) no longer makes (1.49) tend to infinity because S_2 remains positive. Moreover, it is obvious from the definitions of S_1 and S_2 that one cannot choose parameter values that will make $S_1 = S_2 = 0$. Thus L is bounded away from infinity. Taking partial derivatives with respect to β_0, β_1, σ^2_ε, and $\xi_1, \ldots, \xi_n$ and equating the results to zero gives (1.43),

$$-\frac{n}{\hat{\sigma}^2_\varepsilon} + \frac{\lambda}{2\hat{\sigma}^4_\varepsilon}\sum(x_i - \hat{\xi}_i)^2 + \frac{1}{2\hat{\sigma}^4_\varepsilon}\sum(y_i - \hat{\beta}_0 - \hat{\beta}_1\hat{\xi}_i) = 0,$$

and

$$\lambda\left(x_i - \hat{\xi}_i\right) + \hat{\beta}_1\left(y_i - \hat{\beta}_0 - \hat{\beta}_1\hat{\xi}_i\right) = 0, \qquad i = 1, \ldots, n. \tag{1.50}$$

Straightforward calculation yields (1.31a),

$$\hat{\xi}_i = \frac{\lambda x_i + \hat{\beta}_1\left(y_i - \hat{\beta}_0\right)}{\lambda + \hat{\beta}_1^2}, \qquad i = 1, \ldots, n, \tag{1.51}$$

$$\begin{aligned} \hat{\sigma}_\varepsilon^2 &= \frac{1}{2n}\left\{\lambda \sum\left(x_i - \hat{\xi}_i\right)^2 + \sum\left(y_i - \hat{\beta}_0 - \hat{\beta}_1\hat{\xi}_i\right)^2\right\} \\ &= \frac{\lambda}{2n\left(\lambda + \hat{\beta}_1^2\right)} \sum\left(y_i - \hat{\beta}_0 - \hat{\beta}_1\hat{\xi}_i\right)^2, \end{aligned} \tag{1.52}$$

and, from (1.31a),

$$\hat{\sigma}_\varepsilon^2 = \frac{\lambda}{2\left(\lambda + \hat{\beta}_1^2\right)}\left(s_{yy} - 2\hat{\beta}_1 s_{xy} + \hat{\beta}_1^2 s_{xx}\right). \tag{1.53}$$

Substituting (1.51) into (1.50) gives

$$\hat{\beta}_1 = \frac{\left(\lambda + \hat{\beta}_1^2\right)\left(\lambda s_{xy} + \hat{\beta}_1 s_{yy}\right)}{\lambda^2 s_{xx} + 2\lambda\hat{\beta}_1 s_{xy} + \hat{\beta}_1^2 s_{yy}},$$

which simplifies to (1.29).

Thus the solutions of the likelihood equations for β_1 in the linear functional relationship are the same as those of the corresponding linear structural relationship when the error variance ratio, λ, is known. It has been shown that the preceding solutions of the likelihood equations (using (1.30)) indeed attain the maximum value of the likelihood (see, for example, Willassen, 1979; Gleser, 1981).

If one compares the ML estimators for the structural relationship and the functional relationship when the error variance ratio, λ, is known, there are three parameters common to both relationships: β_0, β_1 and σ_ε^2 (or σ_δ^2). The ML estimators of the first two parameters coincide for both cases, but the estimators for the error variance, σ_ε^2, differ by a factor of one-half. This means that at least one of the error variance estimators has to be inconsistent. It is easy to check (see Section 2.1.2) that the functional relationship estimator, (1.53), converges to $\sigma_\varepsilon^2/2$. A consistent estimator is given by (2.2), which was proposed by Lindley (1947).

Case (b): Functional LE estimation when the reliability ratio, κ_ξ, is known.

Here, the true values ξ_i are fixed rather than random and the reliability is not well defined for the functional model. Gleser (1992) introduced a more general reliability definition, which can be used in this situation. The numerical procedures are the same as in case (b) for the structural model (Section 1.3.1). We will not discuss this further here.

Cases (c)–(d): Functional LE estimation when either σ_ε^2 or σ_δ^2 is known.

Suppose, first, that σ_ε^2 is known. The log-likelihood function may be written as

$$L\left(\beta_0, \beta_1, \sigma_\delta^2, \xi_1, \ldots, \xi_n\right) \propto -\frac{n}{2}\log\sigma_\delta^2 - \frac{\lambda}{2\sigma_\delta^2}\sum\left(x_i - \xi_i\right)^2 - \frac{1}{2\sigma_\varepsilon^2}\sum\left(y_i - \beta_0 - \beta_1\xi_i\right)^2.$$

Now one can repeat the arguments given previously to show that this is again unbounded and ML estimates fail to exist. Moberg and Sundberg (1978) discussed the LE estimates for this situation. The likelihood equations give multiple roots, which can be saddle points or local maxima. Moberg and Sundberg partially characterize these roots.

If σ_δ^2 is assumed known, then (1.5) can be rewritten as

$$\xi_i = -\frac{\beta_0}{\beta_1} + \frac{1}{\beta_1}\eta_i$$

and, reversing the roles of x and y, the same argument applies.

Thus, the method of maximum likelihood fails to yield consistent estimators for all parameters of the model if either σ_ε^2 or σ_δ^2 is known. Consistent estimates do exist, however. For example, one can use the ML estimates from the corresponding structural models (see Section 2.1.2).

Case (e): Functional ML estimation when both σ_ε^2 and σ_δ^2 are known.

For the case of both error variances known, the ML estimators are the same as those of known λ, except one need not estimate any error variance.

Case (f): Functional LE estimation when β_0 is known.

For the case where β_0 is known, one again obtains the saddle point results corresponding to (1.44) (Solari, 1969). The LE estimates are:

$$\hat{\beta}_1^2 = \frac{\hat{\sigma}_\varepsilon^2}{\hat{\sigma}_\delta^2} = \frac{\widetilde{s}_{yy}}{\widetilde{s}_{xx}}, \qquad 2\hat{\sigma}_\delta^2 = \widetilde{s}_{xx} - \frac{\widetilde{s}_{xy}}{\hat{\beta}_1}, \qquad 2\hat{\sigma}_\varepsilon^2 = \widetilde{s}_{yy} - \hat{\beta}_1\widetilde{s}_{xy},$$
$$2\hat{\xi}_i = x_i + \frac{y_i - \beta_0}{\hat{\beta}_1}, \tag{1.54}$$

where

$$\widetilde{s}_{xx} = \frac{1}{n}\sum x_i^2, \qquad \widetilde{s}_{yy} = \frac{1}{n}\sum\left(y_i - \beta_0\right)^2, \qquad \widetilde{s}_{xy} = \frac{1}{n}\sum x_i\left(y_i - \beta_0\right).$$

These estimates have all the problems of the estimates (1.44); in particular, they are not consistent and the estimate of β_1 does not make any sense. Again the structural ML estimates for β_0 known provide consistent estimates for this corresponding functional case (see Section 2.1.2).

Comments on maximum likelihood for the functional model.

There has been some confusion in the literature concerning the existence and consistency of ML estimates for the functional model under the variety of error variance assumptions that we have been discussing. As discussed in Section 2.1.2, when the ratio of error variances is known, the ML estimators for the intercept and slope are consistent, but the ML estimator for the unknown error variance is not. On the other hand, when one of the error variances is known, ML estimation fails completely. Under both assumptions, the ratio of the error variances known and one error variance known, the number of observations and the number of parameters are the same. The lack of consistency (of the estimator of error variance) under the first assumption and the breakdown of ML estimation under the second assumption are both related to the fact that ML estimation has problems when the number of incidental parameters increases with sample size. The consistency of the estimators for the intercept and slope in the known error variance ratio case is a matter of good fortune. See the comments in Section 2.1.2.

Example 1.11
Brown (1957) gave nine observations

x :	1.8	4.1	5.8	7.5	9.3	10.6	13.4	14.7	18.9
y :	6.9	12.5	20.0	15.7	24.9	23.4	30.2	35.6	39.1

which were generated from a true linear functional relationship, (1.5)–(1.6), with $\beta_1 = 2$ and $\sigma_\delta^2 = \sigma_\varepsilon^2 = 1$. Thus $\lambda = 1$ and $n = 9$. Furthermore,

$$\sum x_i = 86.1, \qquad \sum y_i = 208.3, \qquad \bar{x} = 9.57, \qquad \bar{y} = 23.14$$

and, rounding to four figures,

$$ns_{xx} = 237.9, \qquad ns_{yy} = 905.4, \qquad ns_{xy} = 451.3.$$

Thus (1.30) gives

$$\hat{\beta}_1 = \frac{(905.4 - 237.9) + \left\{(905.4 - 237.9)^2 + 4 \times 451.3^2\right\}^{1/2}}{2 \times 451.3} = 1.983,$$

and $\hat{\beta}_0 = 23.14 - 1.983 \times 9.57 = 4.163$. The adjusted estimator of σ_ε^2, (2.2), is

$$\widetilde{\sigma}_\varepsilon^2 = \frac{2n}{n-2}\hat{\sigma}_\varepsilon^2 = \frac{9}{7\left(1 + \hat{\beta}_1^2\right)}\left(s_{yy} - 2\hat{\beta}_1 s_{xy} + \hat{\beta}_1^2 s_{xx}\right) = 1.48.$$

The estimator $\hat{\beta}_1$ has performed rather well even with n as low as 9. On the other hand, the estimator of σ_ε^2 seems somewhat unsatisfactory. If the estimator obtained in the corresponding structural relationship is used, then

$$\widehat{\sigma}_\varepsilon^2 = \frac{1}{\left(1 + \hat{\beta}_1^2\right)}\left(s_{yy} - 2\hat{\beta}_1 s_{xy} + \hat{\beta}_1^2 s_{xx}\right) = \frac{7}{9}1.48 = 1.15,$$

which is better than the adjusted estimator.

It is also interesting to compute the least-squares lines of y on x and x on y for comparison. This gives

$$\hat{\beta}_{1R} = \frac{s_{xy}}{s_{xx}} = \frac{451}{238} = 1.90, \quad \text{and} \quad \frac{1}{\hat{\beta}_{1I}} = \frac{s_{yy}}{s_{xy}} = \frac{906}{451} = 2.01.$$

Thus the ME model regression lines are between the y on x and x on y regression lines, as (1.28) indicates. Note that in this example all three lines are reasonably close.

1.3.3 Maximum likelihood estimation for the ultrastructural relationship

The simple normal ultrastructural model is given by (1.5) and (1.6) with

$$\xi_i \sim N\left(\mu_i, \sigma^2\right), \qquad i = 1, \ldots, n.$$

There are $n + 5$ unknown parameters for this model: the 'structural parameters' are β_0, β_1, σ^2, σ_δ^2, and σ_ε^2; and the 'incidental parameters' are $\mu_1, \ldots, \mu_n$. In order to find the ML estimates (if they exist), one must maximize the likelihood function over this $(n + 5)$-dimensional parameter space. This parameter space is not all of $\Re^{n+5}$ ($(n + 5)$-dimensional Euclidean space), but it has a boundary in three of the dimensions because $\sigma^2 \geq 0$, $\sigma_\delta^2 \geq 0$, and $\sigma_\varepsilon^2 \geq 0$. Because of the presence of the incidental parameters, which increase in number with the sample size, ML estimation for the ultrastructural model behaves more like that for the functional model than the structural model.

Case (a): Ultrastructural ML estimation when both λ and ν are known.

Dolby (1976) derived the ML estimates for β_0 and β_1 when λ and $\nu = \sigma^2/\sigma_\delta^2$ are known. Using standard ML techniques a rather tedious calculation results in a quintic equation for $\hat{\beta}_1$,

$$\begin{aligned} h\left(\hat{\beta}_1\right) &= \nu s_{xx}\hat{\beta}_1^5 + \left(3\nu\lambda s_{xx} - \nu s_{yy}\right)\hat{\beta}_1^3 - 2\lambda\left(\nu - 1\right)s_{xy}\hat{\beta}_1^2 \\ &\quad + \left\{2\lambda^2\left(\nu + 1\right)s_{xx} - \lambda\left(\nu + 2\right)s_{xx}\right\}\hat{\beta}_1 - 2\lambda^2(\nu + 1)s_{xy} \\ &= 0, \end{aligned} \tag{1.55}$$

which generally must be solved by iterative numerical methods. There is at least one solution of (1.55) because all odd-order polynomials must have at least one real root. Given the solution, $\hat{\beta}_1$, the estimate for β_0 is given by (1.31a), and the estimate for σ_δ^2 is

$$\hat{\sigma}_\delta^2 = \frac{s_{yy} - 2\hat{\beta}_1 s_{xy} + \hat{\beta}_1^2 s_{xx}}{2\left(\lambda + \hat{\beta}_1^2\right)}.$$

The consistency of the estimate of β_1 is treated in Section 2.1.3.

Case (b): Ultrastructural ML estimation when λ is known.

Dolby (1976) showed that, in the case where only λ is known, there is no solution to the likelihood equations in the interior of the parameter space for the normal ultrastructural model; however, Gleser (1985, Theorem 1) showed that the likelihood is maximized on the $\sigma^2 = 0$ boundary of the parameter set. The parameter, σ^2, is not identifiable in the ultrastructural model. The $\sigma^2 = 0$ ultrastructural model is just the functional model. Thus, the ML estimates for the ultrastructural model are just the ML estimates for the functional model. The argument goes as follows.

For the normal model,

$$(x_i, y_i) \sim N\left\{\left(\mu_i, \beta_0 + \beta_1\mu_i\right), \mathbf{\Omega}\right\},$$

where

$$\mathbf{\Omega} = \sigma_\delta^2 \begin{bmatrix} \nu + 1 & \beta_1\nu \\ \beta_1\nu & \beta_1^2\nu + 1 \end{bmatrix}.$$

The log-likelihood function is

$$L\left(\beta_0, \beta_1, \sigma_\delta^2, \boldsymbol{\mu}, \nu\right) \propto -n \log \sigma_\delta^2 - \frac{n}{2}\left(\lambda + \nu\lambda + \nu\beta_1^2\right) - \frac{Q\left(\beta_0, \beta_1, \boldsymbol{\mu}, \nu\right)}{2\sigma_\delta^2}, \tag{1.56}$$

where $\boldsymbol{\mu} = (\mu_1, \ldots, \mu_n)'$ and

$$Q(\beta_0, \beta_1, \boldsymbol{\mu}, \nu) = \frac{\lambda \sum(x_1 - \mu_i)^2 + \sum(y_i - \beta_0 - \beta_1\mu_i)^2 + \nu \sum(y_i - \beta_0 - \beta_1 x_i)^2}{\lambda + \nu\lambda + \nu\beta_1^2}.$$

If β_0, β_1, σ_δ^2, and ν are fixed, then maximizing the likelihood over $\boldsymbol{\mu}$ is equivalent to minimizing $Q(\beta_0, \beta_1, \boldsymbol{\mu}, \nu)$ over $\boldsymbol{\mu}$. This minimum occurs when

$$\hat{\mu}_i = \frac{\lambda x_i + \beta_1\left(y_i - \beta_0\right)}{\lambda + \beta_1^2}.$$

Now substitute $\hat{\boldsymbol{\mu}} = \left(\mu_1, \ldots, \mu_n\right)'$ for $\boldsymbol{\mu}$ in Q and simplify to obtain (recall equation (1.18))

$$Q\left(\beta_0, \beta_1, \hat{\boldsymbol{\mu}}, \nu\right) = \frac{\sum\left(y_i - \beta_0 - \beta_1 x_i\right)^2}{\lambda + \beta_1^2}. \tag{1.57}$$

Note that (1.57) is independent of ν and

$$L\left(\beta_0, \beta_1, \sigma_\delta^2, \hat{\boldsymbol{\mu}}, \nu\right) \propto -n \log \sigma_\delta^2 - \frac{n}{2}\left(\lambda + \nu\lambda + \nu\beta_1^2\right) - \frac{\sum\left(y_i - \beta_0 - \beta_1 x_i\right)^2}{2\sigma_\delta^2\left(\lambda + \beta_1^2\right)}$$

is decreasing in ν for fixed β_0, β_1, and σ_δ^2, and hence is maximized over ν by $\hat{\nu} = 0$.

Thus one needs to maximize

$$L\left(\beta_0,\ \beta_1,\ \sigma_\delta^2,\ \hat{\boldsymbol{\mu}},\ \nu\right) \propto -n \log \sigma_\delta^2 - \frac{\sum\left(y_i - \beta_0 - \beta_1 x_i\right)^2}{2\sigma_\delta^2\left(\lambda + \beta_1^2\right)}.$$

The solution to the maximization problem above can be easily found as (1.31a),

$$\hat{\mu}_i = \frac{\lambda x_i + \hat{\beta}_1\left(y_i - \hat{\beta}_0\right)}{\lambda + \hat{\beta}_1^2}, \qquad \hat{\nu} = 0, \qquad \hat{\sigma}_\delta^2 = \frac{s_{yy} - 2\hat{\beta}_1 s_{xy} + \hat{\beta}_1^2 s_{xx}}{2\left(\lambda + \beta_1^2\right)},$$

and (1.30).

The inconsistency of the unknown error variance, σ_δ^2 (or σ_ε^2), persists, and no consistent estimator of σ^2 can exist (Gleser, 1985).

Case (c)–(d): Ultrastructural LE estimation when σ_δ^2 and/or σ_ε^2 is known.

Cheng and Van Ness (1991) showed that Gleser's results extend to the case where σ_δ^2 and/or σ_ε^2 is known and σ^2 is not specified, that is, in order to maximize the likelihood function for ultrastructural model, one only need consider the case $\sigma^2 = 0$. Therefore, the ML approach breaks down as it does in the functional case.

This does not mean that consistent estimates do not exist. Consistent method of moments estimators have been given (see Section 2.1.3).

If both σ_δ^2 and σ_ε^2 are known, then the ML estimators are the same as those for the λ known case.

1.4 The ME Model with Correlated Errors

In the previous sections, we required the ME model, (1.2)–(1.3), to have independent measurement errors, δ and ε. This assumption was adopted by most researchers until the 1970s. We now consider the measurement error model when the errors δ and ε are correlated.

1.4.1 The structural model

First look at the structural model. Again suppose that $(\delta_i, \varepsilon_i)$ are jointly normally distributed and that ξ_i and $(\delta_i, \varepsilon_i)$ are independent; (1.22) remains true except for the last term, which becomes

$$\operatorname{cov}(x, y) = \beta_1 \sigma^2 + \sigma_{\delta\varepsilon} = \beta_1 \sigma^2 + \rho_{\delta\varepsilon}\sigma_\delta\sigma_\varepsilon. \tag{1.58}$$

One more parameter, $\sigma_{\delta\varepsilon}$ or $\rho_{\delta\varepsilon}$, has been added, which is respectively the covariance or correlation of δ and ε. There are now seven unknown parameters, so that if β_1 was unidentifiable before, it remains so. If σ_δ^2 is known, σ^2 is identifiable, but $\beta_0, \beta_1, \sigma_\varepsilon^2,$

and $\sigma_{\delta\varepsilon}$ are not. If $\lambda = \sigma^2_\varepsilon/\sigma^2_\delta$ is known, the same is true, while if both σ^2_δ and σ^2_ε are known, β_1^2 is identifiable but β_1 is not. In this last case, the ML estimator of β_1^2 is

$$\hat{\beta}_1^2 = \frac{s_{yy} - \sigma^2_\varepsilon}{s_{xx} - \sigma^2_\delta},$$

provided $s_{xx} > \sigma^2_\delta$ and $s_{yy} \geq \sigma^2_\varepsilon$. Note that, now, the sign of β_1 is not determined by the sign of s_{xy} alone. This is because s_{xy} estimates (1.58), and, if $\sigma_{\delta\varepsilon}$ and $\beta_1$have opposite sign, then s_{xy} and β_1 can have opposite sign.

A commonly used assumption, which can be regarded as a generalization of the case of known ratio of error variance, is that the error covariance matrix is known up to a scalar multiple; that is, it is known that

$$\text{cov}\left(\begin{pmatrix} \delta \\ \varepsilon \end{pmatrix}\right) = \mathbf{\Omega} = \begin{pmatrix} \sigma^2_\delta & \sigma_{\delta\varepsilon} \\ \sigma_{\delta\varepsilon} & \sigma^2_\varepsilon \end{pmatrix} = d^2\mathbf{\Omega}_0, \tag{1.59}$$

where d is an unknown nonzero constant and $\mathbf{\Omega}_0$ is a known positive definite matrix. First observe that (1.59) is equivalent to knowing λ and $\rho_{\delta\varepsilon}$. Thus, it may be assumed, without loss of generality, that the unknown factor d^2 in (1.59) equals σ^2_δ, which implies that

$$\text{cov}\left(\begin{pmatrix} \delta \\ \varepsilon \end{pmatrix}\right) = \mathbf{\Omega} = \sigma^2_\delta \begin{bmatrix} 1 & \rho_{\delta\varepsilon}\lambda^{1/2} \\ \rho_{\delta\varepsilon}\lambda^{1/2} & \lambda \end{bmatrix}. \tag{1.60}$$

One can proceed with the ML estimation directly or one can use a transformation of the data to make the model into an independent errors ME model with known ratio of error variance, $\lambda = 1$. This latter technique was proposed by Gleser (1981, p. 42) with the transformation of (x_i, y_i) to $(x_i^*, y_i^*) = (x_i, y_i)\mathbf{A}$, where $\mathbf{A}$ is the Cholesky decomposition of $\mathbf{\Omega}_0^{-1}$, which is an upper triangular matrix with elements t_{11}, t_{12} and t_{22}. Then (x_i^*, y_i^*) has error covariance matrix $d^2\mathbf{I}_2$, where $\mathbf{I}_2$ is the two-dimensional identity matrix. Let β_0^* and β_1^* be the intercept and slope of this transformed model. Then, using the identities

$$\beta_0^* = t_{22}\beta_0 \quad \text{and} \quad \beta_1^* = t_{12}t_{11}^{-1} + t_{22}t_{11}^{-1}\beta_1,$$

one can recover the estimators of β_0 and β_1.

Another way to obtain ML estimators under the assumption (1.59) is to use the same technique as in Section 1.3 (Jones, 1979). Thus, (1.24) remains unchanged except that (e) becomes

$$(e') \qquad \hat{\beta}_1\hat{\sigma}^2 + \rho_{\delta\varepsilon}\lambda^{1/2}\hat{\sigma}^2_\delta = s_{xy}. \tag{1.61}$$

Further algebraic manipulations yield

$$\begin{aligned} P &= s_{xy} - \rho_{\delta\varepsilon}\lambda^{1/2}s_{xx} = \hat{\beta}_1\hat{\sigma}^2 - \rho_{\delta\varepsilon}\lambda^{1/2}\hat{\sigma}^2, \\ Q &= s_{yy} - \lambda s_{xx} = \hat{\beta}_1^2\hat{\sigma}^2 - \lambda\hat{\sigma}^2, \\ R &= \lambda s_{xy} - \rho_{\delta\varepsilon}\lambda^{1/2}s_{yy} = \lambda\hat{\beta}_1\hat{\sigma}^2 - \rho_{\delta\varepsilon}\lambda^{1/2}\hat{\beta}_1^2\hat{\sigma}^2. \end{aligned}$$

These relations give

$$P\hat{\beta}_1^2 - Q\hat{\beta}_1 - R = 0, \tag{1.62}$$

whose solutions are

$$\hat{\beta}_1 = \frac{Q \pm \left\{Q^2 + 4PR\right\}^{1/2}}{2P}.$$

Note that when $\rho_{\delta\varepsilon} = 0$, the estimator of β_1 should reduce to (1.30), so one should take the positive square root as the estimator, namely,

$$\hat{\beta}_1 = \frac{Q + \left\{Q^2 + 4PR\right\}^{1/2}}{2P}. \tag{1.63}$$

It is easy to see that if only one error variance is known, β_1 is not identifiable; however, if one knows the error covariance in addition to the error variance, then β_1 is identifiable. If σ_δ^2 and $\sigma_{\delta\varepsilon}$ are known, then

$$\hat{\beta}_1 = \frac{s_{xy} - \sigma_{\delta\varepsilon}}{s_{xx} - \sigma_\delta^2} \tag{1.64}$$

is the ML estimator of β_1 provided $s_{xx} - \sigma_\delta^2 > 0$ and $s_{yy} \geq (s_{xy} - \sigma_{\delta\varepsilon})^2/(s_{xx} - \sigma_\delta^2)$. On the other hand, if σ_ε^2 and $\sigma_{\delta\varepsilon}$ are known, then the ML estimator of β_1 is

$$\hat{\beta}_1 = \frac{s_{yy} - \sigma_\varepsilon^2}{s_{xy} - \sigma_{\delta\varepsilon}},$$

provided $s_{yy} > \sigma_\varepsilon^2$ and $s_{xx} \geq (s_{xy} - \sigma_{\delta\varepsilon})^2/(s_{yy} - \sigma_\varepsilon^2)$.

If the covariance matrix, $\mathbf{\Omega}$, is completely known, the model corresponds to the overidentified ME model (case (e) of the structural model). Thus, the estimate of β_1 is (1.63) if additional restrictions are imposed.

1.4.2 The functional model

Now consider ML estimation for the functional relationship when (δ, ε) are normal and $\mathbf{\Omega}$ is known completely. The log-likelihood function is

$$\begin{aligned} L\left(\beta_0, \beta_1, \xi_1, \ldots, \xi_n\right) \;\propto\; & -\frac{1}{2}\sum \left(x_i - \xi_i\,, y_i - \beta_0 - \beta_1\xi_i\right) \\ & \times\mathbf{\Omega}^{-1}\left(x_i - \xi_i\,, y_i - \beta_0 - \beta_1\xi_i\right)', \end{aligned} \tag{1.65}$$

where $\mathbf{\Omega}$ is the covariance matrix of (δ, ε). Taking derivatives of (1.65) with respect to ξ_i, β_0 and β_1 yields, after further algebraic manipulations,

$$\hat{\xi}_i = x_i - \frac{\left(\sigma_{\delta\varepsilon} - \hat{\beta}_1\sigma_\delta^2\right)\left(y_i - \hat{\beta}_0 - \hat{\beta}_1 x_i\right)}{\sigma_\varepsilon^2 - 2\hat{\beta}_1\sigma_{\delta\varepsilon} + \hat{\beta}_1^2\sigma_\delta^2}, \tag{1.66}$$

$$\sum\left(y_i - \hat{\beta}_0 - \hat{\beta}_1 x_i\right) = 0, \tag{1.67}$$

and

$$\sum \left(y_i - \hat{\beta}_0 - \hat{\beta}_1 x_i\right) \hat{\xi}_i = 0. \tag{1.68}$$

Solving (1.66) through (1.68) gives (1.31a) and

$$\hat{\beta}_1^2 \left(\sigma_{\delta\varepsilon} s_{xx} - \sigma_\delta^2 s_{xy}\right) - \hat{\beta}_1 \left(\sigma_\varepsilon^2 s_{xx} - \sigma_\delta^2 s_{yy}\right) - \left(\sigma_{\delta\varepsilon} s_{yy} - \sigma_\varepsilon^2 s_{xy}\right) = 0. \tag{1.69}$$

If one divides (1.69) by σ_δ^2, one obtains (1.61) and the ML estimator of β_1 is (1.63). Looking back, then, we see that for the functional model, we need only know the error covariance matrix up to a scalar multiple. Moreover, the ML estimators of ξ_i, β_0, and β_1 for both cases, knowing $\mathbf{\Omega}$ completely and up to a scalar multiple, are the same. It is left as an exercise (see Exercise 1.8) to show, in the latter case, the ML estimator for the unknown scalar multiple σ_δ^2 of (1.60) is not consistent, which is similar to the case of knowing λ in the classical model with the errors δ and ε independent.

One of the important properties often cited is that the ML estimator of the slope of the ME regression line is between the least-squares regression coefficient of y on x and the reciprocal of the least-squares regression coefficient of x on y. This is no longer true when the measurement errors $(\delta_i, \varepsilon_i)$ are correlated. In other words, when the measurement errors are no longer independent, then (1.28) fails and hence the estimated ME model regression line is not necessarily between the two regression lines. This is true for both the structural and functional models.

1.5 The Equation Error Model

Equation error models are a modification of the standard ME model which are commonly used, for example, in many areas of econometrics (Johnston, 1972, p. 287). For these models the usefulness of the various identifiability assumptions changes from that for the standard ME models.

Consider the situation in which the true values, ξ and η, are not perfectly linearly related and there is an error in the equation, q. In other words, (1.5) is replaced by

$$\eta_i = \beta_0 + \beta_1 \xi_i + q_i, \tag{1.70}$$

where the q_i are independent random variables with mean zero and positive variance, σ_q^2, and q_i is independent of ξ_j for all i and j. Furthermore, it is assumed that $(\delta_i, \varepsilon_i)$ and (q_j, ξ_j) are independent for all i and j. The model (1.70) and (1.3) is called the equation error model or the error-in-the-equation model. Many authors write the equation error model as

$$x_i = \xi_i + \delta_i, \quad y_i = \beta_0 + \beta_1 \xi_i + e_i, \qquad i = 1, \ldots, n, \tag{1.71}$$

where e_i is the combined measurement error of the dependent variable and the equation error, that is, $e_i = \varepsilon_i + q_i$. Thus,

$$\operatorname{cov}\left(\begin{pmatrix} \delta \\ e \end{pmatrix}\right) = \operatorname{cov}\left(\begin{pmatrix} \delta \\ \varepsilon \end{pmatrix}\right) + \begin{pmatrix} 0 & 0 \\ 0 & \sigma_q^2 \end{pmatrix} = \mathbf{\Omega} + \begin{pmatrix} 0 & 0 \\ 0 & \sigma_q^2 \end{pmatrix}.$$

The equation error model looks like it could be absorbed into the model (1.2)–(1.3). The difference is in the way one views q_i. It is assumed that, in practice, one cannot obtain information about q_i, such as σ_q^2, from other experiments or replications. This means that some of the additional assumptions used to obtain identifiability and/or consistent estimates are no longer practical. The typical assumption used for the equation error model is that σ_q^2 is unknown but the measurement errors covariance matrix,

$$\mathbf{\Omega} = \begin{pmatrix} \sigma_\delta^2 & \sigma_{\delta\varepsilon} \\ \sigma_{\delta\varepsilon} & \sigma_\varepsilon^2 \end{pmatrix},$$

is known or estimated from a separate experiment. More information about the equation error model can be found in Section 2.2, Fuller (1987, Chapter 2), and Carroll *et al.* (1995, Chapter 1).

1.5.1 The structural equation error model

In order to find the ML estimator of β_1 for the structural model, first assume that δ and ε are independent. Remember that a new parameter, σ_q^2, has been added, which might mean that at least some of the side assumptions (1.12) no longer imply the identifiability of β_1. Because σ_q^2 can be absorbed into $\sigma_e^2 = \sigma_\varepsilon^2 + \sigma_q^2$, which plays the role of σ_ε^2 in the no-equation-error model, the new parameter, σ_q^2, in some sense, is not really an independent parameter. Thus, some of the conditions in (1.12) lead to the identifiability of β_1 and others do not. Note that (1.24d) becomes

$$(d') \quad s_{yy} = \hat{\beta}_1^2 \hat{\sigma}^2 + \hat{\sigma}_\varepsilon^2 + \hat{\sigma}_q^2.$$

Assumption (1.12a) ($\lambda = \sigma_\varepsilon^2/\sigma_\delta^2$ known) does not lead to the identifiability of β_1. The case of known reliability ratio κ_ξ, (1.12b), results in the same estimators as in the no-equation-error model, namely (1.33). Note that σ_ξ^2 is no longer identifiable, but $\sigma_e^2 = \sigma_\varepsilon^2 + \sigma_q^2$ is, and, in fact, $\hat{\sigma}_e^2$ is the right-hand side of (1.34). If $\hat{\sigma}_e^2$ is negative then $\hat{\beta}_1 = 1/\hat{\beta}_{1I}$. The case (1.12c) ($\sigma_\delta^2$ known) has the same ML estimators as those for the no-equation-error model, that is, (1.35), (1.31c), and $\hat{\sigma}_e^2$ is obtained by replacing $\hat{\sigma}_\varepsilon^2$ in (1.37). The assumption (1.12d) does not make β_1 identifiable because σ_q^2 remains unknown. If both error variances σ_δ^2 and σ_ε^2 are known, then one still does not know σ_e^2 because σ_q^2 remains unknown; therefore, this is not an 'overidentified' situation. The resultant ML estimator of β_1 is the same as that for the σ_δ^2 known case. The only difference is that, with the extra information, σ_ε^2, one can now estimate the equation error variance using

$$\hat{\sigma}_q^2 = \hat{\sigma}_e^2 - \sigma_\varepsilon^2,$$

where $\hat{\sigma}_e^2 = s_{yy} - \hat{\beta}_1 s_{xy}$. Finally, the estimators for the intercept known case are the same as those for the no-equation-error model. Again neither the measurement error variance, σ_ε^2, nor σ_q^2 is identifiable in this case, but their sum σ_e^2 is.

If the measurement errors δ and ε are correlated, then another parameter, $\sigma_{\delta\varepsilon}$, is added. One needs to know either

(i) both σ_δ^2 and $\sigma_{\delta\varepsilon}$, or
(ii) the intercept β_0 and that $Ex \neq 0$.

For case (i), the ML estimator is easily obtained, giving (1.64) provided $s_{xx} - \sigma_\delta^2 > 0$ and $s_{yy} \geq \sigma_\varepsilon^2 + (s_{xy} - \sigma_{\delta\varepsilon})^2/(s_{xx} - \sigma_\delta^2)$. Note that knowledge of σ_ε^2 is also needed if one must verify the latter inequality. The estimator of β_1, under (ii), remains unchanged from the uncorrelated model, because the estimation of β_1 only requires the first moment. None of the rest of the parameters, σ^2, σ_δ^2, $\sigma_{\delta\varepsilon}$, σ_ε^2, nor σ_q^2, is identifiable, however.

1.5.2 The functional equation error model

Next consider the functional equation error model with known

$$\mathbf{\Omega} = \begin{pmatrix} \sigma_\delta^2 & \sigma_{\delta\varepsilon} \\ \sigma_{\delta\varepsilon} & \sigma_\varepsilon^2 \end{pmatrix}$$

and unknown σ_q^2. There are then $n+3$ unknown parameters. Without loss of generality, one can take $\sigma_\delta^2 = \sigma_\varepsilon^2 = 1$ and $\sigma_{\delta\varepsilon} = 0$. Then the log-likelihood function becomes

$$\begin{aligned} L\left(\beta_0, \beta_1, \sigma_q^2, \xi_1, \ldots, \xi_n\right) \quad \propto \quad & -\frac{n}{2}\log\left(1+\sigma_q^2\right) - \frac{1}{2}\sum\left(x_i - \xi_i\right)^2 \\ & -\frac{1}{2\left(1+\sigma_q^2\right)}\sum\left(y_i - \beta_0 - \beta_1\xi_i\right)^2 . \end{aligned} \tag{1.72}$$

Taking derivatives of (1.72) with respect to β_0, β_1, σ_q^2, and ξ_i and equating them to zero yields

$$\begin{aligned} \sum\left(y_i - \hat\beta_0 - \hat\beta_1\hat\xi_i\right) &= 0, \\ \sum\left(y_i - \hat\beta_0 - \hat\beta_1\hat\xi_i\right)\hat\xi_i &= 0, \\ \sum\left(y_i - \hat\beta_0 - \hat\beta_1\hat\xi_i\right)^2 &= n\left(1+\hat\sigma_q^2\right), \end{aligned} \tag{1.73}$$

and

$$\left(1+\hat\sigma_q^2+\hat\beta_1^2\right)\hat\xi_i = \left(1+\hat\sigma_q^2\right)x_i + \left(y_i - \hat\beta_0\right)\hat\beta_1 .$$

Then

$$y_i - \hat\beta_0 - \hat\beta_1\hat\xi_i = \left(1+\hat\sigma_q^2+\hat\beta_1^2\right)^{-1}\left(1+\hat\sigma_q^2\right)\left(y_i - \hat\beta_0 - \hat\beta_1 x_i\right). \tag{1.74}$$

Substitute (1.74) into (1.73) to obtain

$$\left(1+\hat\sigma_q^2+\hat\beta_1^2\right)^{-1}\left\{\frac{1}{n}\sum\left(y_i - \hat\beta_0 - \hat\beta_1 x_i\right)^2\right\} = \left(1+\hat\sigma_q^2\right)^{-1}\left(1+\hat\sigma_q^2+\hat\beta_1^2\right). \tag{1.75}$$

Because

$$y_i - \beta_0 - \beta_1 x_i = \varepsilon_i + q_i - \beta_1 \delta_i,$$

the left-hand side of (1.75) should be estimating unity, but the right-hand side of (1.75) obviously is not estimating unity. Thus, the ML method fails to yield consistent estimators for all parameters in this case. Consistent estimation for the equation error functional model is discussed in Section 2.2.

1.6 The Berkson Model

Berkson (1950) proposed a different regression model with measurement errors, which is appropriate in many applications. Suppose that one is conducting an experiment to determine the relationship between η and ξ, in which x is fixed at a certain series of values, and then the corresponding values of y are measured. For example, in determining the relation between the extension and tension in a spring, one could hang weights x of 10 grams, 20 grams, 30 grams, ..., 100 grams and measure the extensions, y, with a random error ε acting on the true value η. On the other hand, the weights can be also be imperfect, and in attaching a nominal weight of $x = 50$ grams one is, in fact, attaching a weight ξ with error $\delta = 50 - \xi$. Under repetitions of such experiments with different weights, each purporting to be 50 grams, one is really applying a series of true weights ξ_i with errors $\delta_i = 50 - \xi_i$. Thus, the real weights applied are the values of a random variable ξ. The variable, x, is called a **controlled** variable, because its values are fixed in advance, whereas the unknown true values, ξ, are fluctuating.

Suppose that the errors, δ_i, have zero mean. This implies that ξ in the above example has a mean of $50 = x$. Then

$$x_i = \xi_i + \delta_i,$$

where ξ_i and δ_i are perfectly negatively correlated. This is a major difference from the ME model because, in the ME model, ξ_i and δ_i are independent. Thus, the simple linear Berkson model becomes

$$y_i = \beta_0 + \beta_1 \xi_i + \varepsilon_i = \beta_0 + \beta_1 x_i + (\varepsilon_i - \beta_1 \delta_i), \tag{1.76}$$

which is of the same form as (1.10), but radically different. Now x is not random, and neither ε nor δ is correlated with x. Furthermore,

$$E\left(y_i|x_i\right) = E\left(\beta_0 + \beta_1 \xi_i + \varepsilon_i|x_i\right) = \beta_0 + \beta_1 x_i,$$

just as in ordinary linear regression. If it is supposed, as before, that δ_i has the same distribution for all x_i, then (1.76) is a regression equation to which the ordinary least-squares methods can be applied without modification (see Theorem 1.1), and β_0 and β_1 can be estimated and tested without difficulty. This is a regression-type model in which the errors are a function of β_1.

Because least squares provides consistent estimates of β_0 and β_1, these parameters are always identifiable for the Berkson model. However, σ_ε^2 and σ_δ^2 are not identifiable without some side condition. Least squares only provides a consistent estimate of $\text{var}(\varepsilon - \beta_1\delta)$.

There is no doubt that, in many experimental situations, the Berkson model is valid. The variable, x, is often an instrument reading, and the experimenter often tries to hold x to certain preassigned values (not chosen at random, but to cover a specified range adequately). It is known, however, that the instrument is subject to error and does not read the true values, ξ, precisely.

Example 1.12

In many medical and agricultural experiments, the Berkson model might be appropriate.

Suppose one wishes to study the efficacy of a drug at various dosages. One could inject the drug in various amounts, say $x = 0.5\ \text{cm}^3$, $1.0\ \text{cm}^3$, $1.5\ \text{cm}^3$, etc.; but the actual concentrations, ξ, of the drug in the bloodstream would depend on the subject's size, physical activity, physical make-up and so on. If it is blood concentration that is important, then the Berkson model should be considered.

Suppose one wishes to study the effects of a fertilizer on a crop. One could set dials on the fertilizing device at various levels, x, but the actual amount available to the crop, ξ, will depend on many random things such as rainfall, wind patterns, variations in soil composition, and topography.

There are many important differences between the behaviour of estimates for this model and the ME model. For example, the usual estimates of the βs have finite moments for finite sample sizes, which is not true for the usual estimates for the ME model.

The polynomial Berkson model is discussed in Section 6.6.

1.6.1 Least-squares estimation for the Berkson model

For the Berkson model, the ML estimators for β_0 and β_1 can be different from the least-squares estimators when the errors are normally distributed because both the mean and the variance of the y_i are functions of β_1. The usual regression least-squares estimates are also the least-squares estimates for the Berkson model because the Gauss–Markov theorem still holds for the Berkson model (Cheng, 1994):

Theorem 1.1 *Suppose the Berkson model, (1.76), holds with* $E\delta_i = E\varepsilon_i = 0$, $\sigma^2_{\delta_i} = \sigma^2_\delta$, $\sigma^2_{\varepsilon_i} = \sigma^2_\varepsilon$, $\text{cov}(\delta_i, \varepsilon_i) = 0$, *and* $\text{cov}(\delta_i, \delta_j) = \text{cov}(\varepsilon_i, \varepsilon_j) = 0$, $i \neq j$. *Then, the least-squares estimates,*

$$\hat{\beta}_0 = \bar{y} - \hat{\beta}_1\bar{x}. \tag{1.77}$$

and

$$\hat{\beta}_1 = s_{xy}/s_{xx}, \tag{1.78}$$

are uniform minimum variance among all linear unbiased estimates.

Note that $\hat{\beta}_1$ is unbiased for the Berkson model. Because the x_i are not random,

$$\begin{aligned} E\hat{\beta}_1 &= E\frac{\sum(x_i-\bar{x})(y_i-\bar{y})}{\sum(x_i-\bar{x})^2} = \frac{\sum(x_i-\bar{x})E(y_i-\bar{y})}{\sum(x_i-\bar{x})^2} \\ &= \frac{\sum(x_i-\bar{x})E\left(\beta_0+\beta_1x_i+(\varepsilon_i-\beta_1\delta_i)-\beta_0-\beta_1\bar{x}_i-(\bar{\varepsilon}_i-\beta_1\bar{\delta}_i)\right)}{\sum(x_i-\bar{x})^2} \\ &= \beta_1. \end{aligned}$$

1.6.2 Maximum likelihood estimation for the Berkson model

For maximum likelihood, assume that

$$\delta_i \sim N(0,\sigma_\delta^2), \quad \varepsilon_i \sim N(0,\sigma_\varepsilon^2), \qquad i=1,\ldots,n,$$

where δ_i and ε_i are independent; then y_i is normal with mean $\beta_0+\beta_1x_i$ and variance $\sigma_\varepsilon^2+\beta_1^2\sigma_\delta^2$. The log-likelihood function is

$$L(\beta_0,\beta_1,\sigma_\delta^2,\sigma_\varepsilon^2) \propto -\frac{n}{2}\log(\sigma_\varepsilon^2+\beta_1^2\sigma_\delta^2) - \frac{1}{2}\sum\frac{(y_i-\beta_0-\beta_1x_i)^2}{\sigma_\varepsilon^2+\beta_1^2\sigma_\delta^2}.$$

Solving $\partial L/\partial\beta_0=0$ gives (1.77). Measuring x_i and y_i from their means, the log-likelihood function can be simplified to

$$L(\beta_1,\sigma_\delta^2,\sigma_\varepsilon^2) \propto -\frac{n}{2}\log(\sigma_\varepsilon^2+\beta_1^2\sigma_\delta^2) - \frac{1}{2}\sum\frac{(y_i-\beta_1x_i)^2}{\sigma_\varepsilon^2+\beta_1^2\sigma_\delta^2}. \tag{1.79}$$

The estimates of β_1 can now be found for each of the separate cases.

Case 1: Neither σ_δ^2 nor σ_ε^2 known.

From (1.79), $\partial L/\partial\beta_1=0$ yields

$$\hat{\beta}_1\hat{\sigma}_\delta^2(\hat{\sigma}_\varepsilon^2+\hat{\beta}_1^2\hat{\sigma}_\delta^2-s_{yy}+\hat{\beta}_1s_{xy})-\hat{\sigma}_\varepsilon^2(s_{xy}-\hat{\beta}_1s_{xx})=0. \tag{1.80}$$

Both $\partial L/\partial\sigma_\delta^2=0$ and $\partial L/\partial\sigma_\varepsilon^2=0$ lead to the same equation:

$$\hat{\sigma}_\varepsilon^2+\hat{\beta}_1^2\hat{\sigma}_\delta^2 = s_{yy}-2\hat{\beta}_1s_{xy}+\hat{\beta}_1^2s_{xx}. \tag{1.81}$$

Combining (1.80) and (1.81) gives

$$(\hat{\beta}_1s_{xx}-s_{xy})(s_{yy}-2\hat{\beta}_1s_{xy}+\hat{\beta}_1^2s_{xx})=0.$$

Thus, either $\hat{\beta}_1s_{xx}-s_{xy}=0$, which means $\hat{\beta}_1=s_{xy}/s_{xx}$, or $s_{yy}-2\hat{\beta}_1s_{xy}+\hat{\beta}_1^2s_{xx}=0$. By the Cauchy–Schwarz inequality we have $s_{xx}s_{yy}\geq s_{xy}^2$, and the equality holds if and only if, for all i, $y_i=cx_i$ for some constant c. This implies that the equation $s_{yy}-2\hat{\beta}_1s_{xy}+\hat{\beta}_1^2s_{xx}=0$ has no real roots unless equality holds, in which case again $\hat{\beta}_1=s_{xy}/s_{xx}$. Note that only the relation (1.81) is known and there is no estimate for σ_δ^2 or σ_ε^2.

Case 2: Ratio of error variances, λ, known.

First replace σ_ε^2 by $\lambda\sigma_\delta^2$ in (1.79). Solving $\partial L/\partial\beta_1 = 0$ and $\partial L/\partial\sigma_\delta^2 = 0$ simultaneously gives (1.78) and

$$\hat{\sigma}_\delta^2 = \frac{s_{yy} - 2\hat{\beta}_1 s_{xy} + \hat{\beta}_1^2 s_{xx}}{\lambda + \hat{\beta}_1^2}. \tag{1.82}$$

Note that $\hat{\sigma}_\delta^2 \geq 0$.

Case 3: σ_δ^2 known.

Similar arguments to those in case 2 yield (1.78) and

$$\hat{\sigma}_\varepsilon^2 = s_{yy} - 2\hat{\beta}_1 s_{xy} + \hat{\beta}_1^2 s_{xx} - \hat{\beta}_1^2 \sigma_\delta^2, \tag{1.83}$$

provided $s_{xx}(s_{xx}s_{yy} - s_{xy}^2) \geq \sigma_\delta^2 s_{xy}^2$.

Case 4: σ_ε^2 known.

Again, similar arguments to those in case 2 give (1.78) and

$$\hat{\sigma}_\delta^2 = \frac{s_{yy} - 2\hat{\beta}_1 s_{xy} + \hat{\beta}_1^2 s_{xx} - \sigma_\varepsilon^2}{\hat{\beta}_1^2}, \tag{1.84}$$

provided $s_{xx}s_{yy} - s_{xy}^2 \geq \sigma_\varepsilon^2 s_{xx}$.

Case 5: Both of the error variances known.

Setting $\partial L/\partial\beta_1 = 0$ in (1.79), $\hat{\beta}_1$ is a root of the following equation

$$\hat{\beta}_1^3 \sigma_\delta^4 + \hat{\beta}_1^2 \sigma_\delta^2 s_{xy} + \hat{\beta}_1(\sigma_\delta^2\sigma_\varepsilon^2 + \sigma_\varepsilon^2 s_{xx} - \sigma_\delta^2 s_{yy}) - \sigma_\varepsilon^2 s_{xy} = 0. \tag{1.85}$$

Unfortunately, $\hat{\beta}_1 = s_{xy}/s_{xx}$ is not a solution to (1.85) unless σ_δ^2 and σ_ε^2 satisfy (1.81) with $\hat{\sigma}_\delta^2$ and $\hat{\sigma}_\varepsilon^2$ being replaced by σ_δ^2 and σ_ε^2, respectively.

Example 1.13
We now generate an artificial set of data to demonstrate the use of the methods of this section. We performed the following experiment:

1. Generate 20 values, $x_1, \ldots, x_{20}$: $x_i = 2i$.
2. Generate 20 independent values, $\delta_1, \ldots, \delta_{20}$, from $\delta \sim N(0, \sigma_\delta^2)$.
3. Generate 20 independent values, $\varepsilon_1, \ldots, \varepsilon_{20}$, from $\varepsilon \sim N(0, \sigma_\varepsilon^2)$.
4. Calculate the 20 values $\xi_1, \ldots, \xi_{20}$, where $\xi_i = x_i - \delta_i$.
5. Form the data

$$\begin{pmatrix} x_i \\ y_i \end{pmatrix} = \begin{pmatrix} x_i \\ \beta_0 + \beta_1\xi_i + \varepsilon_i \end{pmatrix}, \qquad i = 1, \ldots, 20.$$

Thus, the x_i are fixed and known, but the ξ_i are calculated from the x_i and the δ_i and are assumed unknown. Thus, these data perfectly fit the linear Berkson model. For $\beta_0 = 1$, $\beta_1 = 2$, and $\sigma_\delta^2 = \sigma_\varepsilon^2 = 16$, the data are given in Table 1.1.

Table 1.1 Twenty Berkson model data: $\beta_0 = 1,\ \beta_1 = 2,$ and $\sigma_\delta^2 = \sigma_\varepsilon^2 = 16$

	x	y		x	y		x	y
1	2	17.31	8	16	33.59	15	30	46.14
2	4	6.77	9	18	40.77	16	32	64.70
3	6	4.32	10	20	46.06	17	34	69.49
4	8	7.57	11	22	43.78	18	36	58.84
5	10	18.13	12	24	54.79	19	38	86.85
6	12	34.45	13	26	50.06	20	40	84.17
7	14	44.80	14	28	70.81			

For these data, we calculate

$$\bar{x} = 21, \qquad \bar{y} = 44.17,$$

$$s_{xx} = 133, \qquad s_{yy} = 575.36, \qquad s_{xy} = 258.79,$$

which gives

$$\hat{\beta}_1 = s_{xy}/s_{xx} = 1.946 \quad \text{and} \quad \hat{\beta}_0 = \bar{y} - \hat{\beta}_1\bar{x} = 3.30.$$

These are reasonable estimates for this sample size.

If we assume that $\lambda = 1$, is known, then by (1.82),

$$\hat{\sigma}_\delta^2 = \frac{575.36 - 2 \times 1.946 \times 258.79 + 1.946^2 \times 133}{1 + 1.946^2} = 15.00.$$

The structural ME model for λ known gives $\hat{\beta}_1 = 2.17$ for these data, slightly less accurate than the Berkson estimate. We know that the ME model estimate will be larger than the Berkson model estimate (least-squares estimate) by (1.20).

If we assume that $\sigma_\delta^2 = 16$ is known, then (1.83) yields $\hat{\sigma}_\varepsilon^2 = 11.23$. The structural ME model for σ_δ^2 known gives $\hat{\beta}_1 = 2.21$, which is also less accurate than the Berkson model estimate.

If we assume that $\sigma_\varepsilon^2 = 16$ is known, then (1.84) yields $\hat{\sigma}_\delta^2 = 14.74$. The structural ME model for σ_ε^2 known gives $\hat{\beta}_1 = 2.16$, which is again a less accurate estimate than the Berkson model estimate.

1.6.3 The Berkson model with equation error

The Berkson model can also be written with equation error. In this case, (1.76) becomes

$$y_i = \beta_0 + \beta_1\xi_i + q_i + \varepsilon_i = \beta_0 + \beta_1 x_i + (q_i + \varepsilon_i - \beta_1\delta_i),$$

where the q_i are independent and have mean zero and positive variance, σ_q^2. Furthermore q_i is independent of ε_j and δ_j for all j. Again, least-squares methods can be applied without modification. Maximum likelihood estimates for β_0 and β_1 can also be obtained for the normal linear Berkson model (see Exercise 1.15). If estimates of the error variances are needed, then one must be given appropriate side conditions.

1.6.4 Further comments

If the measurement errors δ and ε are correlated (with or without equation error), the least-squares method still applies to Berkson model (for the ML estimates of β_0 and β_1 for the normal Berkson model, see Exercise 1.16). Moreover, even if the values at which x is controlled are themselves random variables (i.e. determined by some process of random selection), the preceding analysis applies if the errors δ and ε are uncorrelated with x (but ξ and δ are correlated, of course). This latter assumption is usually fulfilled, but the former (δ uncorrelated with x) can be a problem. In terms of the spring example, suppose that one randomly selected the available weights, and used these for the experiment. The requirement that δ be uncorrelated with x now implies, for example, that the larger weights should not tend to have larger or smaller errors of determination in their nominal values than do the smaller weights. Whether this is so will depend on the situation.

It is comforting, after our earlier discussions of the difficulties with situations involving measurement errors, that the standard least-squares analysis works for this common experimental situation. This fact illustrates the point, which cannot be overly stressed, that a thorough analysis of the sources of error and of the nature of the observational process is essential in finding the correct inferential methods, and, can, as in this case, lead to a simple solution of an apparently difficult problem.

1.7 Maximum Likelihood Estimation of Transformed Data: Elimination of Nuisance Parameters

We have seen that ML estimation for the functional and ultrastructural relationships is not satisfactory. This is due to the presence of incidental parameters whose number increases with the sample size. More generally, in parametric problems with many nuisance parameters (which need not necessarily increase with sample size), the direct use of ML estimation of the parameters of interest is known to be unreliable because the ML estimator can have a distribution concentrated far from the true value (see, for example, Cox, 1993). This problem can be handled in several ways: for example, by using empirical Bayes estimation, Bayesian estimation, or by conditioning or marginalization, which produces a likelihood function depending primarily on the parameters of interest.

This section treats the ML estimation of transformed data in which the nuisance parameters are eliminated. Generally speaking, for a vector of observed random variables having density depending on a vector parameter $\boldsymbol{\theta} = (\boldsymbol{\theta}_1, \boldsymbol{\theta}_2)$, where $\boldsymbol{\theta}_1$ is the (vector) parameter of interest and $\boldsymbol{\theta}_2$ is the (vector) nuisance parameter, one can find another vector $\mathbf{Z} = \mathbf{Z}(\mathbf{X}, \boldsymbol{\theta}_1)$ whose density depends only on $\boldsymbol{\theta}_1$ and not on $\boldsymbol{\theta}_2$.

One can then work on the likelihood function of the transformed data $\mathbf{Z}$ instead of the original data $\mathbf{X}$. The idea of using a function of the observations rather than the observations themselves as the source of the likelihood functions seems to be due to Bartlett (1936) (see also Sprent, 1976). Nuisance parameters are parameters that are not of major interest and need not necessarily be incidental.

Example 1.14

Edwards (1992, p. 116) gave an example of the use of transformed data. Consider a random sample $\mathbf{X} = (X_1, \ldots, X_n)$ from a normal distribution with mean μ and variance σ^2. If our interest is in μ, then the nuisance parameter is σ^2 and vice versa. In the former case, one can use $t = (n-1)^{\frac{1}{2}} s^{-1}(\bar{x} - \mu)$, with $s^2 = n^{-1} \sum (x_i - \bar{x})^2$, which is distributed as Student's t with $n-1$ degrees of freedom. The resulting ML estimator of μ based on the distribution of t is $\bar{x}$ (that is, $\hat{\mu} = \bar{x}$ satisfies $\partial f_t / \partial \mu = 0$), which is the same as the ML estimator based on the original data $\mathbf{X}$. For the latter case, however, note that $v = ns^2/\sigma^2$ is distributed as chi-square with $n-1$ degrees of freedom. The resulting ML estimator of σ^2 based on the distribution of v is $ns^2/(n-1)$, which is unbiased, in contrast with the biased ML estimator s^2 derived from the original data $\mathbf{X}$.

Sprent (1976) and Cox (1993) investigated the transformed data approach for the functional relationship with uncorrelated measurement errors. In fact, their approach also works with correlated errors. Note that

$$z_i \equiv \frac{y_i - \beta_0 - \beta_1 x_i}{(\sigma_\varepsilon^2 - 2\beta_1 \sigma_{\delta\varepsilon} + \beta_1^2 \sigma_\delta^2)^{1/2}}$$

is normally distributed with mean zero and unit variance. Assuming the error covariance matrix is known, the log-likelihood function of $(z_1, \ldots, z_n)$ is

$$L(\beta_0, \beta_1) \propto -\frac{1}{2\left(\sigma_\varepsilon^2 - 2\beta_1 \sigma_{\delta\varepsilon} + \beta_1^2 \sigma_\delta^2\right)} \sum \left(y_i - \beta_0 - \beta_1 x_i\right)^2.$$

The resulting ML estimators are (1.31a) and $\hat{\beta}_1$, which satisfies (1.69), and the solution with positive square root achieves the maximum. In other words, the ML estimator of β_1 coincides with (1.63).

Supposing the error covariance matrix is known up to a scalar multiple, one may assume, without loss of generality, that the unknown factor is σ_δ^2. Let

$$u_i \equiv \frac{y_i - \beta_0 - \beta_1 x_i}{\left(\lambda - 2\beta_1 \rho_{\delta\varepsilon} \lambda^{1/2} + \beta_1^2\right)^{1/2}},$$

which is normally distributed with mean zero and variance σ_δ^2. Thus, the log-likelihood function of $(u_1, \ldots, u_n)$ becomes

$$L\left(\beta_0, \beta_1, \sigma_\delta^2\right) \propto -\frac{n}{2\sigma_\delta^2} - \frac{1}{2\sigma_\delta^2} \frac{\sum \left(y_i - \beta_0 - \beta_1 x_i\right)^2}{\lambda - 2\beta_1 \rho_{\delta\varepsilon} \lambda^{1/2} + \beta_1^2}.$$

It is easily seen that the ML estimators of β_0 and β_1 are the same as those derived from z_i, and

$$\hat{\sigma}_\delta^2 = \frac{1}{n} \frac{\sum \left(y_i - \hat{\beta}_0 - \hat{\beta}_1 x_i\right)^2}{\lambda - 2\hat{\beta}_1 \rho_{\delta\varepsilon} \lambda^{1/2} + \hat{\beta}_1^2}. \tag{1.86}$$

Note that (1.86) is a consistent estimator of σ_δ^2, in contrast with the inconsistent ML estimator of the untransformed data (cf. (1.52)).

Although both Sprent (1976) and Cox (1993) focused on the functional relationship, their approach actually works for the structural and ultrastructural relationships as well. There are a few points that merit our attention, however. First of all, the above approach works only for the case when the error covariance matrix is known at least up to a scalar multiple. It does not work for the case where one error variance is known; neither does it work for the equation error model. Second, there can be more than one way to select $\mathbf{Z}$; for example, $z_i' = y_i - \beta_0 - \beta_1 x_i$, which is normal with mean zero and variance $\sigma_\varepsilon^2 - 2\beta_1 \sigma_{\delta\varepsilon} + \beta_1^2 \sigma_\delta^2$, or $z_i'' = (y_i - \beta_0 - \beta_1 x_i)^2 / (\sigma_\varepsilon^2 - 2\beta_1 \sigma_{\delta\varepsilon} + \beta_1^2 \sigma_\delta^2)$, which is chi-square with one degree of freedom. Neither the ML estimation of z_i' nor that of z_i'' results in a sensible (for example, consistent) estimator of β_1.

Cox (1993) proposed that the ML estimating equation of the transformed data for the parameter of interest should be unbiased. He also gave conditions that result in an unbiased estimating equation; that is, the expectation of the defining equation is zero. For the measurement error model, only $\mathbf{Z}$, among $\mathbf{Z}$, $\mathbf{Z}'$ and $\mathbf{Z}''$, satisfies such conditions.

1.8 Bibliographic Notes and Discussion

We close this chapter with some bibliographic remarks. There is a vast literature on ME models. The aim here is not to give a complete bibliographic history but only to mention some additional topics and re-emphasize some items already mentioned.

The first paper on ME models (Adcock, 1877) concerns the least-squares fitting of a straight line when both variables are subject to error. Adcock's (1877; 1878) least-squares method is now known as orthogonal regression because he assumed that the error variances were equal (see Finney, 1996). Kummel (1879) extended Adcock's result (with a correction to Adcock's formulation) to the case where the ratio of error variances is known, and Pearson (1901) extended Adcock's solution to the multivariate case, although Pearson seemed to be unaware of the previous work by Adcock and Kummel. Basically, all these papers treated the problem by minimizing the perpendicular and/or the weighted distance between the observations and the line (surface). As we have seen in Section 1.3, the resulting estimator of the slope is the ML estimator under a normality assumption when the ratio of error variances is known. Orthogonal regression has been rediscovered many times in a variety of areas. Many people equate ME models and orthogonal regression, but the statistical model that corresponds to orthogonal regression is a very special case of ME model: the no-equation-error ME model with known ratio of error variances and scaled so that this ratio is one.

1.8.1 Error in the equation

A diversity of terminology evolved, namely, errors-in-variables (EV) models, regression with errors in the independent variables, functional or structural models, and measurement error models. The term *EV models* seems to be more popular in econometrics while the other terms are used most often in the statistical literature. Both Lindley (1947) and Kendall (1951) regarded *functional relationships* and *structural relationships* as referring to models without equation error. If the model had equation error, they referred to it as the *regression situation* (Madansky, 1959, p. 175). Although both Lindley and Kendall realized that equation error can occur in practice, they did not apply the notions of functional or structural relationships to the equation error situation. For a long while, most statistical literature on ME models was not concerned with equation error. For example, Madansky (1959), Moran (1971), Kendall and Stuart (1979), Anderson (1976; 1984), Sprent (1990) and Cheng and Van Ness (1994) were all focused on models without equation error. In this book, we use the terms 'functional' and 'structural' only to distinguish the nature of ξ: random or fixed. The models might or might not have equation error.

Many practitioners encounter ME models that do have equation error. For example, Johnston (1972, p. 287) stated that for many purposes in econometric analysis the exact (no-equation-error) model is not appropriate. Therefore, the classical functional and structural relationships, discussed by Kendall and Lindley, do not seem to cover all the situations which arise in econometrics. In many applications econometricians and others envision a model where the exact relationship between the true variables is unknown and, to allow for this, an error term is added to the relationship between the true variables:

$$\eta_i = \beta_0 + \beta_1 \xi_i + q_i .$$

Thus, some authors consider that there are two essentially different classes of models, one without equation error and one with equation error. For model estimation purposes, however, it will be seen in Chapter 4 that this distinction has no effect on intercept and slope parameter estimation for the nonnormal model (without side conditions) and, as seen in this chapter, only has the effect of eliminating certain side conditions as being practically reasonable for providing consistent estimates of the intercept and slope in the normal model. Those side conditions which lead to consistent estimates in the normal equation error model lead to the same estimates of the intercept and slope as in the no-equation-error model. The slight difference in the estimation of the model parameters in these two classes of models is only in the estimation of the error variances. In the equation error model without additional information on the error variance, σ_ε^2, one can only estimate the sum of the error variances, $\sigma_\varepsilon^2 + \sigma_q^2$, and not each individual error variance.

In the ordinary regression model,

$$y_i = \beta_0 + \beta_1 x_i + e_i ,$$

the errors e_i, indeed, can be a combination of measurement error and equation error. Because the equation error is most likely independent of the measurement error, a combination of these two errors does not fundamentally change the model and the standard estimation approaches remain unchanged. Thus, equation error does not come up in the classical regression literature. The ME model, on the other hand, uses additional assumptions about the error structure in order to make the model identifiable or to allow for consistent estimation of the parameters. The introduction of equation error does affect these models because it changes the types of additional assumptions on the error structure that make sense in practice. For example, one effect of equation error in ME models is that it is unlikely that orthogonal regression can be used because the assumption that the ratio of the (total) error variances is known does not seem practically reasonable because the ratio of the total error variances is

$$\lambda_t = \frac{\sigma_\varepsilon^2 + \sigma_q^2}{\sigma_\delta^2}$$

and the typical assumption is that σ_q^2 and $\sigma_q^2/\sigma_\delta^2$ are unknown.

Carroll and Ruppert (1996) pointed out the seemingly wide misuse of orthogonal regression. In our opinion, the ME model, unlike the classical regression model, has many different types of assumptions and one should take care in selecting the correct ME model for an application and then use the appropriate estimation procedures.

1.8.2 Maximum likelihood estimation

Maximum likelihood estimation in the structural model does not pose any problem when the model is identifiable because the number of parameters is fixed. A problem arises, however, in the functional and ultrastructural models because the number of unknown incidental parameters, the ξ_i or the μ_i, increases with the sample size. There is no general theory guaranteeing the existence and suitable large sample properties for ML estimators under these circumstances and one needs to check ML estimation on a case-by-case basis in this situation. Some cases result in satisfactory estimates whereas others do not. Examples of the two situations are the case when the error covariance matrix is known up to a proportionality factor and the case when one error variance is known (see Section 1.3). We emphasize again that this phenomenon is not a problem of degrees of freedom (see Sections 1.3 and 2.1).

For the completely unknown error variances case of the normal functional model, Dent (1935) gave the solution of the likelihood equations (1.44), which is called the *geometric mean functional relationship estimator* of the slope. This is because $\hat{\beta}_1$ is the geometric mean of the least-squares coefficient for the regression of y on x and the reciprocal of that for x on y. Anderson and Rubin (1956) proved that the likelihood function is actually unbounded, and it was shown by Solari (1969) that the geometric mean estimator is not the ML estimator but only a gives a saddle point of the likelihood equations. However, this estimate is still mentioned and used in certain applications; see Jolicoeur (1975) and Ricker (1975).

Barker *et al.* (1988) showed that the geometric mean estimate can be interpreted as the estimate that minimizes an error cost functional based on the sum of the triangular areas formed by connecting the measured data points to the estimated line with lines

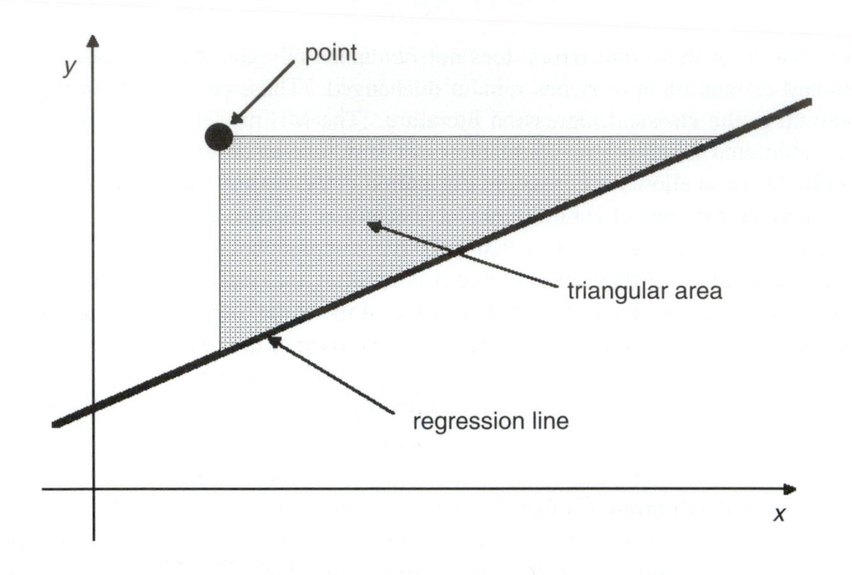

Figure 1.3: Triangular areas for the cost function.

parallel to the coordinate axes (see Figure 1.3). Such a least triangles approach offers a seemingly appealing geometric interpretation of the estimate. Despite this property, however, the geometric mean estimate is neither consistent nor does it possess other desirable statistical properties. We do not recommend its use for the ME model.

1.8.3 The Berkson model

After the appearance of Berkson's (1950) paper, there was some confusion and dispute among statisticians concerning model fitting. Lindley (1953) helped settle this dispute with the following observations. A key assumption in the standard ME model is that the true variable ξ_i is independent of the error $(\delta_i, \varepsilon_i)$. Berkson called such x_i *uncontrolled* observations. If the measured quantity is fixed but the real value differs by an error from the measurement, then

$$\xi_i = x_i - \delta_i. \tag{1.87}$$

That is, the true value is the observed (fixed beforehand) value minus a random error. The minus sign is taken for consistency with our previous ME model notation. The major difference now is that ξ_i and δ_i are no longer independent. Berkson called such x_i *controlled* observations. Controlled observations occur in many experiments and the associated model is called the controlled variable model or the Berkson model. Some authors, such as Carroll *et al.* (1995), wish to distinguish the Berkson model and the usual ME model and write the Berkson model in the form

$$\xi_i = x_i + \delta_i, \tag{1.88}$$

instead of (1.87). The main reason for this is that in the usual ME model the conditional distribution of x given ξ is modelled, while in the Berkson model the conditional distribution of ξ given x is modelled. This corresponds to the fact that ξ is independent of δ in the usual ME model while x is independent of δ in the Berkson model. Note that the measurement error, δ_i, is often symmetric about zero (as in the normal case) so that the sign of the error in (1.88) is not essential.

Least squares has been applied to the Berkson model, and this in essence means that one just ignores the measurement error. The ML approach for the normal Berkson model without equation error was studied by Cheng (1994), and this also leads to least squares except when both error variances are known (Section 1.6). Another important issue is that the least-squares estimate is unbiased in Berkson model only in the linear case. Any slight deviation from linearity, such as a quadratic model, will result in biased and inconsistent estimates from the least-squares method. In other words, one cannot ignore the measurement error in the Berkson model if the model is nonlinear. For more information, see Section 6.6.

1.8.4 Total least squares

In some areas, such as numerical analysis, the field of *total least squares*, first introduced by Golub and Van Loan (1981), is very active. It has grown rapidly in many directions and includes many applications in engineering and computer science. Much of the work was summarized in Van Huffel and Vandewalle (1991) and, more recently, in Van Huffel (1997). The statistical model that corresponds to total least squares is the no-equation-error ME model when the error covariance matrix is known up to a proportionality factor (see Cheng and Van Ness, 1997a; Amemiya, 1997). In other words, total least squares is a particular case of an ME model and, in its simplest version, is just orthogonal regression. As just mentioned, total least squares is not an appropriate approach for other cases of the ME model and, in particular, not for the equation error model.

1.9 Exercises

1.1 *Show that each of the six additional assumptions in Section 1.2 makes the normal structural model identifiable.*

1.2 *Show using geometric arguments that (1.18) sums the squares of the orthogonal distances to the regression line.*

1.3 *Make up an artificial data set of four points for which standard regression, inverse regression, and orthogonal regression all give different regression lines.*

1.4 *Make up an artificial noncollinear data set of four points for which standard regression and orthogonal regression give the same regression line.*

1.5 *Show that the ML estimator $\hat{\beta}_1$, given by (1.30), is a monotone function of λ.*

1.6 *(Birch, 1964; Kendall and Stuart, 1979) When the conditions $s_{xx} > \sigma_\delta^2$ and (1.27b) are not satisfied in case (c) of Section 1.4.1, show that the likelihood function is maximized either when $\hat{\sigma}_\varepsilon^2 = 0$ or when $\hat{\sigma}^2 = 0$. Which of these two results gives the overall ML estimate? Show that the ML estimate for β_1 is*

$$\hat{\beta}_1 = s_{yy} / s_{xy}.$$

1.7 *If $s_{xy} \neq 0$, prove that the sample ordinary regression line and the sample inverse regression line (and the orthogonal regression line) are the same, when the axes are interchanged, if and only if all the data, $\{(x_1, y_1)', \ldots, (x_n, y_n)'\}$, are collinear.*

1.8 *Show that, for the functional equation error model with $\mathbf{\Omega}$ known up to a scalar multiple σ_δ^2, the ML estimator of that scalar multiple in (1.60) is not consistent.*

1.9 *For ML estimation in the structural model, one assumes that (x_i, y_i) is normal and then the claim is made that this is equivalent to the assumption that $(\xi_i, \delta_i, \varepsilon_i)$ is normal. (Note that the prior assumption is that the $(\delta_i, \varepsilon_i)$ are independent and identically distributed and independent of ξ_i and that δ_i and ε_i may be correlated.) Is this statement true? Justify it or disprove it.*

1.10 *(Bowden and Turkington, 1984, p. 3) Solari (1969) showed that, in the no-intercept functional model under the normality assumption, the solution of the likelihood equation is a saddle point. The failure of ML estimation in this case is not an identifiability problem but instead is due to the presence of incidental parameters. Use the result of Bowden (1973) to show that the parameter $\boldsymbol{\theta} = (\beta_1, \sigma_\delta^2, \sigma_\varepsilon^2; \xi_i, i = 1, \ldots, n)'$ is identifiable.*

1.11 *(Barker et al., 1988) Show that the geometric mean estimate (Section 1.8.2) minimizes the sum of the triangular areas of Figure 1.3.*

1.12 *If σ_δ^2 is known, then $\hat{\beta}_1$ is (1.35). Now, if σ_δ^2 is estimated by (1.31b), use (1.35) and (1.31b) to show that $\hat{\beta}_1$ is (1.30).*

1.13 *The data in Table 1.2 are given in Kendall and Stuart (1979, p. 311) and Stuart and Ord (1991, p. 981).*

Table 1.2 *Wheat and potato yields in 48 counties in England in 1936*

Co.	Wheat	Potatoes	Co.	Wheat	Potatoes	Co.	Wheat	Potatoes
1	16.0	5.3	17	15.7	4.9	33	13.8	6.5
2	16.0	6.6	18	14.3	5.1	34	14.4	6.2
3	16.4	6.1	19	13.8	5.5	35	13.4	5.2
4	20.5	5.5	20	12.8	6.7	36	11.2	6.6
5	18.2	6.9	21	12.0	6.5	37	14.4	5.8
6	16.3	6.1	22	15.6	5.2	38	15.4	6.3
7	17.7	6.4	23	15.8	5.2	39	18.5	6.3
8	15.3	6.3	24	16.6	7.1	40	16.4	5.8
9	16.5	7.8	25	14.3	4.9	41	17.0	5.9
10	16.9	8.3	26	14.4	5.6	42	16.9	6.5
11	21.8	5.7	27	15.2	6.4	43	17.5	5.8
12	15.5	6.2	28	14.1	6.9	44	15.8	5.7
13	15.8	6.0	29	15.4	5.6	45	19.2	7.2
14	16.1	6.1	30	16.5	6.1	46	17.7	6.5
15	18.5	6.6	31	14.2	5.7	47	15.2	5.4
16	12.7	4.8	32	13.2	5.0	48	17.1	6.3

The yields of wheat (in hundredweight (cwt = 112 lb.)) and of potatoes (in tons (1 ton = 20 cwt) per acre) are given for 48 counties in England for the year 1936. Let x be per acre wheat yield and y be per acre potato yield.

(a) *Do a scatter plot of the data.*
(b) *Perform a simple linear regression of y on x and plot the resulting regression line on the scatter plot.*
(c) *Perform a simple linear regression of x on y and plot the resulting regression line on the scatter plot.*
(d) *Since these two variables are of the same type, assume that the error in measuring the variables is proportional to the average size of the variables, that is, that $\sigma_y/\sigma_x \cong \bar{y}/\bar{x}$, and take $\lambda \cong 0.15$. Estimate the simple linear ME model line and plot it on the scatter plot.*
(e) *Suppose $\lambda = 1$ in part (d). What happens? Explain.*
(f) *Suppose you rescale so that both variables are measured in tons per acre and take $\lambda = 1$. What happens? Explain.*
(g) *Under what sampling conditions would one use a structural model and when would one use a functional model?*
(h) *Comment on the use of these models for these data.*

1.14 *Do a simulation study of the Berkson model similar to that of Exercise 3.4, but use Berkson's model to generate the data with*

$$x_i = \frac{20\,i}{n} \quad \text{and} \quad \xi_i = x_i - \delta_i.$$

1.15 *Obtain the ML estimator for β_1 for the Berkson model with equation error. Assume the normal model with independent errors.*

1.16 *Obtain the ML estimator for β_1 for the Berkson model without equation error, but with correlated errors, ε and δ. Assume the normal model.*

2

Properties of Estimates and Predictors

This chapter covers the properties of the estimates of the ME model parameters. It includes discussions of their moments, consistency, asymptotic distributions, and confidence intervals. It also discusses the properties of the predictors of y given x and the use of ME models in the calibration problem. It will be seen that the ME model estimators have some unusual properties – some of them quite unlike those of ordinary regression estimators.

2.1 Asymptotic Properties of ME Model Parameter Estimates

2.1.1 The structural ME model

For the structural model, the number of unknown parameters stays fixed as the sample size increases. This is important because most general asymptotic theorems on maximum likelihood estimates require that the dimension of the parameter vector remain constant as the sample size increases. Thus, by the general properties of ML estimates (see, for example, Stuart and Ord, 1991, Chapter 18; or Lehmann, 1983, Chapter 6), the ML estimates of Section 1.3.1 are consistent and asymptotically normal when they exist. These details of these results are subsumed under the theorems for the ultrastructural model in Section 2.1.3 because the method of moments estimates used for the ultrastructural model there are just the ML estimates for the structural model and the structural model is just a special case of the ultrastructural model. For example, if λ is known, Theorem 2.1 applies. In the notation of that theorem, the structural model is the special case of the ultrastructural model with

$$E\xi = \mu = \mu^*, \qquad s^*_{\mu\mu} = 0, \qquad \theta = \sigma_\varepsilon^2 + \beta_1^2\sigma_\delta^2$$

and

$$\psi = \sigma^{-4}\{\theta(\sigma^2 + \sigma_\delta^2) - \beta_1^2\sigma_\delta^4\} = \frac{\theta}{\sigma^2} + \frac{\sigma_\varepsilon^2\sigma_\delta^2}{\sigma^4} = \frac{\sigma_\varepsilon^2\sigma^2 + \sigma_\varepsilon^2\sigma_\delta^2 + \beta_1^2\sigma_\delta^2\sigma^2}{\sigma^4}.$$

Thus, the vector of ML estimates has the property that

$$n^{1/2}\begin{pmatrix}\hat\beta_0\\ \hat\beta_1\\ \hat\sigma_\delta^2\end{pmatrix} \xrightarrow{d} N\left(\begin{pmatrix}\beta_0\\ \beta_1\\ \sigma_\delta^2\end{pmatrix}, \boldsymbol{\Gamma}\right) \qquad \text{as } n \to \infty,$$

where '$\xrightarrow{d}$' indicates convergence in distribution, and $N(\mathbf{m}, \boldsymbol{\Gamma})$ is the three-dimensional normal distribution with mean vector, $\mathbf{m}$, and covariance matrix,

$$\boldsymbol{\Gamma}(\hat\beta_0, \hat\beta_1, \hat\sigma_\delta^2) \equiv \begin{bmatrix} \theta+\mu^2\psi & -\mu\psi & 0\\ -\mu\psi & \psi & 0\\ 0 & 0 & 2\sigma_\delta^4\end{bmatrix}.$$

Thus one can conclude that $\hat\sigma_\delta^2$ is asymptotically independent of $\hat\beta_0$ and $\hat\beta_1$ and that the asymptotic covariance of $\hat\beta_0$ and $\hat\beta_1$ is $-\mu\psi$.

2.1.2 The functional ME model

By the general properties of sample moments (see, for example, Stuart and Ord, 1994, Chapter 10); the sample variances and covariances, s_{xx}, s_{yy}, and s_{xy}, converge in probability to their expectations. Thus, for example, for the functional relationship

$$\begin{aligned} s_{xx} &\xrightarrow{p} s_{\xi\xi}^* + \sigma_\delta^2,\\ s_{yy} &\xrightarrow{p} \beta_1^2 s_{\xi\xi}^* + \sigma_\varepsilon^2,\\ s_{xy} &\xrightarrow{p} \beta_1 s_{\xi\xi}^*, \end{aligned} \tag{2.1}$$

where (1.48) is assumed. The convergence in probability in (2.1) can be made strong convergence under slightly different regularity conditions.

Consistency of $\hat\beta_0$ and $\hat\beta_1$ when λ is known.

The intercept and slope parameter estimates are consistent for the functional model when λ is known (Section 1.3.2). Substituting (2.1) into (1.30) gives

$$\hat\beta_1 \xrightarrow{p} \frac{\beta_1^2 s_{\xi\xi}^* - \lambda s_{\xi\xi}^* + [\{\beta_1^2 s_{\xi\xi}^* + \lambda\sigma_\delta^2 - \lambda(s_{\xi\xi}^* + \sigma_\delta^2)\}^2 + 4\lambda\beta_1^2(s_{\xi\xi}^*)^2]^{\frac{1}{2}}}{2\beta_1 s_{\xi\xi}^*} = \beta_1.$$

The consistency of $\hat\beta_0$ follows immediately from (1.31a) and the consistency of $\hat\beta_1$, $\bar{x}$, and $\bar{y}$. For the asymptotic distributions, see the theorems in Section 2.1.3.

Inconsistency of $\hat\sigma_\varepsilon^2$ when λ is known.

Substituting (2.1) into (1.53) yields

$$\hat\sigma_\varepsilon^2 \xrightarrow{p} \frac{\lambda}{2(\lambda+\beta_1^2)}(\beta_1^2 s_{\xi\xi}^* + \sigma_\varepsilon^2 - 2\beta_1^2 s_{\xi\xi}^* + \beta_1^2(s_{\xi\xi}^* + \sigma_\delta^2)) = \frac{\sigma_\varepsilon^2}{2},$$

which implies that $\hat{\sigma}_\varepsilon^2$ is inconsistent. Lindley (1947, p. 238) considered this a problem of *degrees of freedom*. He explained that the ML estimator for the variance σ^2 of a normal sample with unknown mean μ and variance σ^2 gives the factor n instead of $n-1$. Thus, he reasoned that the inconsistency of the ML estimator for σ_ε^2 is a reflection of the small sample bias of ML estimators in general. In the present situation, it is therefore necessary to adjust for the degrees of freedom. The correct factor is not the number of observations, $2n$, but the degrees of freedom, which is $2n-(n+2)=n-2$. A consistent estimator is therefore

$$\begin{aligned}
\text{(a) } \tilde{\sigma}_\varepsilon^2 &= \tfrac{2n}{n-2}\hat{\sigma}_\varepsilon^2 = \frac{\lambda n}{(n-2)(\lambda+\hat{\beta}_1^2)}(s_{yy}-2\hat{\beta}_1 s_{xy}+\hat{\beta}_1^2 s_{xx}),\\
\text{(b) } \tilde{\sigma}_\delta^2 &= \tfrac{2n}{n-2}\hat{\sigma}_\delta^2 = \frac{n}{(n-2)(\lambda+\hat{\beta}_1^2)}(s_{yy}-2\hat{\beta}_1 s_{xy}+\hat{\beta}_1^2 s_{xx}).
\end{aligned} \tag{2.2}$$

This adjusted estimator has been adopted by most researchers.

As discussed in Section 1.3.2, the inconsistency of ML estimate, $\hat{\sigma}_\varepsilon^2$, for the functional model is a result of the fact that the number of unknown parameters in the model grows with the sample size. With each new observation, (x_i, y_i), a new unknown parameter, ξ_i, is introduced. Thus, the standard theorems giving asymptotic (as $n \to \infty$) results for ML estimation do not apply to the functional and ultrastructural ML estimates.

Consistent estimates when either σ_ε^2 or σ_δ^2 is known.

Section 1.3.2 states that maximum likelihood fails to provide consistent estimates for the functional model when either σ_ε^2 or σ_δ^2 is known. Consistent estimates do exist, however. For example, the ML estimates ((1.35), (1.37b), (1.38), and (1.40)) for the structural model with either σ_ε^2 or σ_δ^2 known provide consistent estimates for the corresponding functional model. Moberg and Sundberg (1978, p. 63) pointed out the remarkable fact that these structural estimates are consistent even in the functional case. These estimates can be considered method of moments estimates for the functional model.

For example, consider (1.35) and (1.37b); from (2.1),

$$\hat{\beta}_1 = \frac{s_{xy}}{s_{xx}-\sigma_\delta^2} \xrightarrow{p} \frac{\beta_1 s_{\xi\xi}^*}{s_{\xi\xi}^*} = \beta_1,$$

and

$$\hat{\sigma}_\varepsilon^2 = s_{yy}-\hat{\beta}_1 s_{xy} \xrightarrow{p} \beta_1^2 s_{\xi\xi}^* + \sigma_\varepsilon^2 - \beta_1^2 s_{\xi\xi}^* = \sigma_\varepsilon^2.$$

For the asymptotic distributions, see the theorems in Section 2.1.3.

Consistent estimates when both σ_ε^2 and σ_δ^2 are known.

Here λ is known and one can use the previous results for the λ known case. Now, however, all estimates are consistent because neither σ_ε^2 nor σ_δ^2 have to be estimated.

Consistent estimates when β_0 is known.

Section 1.3.2 states that maximum likelihood fails to provide consistent estimates for the functional model when β_0 is known. Again, consistent estimates do exist. The ML estimates ((1.37b), (1.40), and (1.41)) for the structural model with β_0 known are also consistent method of moments estimates for the corresponding functional model.

For example, consider (1.41); from (2.1),

$$\hat{\beta}_1 = \frac{\bar{y} - \beta_0}{\bar{x}} \xrightarrow{p} \frac{\beta_0 + \beta_1 \xi^* - \beta_0}{\xi^*} = \beta_1.$$

2.1.3 The ultrastructural ME model

The ultrastructural model has all the incidental parameter problems of the functional model; it also has an additional structural parameter, σ^2. The conditions (1.48) are replaced by the assumption that

$$\mu^* = \lim_{n\to\infty} \bar{\mu}_n = \lim_{n\to\infty} \frac{1}{n} \sum \mu_i \quad \text{and} \quad s^*_{\mu\mu} = \lim_{n\to\infty} \frac{1}{n} \sum (\mu_i - \bar{\mu}_n)^2 \tag{2.3}$$

exist.

Consistency of $\hat{\beta}_1$ when both λ and ν are known.

Patefield (1978) showed that the consistency of the quintic (1.55) depends on the sequence $\{\mu_i\}$. Suppose there exists a constant γ such that

$$\lim_{n\to\infty} n^{-\gamma} \bar{\mu}_n = \mu_*, \qquad \lim_{n\to\infty} n^{-2\gamma-1} \sum (\mu_i - \mu^*)^2 = \nu_*. \tag{2.4}$$

If the μ_i are equally spaced, then $\gamma = 1$; and if the μ_i have finite range, then the $\{\mu_i, i = 1, \ldots, \infty\}$ have finite-sample mean and variance and $\gamma = 0$. If $\gamma > 0$, then there is a solution of (1.55) that is consistent; whereas, if $\gamma \leq 0$, Patefield showed that there is no consistent solution of (1.55).

Consistent estimates when λ is known.

The ML estimates are just the ML estimates of the corresponding functional model. Thus maximum likelihood provides consistent estimates of β_0 and β_1, but not of σ^2_ε (or σ^2_δ). The adjusted estimator (2.2) is again consistent. The following results are proved in Gleser (1985).

Theorem 2.1 *Assume the ultrastructural model with λ known and (2.3) true. If $\sigma^2 + s^*_{\mu\mu} > 0$, then the estimates (1.30), (1.31a), and (2.2b) are strongly consistent, asymptotically jointly normal with the covariance of the asymptotic distribution of $n^{1/2}(\hat{\beta}_0 - \beta_0, \hat{\beta}_1 - \beta_1, \tilde{\sigma}^2_\delta - \sigma^2_\delta)'$ equal to*

$$\mathbf{\Gamma}(\hat{\beta}_0, \hat{\beta}_1, \tilde{\sigma}^2_\delta) \equiv \begin{bmatrix} \theta + (\mu^*)^2\psi & -\mu^*\psi & 0 \\ -\mu^*\psi & \psi & 0 \\ 0 & 0 & 2\sigma^4_\delta \end{bmatrix},$$

where

$$\theta = \sigma_\varepsilon^2 + \beta_1^2\sigma_\delta^2, \; \psi = \tau^{-2}\{\theta(\tau + \sigma_\delta^2) - \beta_1^2\sigma_\delta^4\} \quad \text{and} \quad \tau = \sigma^2 + s_{\mu\mu}^*.$$

Theorem 2.2 *Assume the ultrastructural model with λ known and (2.3) true. Let $\mathcal{C}$ be the class of all estimators $(\check{\beta}_0, \check{\beta}_1, \check{\sigma}_\delta^2)$ of $(\beta_0, \beta_1, \sigma_\delta^2)$ that are consistent and asymptotically normal, with the covariance matrix of their asymptotic distribution depending on the sequence $\{\mu_i, \; i = 1, 2, \ldots\}$ only through the limits μ^* and $s_{\mu\mu}^*$. Within $\mathcal{C}$ the estimators, $(\hat{\beta}_0, \hat{\beta}_1, \tilde{\sigma}_\delta^2)$, are best asymptotically normal; that is, for any 3-vector, **t**,*

$$\mathbf{t}\boldsymbol{\Gamma}(\hat{\beta}_0, \hat{\beta}_1, \tilde{\sigma}_\delta^2)\mathbf{t}' \leq \mathbf{t}\boldsymbol{\Gamma}(\check{\beta}_0, \check{\beta}_1, \check{\sigma}_\delta^2)\mathbf{t}'$$

where $(\check{\beta}_0, \check{\beta}_1, \check{\sigma}_\delta^2) \in \mathcal{C}$ and $\boldsymbol{\Gamma}(\check{\beta}_0, \check{\beta}_1, \check{\sigma}_\delta^2)$ is the covariance of the asymptotic distribution of

$$n^{1/2}[(\check{\beta}_0, \check{\beta}_1, \check{\sigma}_\delta^2) - (\beta_0, \beta_1, \sigma_\delta^2)].$$

Consistent estimates when either σ_ε^2 or σ_δ^2 is known.

As in the previous case, maximum likelihood leads to the functional model ($\sigma^2 = 0$, see Section 1.3.3), therefore ML estimation breaks down when either σ_ε^2 or σ_δ^2 is known. Cheng and Van Ness (1991) presented consistent method of moments estimators for the σ_δ^2 known case, which are the structural ML estimates $\hat{\beta}_0, \hat{\beta}_1$, and $\hat{\sigma}_\varepsilon^2$ given by (1.31a), (1.35), and (1.37) provided $s_{xx} > \sigma_\delta^2$ and $s_{yy} \geq s_{xy}^2/\left(s_{xx} - \sigma_\delta^2\right)$. Cheng and Van Ness (1994) also gave consistent method of moments estimators for the σ_ε^2 known case, which are the structural ML estimates $\hat{\beta}_0, \hat{\beta}_1$, and $\hat{\sigma}_\varepsilon^2$ given by (1.31a), (1.38), and (1.40) provided $s_{yy} > \sigma_\varepsilon^2$ and $s_{xx} \geq s_{xy}^2/\left(s_{yy} - \sigma_\varepsilon^2\right)$. Cheng and Van Ness (1991; 1994) also proved the following results.

Theorem 2.3 *Assume the ultrastructural model with σ_δ^2 known and (2.3) true. If $\tau \equiv \sigma^2 + s_{\mu\mu}^* > 0$, then the estimates $\hat{\beta}_0, \hat{\beta}_1$, and $\hat{\sigma}_\varepsilon^2$ given by (1.31a), (1.35), and (1.37) are strongly consistent, asymptotically jointly normal with the covariance of the asymptotic distribution of $n^{1/2}(\hat{\beta}_0 - \beta_0, \hat{\beta}_1 - \beta_1, \hat{\sigma}_\varepsilon^2 - \sigma_\varepsilon^2)'$ equal to*

$$\boldsymbol{\Gamma}_1(\hat{\beta}_0, \hat{\beta}_1, \hat{\sigma}_\varepsilon^2) \equiv \begin{bmatrix} \theta + (\mu^*)^2\psi_1 & -\mu^*\psi_1 & -\mu^*\rho \\ -\mu^*\psi_1 & \psi_1 & \rho \\ -\mu^*\rho & \rho & 2\theta^2 \end{bmatrix}, \tag{2.5}$$

where

$$\psi_1 = \tau^{-2}\{\theta(\tau + \sigma_\delta^2) + \beta_1^2\sigma_\delta^4\} \quad \text{and} \quad \rho = -2\tau^{-1}\beta_1\sigma_\delta^2\theta.$$

Theorem 2.4 *Theorem 2.2 is true with σ_δ^2 known instead of λ known and the method of moments estimates replaced by $(\hat{\beta}_0, \hat{\beta}_1, \hat{\sigma}_\varepsilon^2)$ given by (1.31a), (1.35), and (1.37) and the covariance matrix replaced by (2.5).*

Theorem 2.5 *Assume the ultrastructural model with σ_ε^2 known and (2.3) true. If $\tau > 0$, then the estimates $\hat{\beta}_0, \hat{\beta}_1$, and $\hat{\sigma}_\delta^2$ given by (1.31a), (1.38), and (1.40) are* *strongly consistent, asymptotically jointly normal with the covariance of the asymptotic distribution of* $n^{1/2}(\hat{\beta}_0 - \beta_0, \hat{\beta}_1 - \beta_1, \hat{\sigma}_\delta^2 - \sigma_\delta^2)'$ *equal to*

$$\boldsymbol{\Gamma}_2(\hat{\beta}_0, \hat{\beta}_1, \hat{\sigma}_\varepsilon^2) \equiv \begin{bmatrix} \theta + (\mu^*)^2\psi_2 & -\mu^*\psi_2 & -\mu^*\rho_1 \\ -\mu^*\psi_2 & \psi_2 & \rho_1 \\ -\mu^*\rho_1 & \rho_1 & 2\beta_1^{-4}\theta^2 \end{bmatrix}, \tag{2.6}$$

where

$$\psi_2 = \tau^{-2}\beta_1^{-2}\{(\beta_1^2\tau + \sigma_\varepsilon^2)\theta + \sigma_\varepsilon^4\} \quad \textit{and} \quad \rho_1 = 2\beta_1^{-3}\tau^{-1}\sigma_\varepsilon^2\theta.$$

Theorem 2.6 *Theorem 2.2 is true with* σ_ε^2 *known instead of* λ *known and the method of moments estimates replaced by* $(\hat{\beta}_0, \hat{\beta}_1, \hat{\sigma}_\varepsilon^2)$ *given by (1.31a), (1.38), and (1.40) and the covariance matrix replaced by (2.6).*

The results of these six theorems are also applicable to the corresponding structural models ($s_{\mu\mu}^* = 0$) and functional models ($\sigma^2 = 0$). In the ultrastructural and structural models, the asymptotic covariance matrices hold for the corresponding estimates when ξ_i is not normal provided that σ^2 exists.

Note that the asymptotic covariance matrices in Theorems 2.1–2.3 do not require ξ to be normally distributed, only that (δ, ε) be normal.

Consistent estimates when both σ_ε^2 and σ_δ^2 are known.

Here, again, λ is known. The previous results for the λ known case apply. All estimates are consistent because neither σ_ε^2 nor σ_δ^2 has to be estimated.

Consistent estimates when β_0 is known.

This is the same situation as with the functional model. Maximum likelihood fails to provide consistent estimates when β_0 is known, but again the ML estimates ((1.37), (1.40), and (1.41)) for the structural model with β_0 known provide consistent method of moments estimates for the corresponding ultrastructural model.

For example, consider (1.41); from (2.3),

$$\hat{\beta}_1 = \frac{\bar{y} - \beta_0}{\bar{x}} \xrightarrow{p} \frac{\beta_0 + \beta_1\mu^* - \beta_0}{\mu^*} = \beta_1.$$

Summary table for the standard ME model.

The estimators for the ME model without equation error are summarized in Table 2.1. There are, of course, many possible method of moments estimators, Table 2.1 gives the most common ones. The method of moments estimator for β_1 has several reasonable possibilities; the chosen one matches the λ known case; under normality, it has smaller asymptotic variance than $\hat{\beta}_2$ or $\hat{\beta}_3$ (see Fuller, 1987, p. 33). For the functional model,

(1.48) is assumed; and for the ultrastructural model, (2.3) is assumed. The ML estimates are based on the normal model.

Table 2.1 Measurement error model structural parameter estimators

Known parameter	Functional & Ultrastructural Model	
	Maximum likelihood	Method of moments
λ	$\hat\beta_{0,1}, \hat\beta_{1,1}, \hat\sigma^2_{\delta,1}$	$\hat\beta_{0,1}, \hat\beta_{1,1}, 2\hat\sigma^2_{\delta,1}$
$\sigma^2_\varepsilon, \sigma^2_\delta$	$\hat\beta_{0,1}, \hat\beta_{1,1}$	$\hat\beta_{0,1}, \hat\beta_{1,1}$
σ^2_δ	breakdown	$\hat\beta_{0,2}, \hat\beta_{1,2}, \hat\sigma^2_{\varepsilon,1}$
σ^2_ε	breakdown	$\hat\beta_{0,3}, \hat\beta_{1,3}, \hat\sigma^2_{\delta,2}$
β_0	saddle point breakdown	$\hat\beta_{1,4}, \hat\sigma^2_{\delta,3}, \hat\sigma^2_{\varepsilon,2}$

Known parameter	Structural Model	
	Maximum likelihood	Method of moments
λ	$\hat\beta_{0,1}, \hat\beta_{1,1}, \hat\mu, 2\hat\sigma^2_{\delta,1}, \hat\sigma^2_1$	$\hat\beta_{0,1}, \hat\beta_{1,1}, \hat\mu, 2\hat\sigma^2_{\delta,1}, \hat\sigma^2_1$
$\sigma^2_\varepsilon, \sigma^2_\delta$	$\hat\beta_{0,1}, \hat\beta_{1,1}, \hat\mu, \hat\sigma^2_5$	$\hat\beta_{0,1}, \hat\beta_{1,1}, \hat\mu, \hat\sigma^2_1$
σ^2_δ	$\hat\beta_{0,2}, \hat\beta_{1,2}, \hat\mu, \hat\sigma^2_{\varepsilon,1}, \hat\sigma^2_2$	$\hat\beta_{0,2}, \hat\beta_{1,2}, \hat\mu, \hat\sigma^2_{\varepsilon,1}, \hat\sigma^2_2$
σ^2_ε	$\hat\beta_{0,3}, \hat\beta_{1,3}, \hat\mu, \hat\sigma^2_{\delta,2}, \hat\sigma^2_3$	$\hat\beta_{0,3}, \hat\beta_{1,3}, \hat\mu, \hat\sigma^2_{\delta,2}, \hat\sigma^2_3$
β_0	$\hat\beta_{1,4}, \hat\mu, \hat\sigma^2_{\delta,3}, \hat\sigma^2_{\varepsilon,2}, \hat\sigma^2_4$	$\hat\beta_{1,4}, \hat\mu, \hat\sigma^2_{\delta,3}, \hat\sigma^2_{\varepsilon,2}, \hat\sigma^2_4$

The following definitions are required for Table 2.1:

$$\hat\beta_{0,i} = \bar y - \hat\beta_{1,i}\bar x, \qquad i = 1, 2, 3;$$

$$\hat\beta_{1,1} = \frac{s_{yy} - \lambda s_{xx} + ((s_{yy} - \lambda s_{xx})^2 + 4\lambda s_{xy}^2)^{1/2}}{2s_{xy}}, \qquad s_{xy} \neq 0;$$

$$\hat\beta_{1,2} = s_{xy}/(s_{xx} - \sigma^2_\delta), \qquad \text{assuming that } s_{xx} > \sigma^2_\delta \text{ and } s_{yy} \geq s^2_{xy}/(s_{xx} - \sigma^2_\delta);$$

$$\hat\beta_{1,3} = (s_{yy} - \sigma^2_\varepsilon)/s_{xy}, \qquad \text{assuming that } s_{yy} > \sigma^2_\varepsilon, \text{ and } s_{xx} \geq s^2_{xy}/\left(s_{yy} - \sigma^2_\varepsilon\right);$$

$$\hat\beta_{1,4} = \left(\bar y - \beta_0\right)/\bar x, \qquad \text{assuming that } \bar x \neq 0;$$

$$\hat\mu = \bar x;$$

$$\hat\sigma^2_{\delta,1} = \frac{s_{yy} - 2\hat\beta_{1,1}s_{xy} + \hat\beta^2_{1,1}}{2(\lambda + \hat\beta^2_{1,1})};$$

$$\hat\sigma^2_{\delta,2} = s_{xx} - s_{xy}/\hat\beta_{1,3}; \qquad \hat\sigma^2_{\delta,3} = s_{xx} - s_{xy}/\hat\beta_{1,4};$$

$$\hat\sigma^2_{\varepsilon,1} = s_{yy} - \hat\beta_{1,2}s_{xy}; \qquad \hat\sigma^2_{\varepsilon,2} = s_{yy} - \hat\beta_{1,4}s_{xy};$$

$$\hat\sigma^2_i = s_{xy}/\hat\beta_{1,i}, \qquad i = 1, \ldots, 4$$

(assuming that the $\hat\sigma^2$ involved is nonnegative); and

$$\hat\sigma^2_5 = \frac{s_{yy} + \lambda s_{xx} - 2\sigma^2_\varepsilon + ((s_{yy} - \lambda s_{xx})^2 + 4\lambda s^2_{xy})^{1/2}}{2(\lambda + \hat\beta^2_{1,1})}.$$

Recall that the ML estimates for the structural model require that the estimated variances be nonnegative and that the denominators of the estimates be nonzero. These restrictions were covered in Section 1.3.

2.2 Asymptotic Properties of Equation Error Model Estimates

The equation error model, defined in Section 1.5, introduces another type of error which allows factors, other than the ξ_i, to be responsible for the variation in η_i, that is, the true values η_i and ξ_i are not perfectly related. From (1.70),

$$\eta_i = \beta_0 + \beta_1 \xi_i + q_i, \qquad i = 1, \ldots, n \tag{2.7}$$

where the q_i are the 'equation errors'. It is assumed that (ξ_i, q_i) are independent of $(\delta_j, \varepsilon_j)$ for all i and j. The existence of equation error affects the practical usefulness of the side conditions used to make the parameters identifiable and/or consistently estimable.

Define $\lambda_q = \sigma_q^2/\sigma_\delta^2$. As mentioned in Chapter 1, the assumption that λ_q is known does not seem reasonable because it does not seem to be possible in practice to obtain the information about $\sigma_q^2/\sigma_\delta^2$. On the other hand, $\sigma_\varepsilon^2/\sigma_\delta^2$ might be available; for example, if x and y are both measured on the same scale, then it might be reasonable to assume that $\lambda = \sigma_\varepsilon^2/\sigma_\delta^2 = 1$ (cf. Carroll *et al.*, 1995, p. 32). Unfortunately, knowing only λ does not make β_1 identifiable in the equation error model. Similarly, assuming knowledge of σ_ε^2 is not useful because although information might be obtained about σ_ε^2 via replications of y, β_1 still cannot be estimated due to the unknown equation error variance σ_q^2.

On the other hand, assuming knowledge of σ_δ^2 is realistic in practice, because prior knowledge of σ_δ^2 can be obtained from replications of x, and this assumption makes β_1 identifiable. Moreover, this assumption is widely used in *nonlinear* ME models (see Chapter 6; and Carroll *et al.*, 1995). Recalling (1.71), with this additional assumption, the equation error model can be written as a corresponding ME model also with σ_δ^2 known. For the structural relationship, assuming that q_i is normal, the ML estimators for the parameters are the same as the corresponding ones for the no-equation-error model except that $\hat{\sigma}_\varepsilon^2$ is replaced by $\hat{\sigma}_e^2$. Note that neither σ_ε^2 nor σ_q^2 is identifiable but their sum, $\sigma_e^2 = \text{var}(\varepsilon + q)$, is. For the functional relationship, ML estimation breaks down as it does in the no-equation-error model, but the method of moments estimators can be used with $\hat{\sigma}_\varepsilon^2$ replaced by $\hat{\sigma}_e^2$. It is worth pointing out that the σ_δ^2 known case of the classical ME model can be treated as a *degenerate* equation error model in which $\sigma_q^2 = 0$.

As for the case when both σ_δ^2 and σ_ε^2 are known, the ML estimators of the parameters in the structural relationship are the same as those for the σ_δ^2 known case, except that now σ_q^2 can be estimated. Note that this case is quite different from the corresponding one for no-equation-error structural relationship. In the equation error model, knowing both σ_δ^2 and σ_ε^2 does not lead to an overidentified model. For the functional relationship, ML estimation breaks down, which is, again, quite different from the corresponding no-equation-error functional model. Here again, the ML estimators in the structural relationship can be used and treated as method of moments estimators (with $\hat{\sigma}_\varepsilon^2$ replaced by $\hat{\sigma}_e^2$).

Finally, for the β_0 known case, the ML estimators for the parameters are the same as the corresponding ones for the no-equation-error model except that only σ_e^2 can be estimated and not σ_ε^2 and σ_q^2 separately. Again ML estimation results in a saddle point for the functional relationship.

Table 2.2 Equation error model structural parameter estimators

Known parameter	Functional & Ultrastructural Model	
	Maximum likelihood	Method of moments
λ	unidentified	unidentified
$\sigma_\varepsilon^2, \sigma_\delta^2$	breakdown	$\hat{\beta}_0, \hat{\beta}_{1,1}, \hat{\sigma}_q^2$
σ_δ^2	breakdown	$\hat{\beta}_0, \hat{\beta}_{1,1}, \hat{\sigma}_{e,1}^2$
σ_ε^2	unidentified	unidentified
β_0	saddle point breakdown	$\hat{\beta}_{1,2}, \hat{\sigma}_\delta^2, \hat{\sigma}_{e,2}^2$

Known parameter	Structural Model	
	Maximum likelihood	Method of moments
λ	unidentified	unidentified
$\sigma_\varepsilon^2, \sigma_\delta^2$	$\hat{\beta}_0, \hat{\beta}_{1,1}, \hat{\mu}, \hat{\sigma}_1^2, \hat{\sigma}_q^2$	$\hat{\beta}_0, \hat{\beta}_{1,1}, \hat{\mu}, \hat{\sigma}_1^2, \hat{\sigma}_q^2$
σ_δ^2	$\hat{\beta}_0, \hat{\beta}_{1,1}, \hat{\mu}, \hat{\sigma}_{e,1}^2, \hat{\sigma}_1^2$	$\hat{\beta}_0, \hat{\beta}_{1,1}, \hat{\mu}, \hat{\sigma}_{e,1}^2, \hat{\sigma}_1^2$
σ_ε^2	unidentified	unidentified
β_0	$\hat{\beta}_{1,2}, \hat{\mu}, \hat{\sigma}_\delta^2, \hat{\sigma}_{e,2}^2, \hat{\sigma}_2^2$	$\hat{\beta}_{1,2}, \hat{\mu}, \hat{\sigma}_\delta^2, \hat{\sigma}_{e,2}^2, \hat{\sigma}_2^2$

The estimators for the equation error model are summarized in Table 2.2, for which the following definitions are required:

$$\hat{\beta}_0 = \bar{y} - \hat{\beta}_{1,1}\bar{x};$$

$$\hat{\beta}_{1,1} = s_{xy}/(s_{xx} - \sigma_\delta^2), \qquad \text{assuming that } s_{xx} > \sigma_\delta^2 \text{ and } s_{yy} \geq s_{xy}^2/(s_{xx} - \sigma_\delta^2);$$

$$\hat{\beta}_{1,2} = (\bar{y} - \beta_0)/\bar{x}, \qquad \text{assuming that } \bar{x} \neq 0;$$

$$\hat{\mu} = \bar{x};$$

$$\hat{\sigma}_\delta^2 = s_{xx} - s_{xy}/\hat{\beta}_{1,2};$$

$$\hat{\sigma}_{e,1}^2 = s_{yy} - \hat{\beta}_{1,1}s_{xy}; \qquad \hat{\sigma}_{e,2}^2 = s_{yy} - \hat{\beta}_{1,2}s_{xy};$$

$$\hat{\sigma}_i^2 = s_{xy}/\hat{\beta}_i, \qquad i = 1, 2;$$

$$\hat{\sigma}_q^2 = \hat{\sigma}_{e,1}^2 - \sigma_\varepsilon^2,$$

assuming that all the estimated variances involved are nonnegative.

2.3 Finite-Sample Properties

We have seen that many of the ML estimators and/or the method of moments estimators derived in Chapter 1 have reasonable asymptotic properties. As shall be seen, however, the finite-sample properties, which are the most important to the practitioner, are not so good. Properties such as unbiasedness, minimum variance, and minimum mean squared error do not hold for $\hat{\beta}_0$ and $\hat{\beta}_1$, in contrast to the ML (least-squares) estimators in ordinary regression. In fact, an unfortunate property of $\hat{\beta}_1$, shared by all cases except the reliability ratio known case, is that its first moment does not exist. More specifically, both the positive and negative parts of $E\hat{\beta}_1$ are infinite in these cases, which implies that $E|\hat{\beta}_1|$ is infinite (see Exercises 2.1–2.3). This lack of any moments of an estimator is quite uncommon in statistics and it might seem to some readers that this is a fatal flaw of the ME model. In fact, all it means is that the distribution of the estimate, say $\hat{\beta}_1$, has 'fat' tails. By the results of the previous section, for n large, this distribution must look something like a normal distribution but with somewhat thicker tails. One can still obtain such things as confidence intervals for the parameter values (see the next section).

The nonexistence of the first moment of $\hat{\beta}_1$ need only be shown for the functional model because the structural model will follow from the fact that, if the conditional expectation is infinite, then the overall expectation is infinite,

$$E(|\hat{\beta}_1| \mid \xi) = \infty \Rightarrow E(|\hat{\beta}_1|) = \infty.$$

For the functional model, this was first shown for the λ known case (see, for example, Anderson, 1976). It is therefore true when both error variances are known because the estimator is the same. For σ_δ^2 known, Fuller (1987, p. 28, Problem 13) gave an argument for the structural model via conditional expectation. The argument for the functional model is similar (see Exercise 2.1). For σ_ε^2 known and the intercept known it can also be shown that the expectation is not finite (see Exercises 2.2 and 2.3). Note that when the reliability ratio is known then the first moment exists because $\hat{\beta}_1$ is the inverse of the reliability ratio, a known constant, multiplied by the ordinary least squares estimator and the latter has finite first moment. The assumption of normality is not necessary for the proofs that $E\hat{\beta}_1$ does not exist. This holds for very general distributional assumptions (see Exercises 2.1–2.3; and Cheng and Van Ness, 1997a).

Of course, for any finite sample size, n, the estimator, $\hat{\beta}_1$, also has infinite variance, although it possesses an asymptotic distribution with both finite mean and variance. This means that the distribution of $\hat{\beta}_1$ has tails too heavy for the first and second moments to exist, but, as the sample size tends to infinity, this distribution's tails become thinner and thinner until it converges to a distribution which does have finite moments. The popular comparison of expected mean squared error of estimators is not applicable in finite-sample situations for these estimates because the estimator has neither a mean nor a variance and the mean squared error is infinite. One possible way to compare competing estimators is to use (2.8).

Many practitioners as well as researchers use the ordinary least-squares estimator $\hat{\beta}_{1R} = s_{xy}/s_{xx}$ even though they realize they are dealing with the ME model. Though this estimator possesses finite mean and variance, it is both biased and inconsistent. But the least-squares estimator is familiar and easy to compute from readily available software. On the other hand, the ML estimator for the ME model has been shown to be asymptotically best. Thus the large sample comparisons favour the ME model estimators, but this does not necessarily imply that these estimators do better for finite sample sizes. If the expected mean squared errors of these two estimators are compared, the ML estimator loses because it has an infinite mean squared error. Anderson (1976) compared the asymptotic distributional concentrations of the least-squares and ML estimators of the slope β_1 in the functional relationship when the errors are normally distributed with diagonal covariance matrix, $\sigma_\delta^2 \mathbf{I}$. He showed that, subject to the accuracy of the distributional expansions used for the least-squares estimator $\hat{\beta}_{1R}$ and the ML estimator $\hat{\beta}_1$ of (1.30), asymptotically

$$P\{\mid \hat{\beta}_1 - \beta_1 \mid \leq t\} \leq P\{\mid \hat{\beta}_{1R} - \beta_1 \mid \leq t\} \tag{2.8}$$

for all $t > 0$, except when the magnitude of the slope β_1 is small. Thus the ML estimator is superior asymptotically to the least-squares estimator in the sense that the former is closer to the true parameter in probability. For cases other than λ known, no similar result has been obtained except for the both error variances known case, which shares the same estimator with the λ known case.

There are very few finite-sample properties of $\hat{\beta}_1$ given in the literature due to the complexity of the situation. It would be interesting and useful to know more about the exact distribution of $\hat{\beta}_1$. In the case where the ratio of error variances is known, Mariano (1969) derived the exact distribution of $\hat{\beta}_1$ for the functional model (see also Anderson and Sawa, 1982), whereas Wong (1989) derived the exact distribution in the structural model in the case where both error variances are known. In both cases the distributions are a function of the reliability ratio and, unfortunately, very complicated. It is very difficult to extract much useful information such as confidence intervals from these distributions (see the next section). The exact distributional theory of other cases is still unknown.

In practice it seems quite reasonable that one would know a bound K on $|\beta_1|$. In almost any application, it is known that the slope is less than some large number. In this case, one can modify any of the previous estimates $\hat{\beta}_1$ to

$$\hat{\beta}_{1,K} = \min(|\hat{\beta}_1|, K)\,\mathrm{sign}(\hat{\beta}_1). \tag{2.9}$$

This new estimate obviously has finite moments of all orders and, if $|\beta_1|$ is indeed less than K, then this new estimate is consistent by an argument using the triangle inequality (Cheng and Van Ness, 1994). Thus, there are quite reasonable ME model estimates available which do have finite moments and finite mean squared errors for finite sample sizes. Estimates of the form (2.9) have not been studied in detail.

2.4 Implications Regarding Confidence Regions

Heretofore, we have only discussed the point estimation of the parameters; however, it is also very important to assess the accuracy of these estimates. The focus of this section is on confidence intervals for β_1 with a brief discussion of joint confidence regions for (β_0, β_1).

2.4.1 The Gleser–Hwang effect

Before proceeding, an interesting but rather unsettling result due to Gleser and Hwang (1987) needs to be introduced. This result, applied to ME models, says that every confidence set for (β_0, β_1) with nontrivial confidence level $1 - \gamma$ (i.e. $0 \leq \gamma < 1$) will be unbounded with positive probability. This implies that such confidence sets have infinite expected size. Conversely, any bounded confidence set for (β_0, β_1) will have confidence level equal to zero with probability one. Moreover, for any scalar function $g(\beta_0, \beta_1)$ with finite range, $[a, b]$, any confidence interval for $g(\beta_0, \beta_1)$ which has confidence level greater than one-half must have positive probability of containing the entire range. Consequently, if one tries to reparameterize (β_0, β_1) into new parameters having a bounded parameter space, the confidence set is still not very satisfactory. We refer to this result as the zero confidence level effect or the Gleser–Hwang effect.

To explain the idea behind this effect, consider the structural model with identity error covariance matrix. Suppose there are data, $\mathbf{Z}$, from which one is to find a confidence region $C(\mathbf{Z})$. Let $\boldsymbol{\theta}$ be a parameter vector; then the confidence level is defined by

$$1 - \gamma = \inf_{\boldsymbol{\theta}}[P_{\boldsymbol{\theta}}\{\boldsymbol{\theta} \in C(\mathbf{Z})\}]. \tag{2.10}$$

The structural ME model can be made arbitrarily close to the model

$$x_i = \mu + \delta_i\,, \qquad y_i = \beta_0 + \beta_1\mu + \varepsilon_i\,, \qquad 1 = 1, \ldots, n, \tag{2.11}$$

by letting σ^2 approach zero. The (x_i, y_i) from (2.11) are independent and identically distributed with identity covariance matrix. This degenerate model is not identifiable because there are infinitely many (β_0, β_1) that satisfy $\mu_y = \beta_0 + \beta_1\mu_x$. The infimum in (2.10) takes $\boldsymbol{\theta}$ arbitrarily close to the unidentified model, (2.11). This is the basic idea behind the Gleser–Hwang effect.

It is important to note that if one a priori restricts the parameters to a closed subset not including $\sigma^2 = 0$ (or $\sum_{i=1}^{n}(\xi_i - \bar{\xi})^2 = 0$ in the functional model), then the Gleser–Hwang effect disappears. This might be quite reasonable in practice. For example, one might be certain that σ^2 is bounded below by some small number, $b > 0$, and that the error variances are bounded above by some large number. Unless these bounds are reasonably restrictive, however, the confidence regions might still be quite large. One needs the signal to noise ratio to be sufficiently large (see Section 2.4.2) to ensure reasonable confidence intervals.

2.4.2 Asymptotic confidence regions

In practice, perhaps the most common interval estimates for ME models, when the models are identifiable, are constructed using consistent estimators that are asymptotically normal (see Section 2.1.3; see also, for example, Fuller, 1987, Chapter 1). One can obtain the asymptotic single or joint confidence regions for the parameters by using the asymptotic covariance matrix. For example,

$$\hat{\beta}_1 \pm z_{\gamma/2}\hat{\sigma}_{\hat{\beta}_1} n^{-1/2} \tag{2.12}$$

is a large sample $1-\gamma$ confidence interval for β_1, in which z_ν is the $100(1-\nu)$ percentile of the standard normal distribution and $\hat{\sigma}_{\hat{\beta}_1}$ is a consistent estimator of the standard deviation of the large sample distribution of $\hat{\beta}_1$ given earlier in Section 2.1. Although the coverage probability of interval (2.12) might approach $1-\gamma$ as n becomes large for a fixed value of the parameter vector, it does not do so uniformly over the parameter space. Any use of such an asymptotic confidence interval requires knowledge of two things: the actual coverage probability and the expected length of the confidence interval. Because the confidence interval (2.12) is always bounded it has zero confidence level for any fixed n. Consequently, as n tends to infinity, the confidence level remains zero and does not tend to the nominal confidence level, $1-\gamma$. This casts serious doubts on the appropriateness of the use of (2.12) or any other confidence interval of finite length.

Example 2.1

As an example of the computation of an asymptotic confidence interval for β_1, consider the structural model with λ known. Theorem 2.1 gives the asymptotic variance as

$$\begin{aligned}
\sigma^2_{\hat{\beta}_1} &= \frac{1}{\sigma^4}\{(\sigma^2_\varepsilon+\beta_1^2\sigma^2_\delta)(\sigma^2+\sigma^2_\delta)-\beta_1^2\sigma^4_\delta\} \\
&= \frac{1}{\sigma^2}(\sigma^2_\varepsilon+\beta_1^2\sigma^2_\delta)+\frac{\sigma^2_\varepsilon\sigma^2_\delta}{\sigma^4}=\frac{\sigma^2_\delta}{\sigma^2}(\lambda+\beta_1^2)+\lambda\frac{\sigma^4_\delta}{\sigma^4}. \qquad (2.13)
\end{aligned}$$

In order to obtain an estimate $\hat{\sigma}_{\hat{\beta}_1}$ for use in (2.12), one can substitute the estimates given in equations (1.30) and (1.31) into (2.13) and take the square root.

Problems arise in the structural relationship when σ^2 is very close to 0, in the functional relationship when all ξ_i are almost the same, and in the ultrastructural relationship when $\sigma^2+\sum(\mu_i-\bar{\mu}_n)^2/n$ is small. As Gleser (1987) points out, it is intuitively clear that when the ξ_i do not vary very much, it is difficult to fit a straight line through the points $(\xi_i, \beta_0+\beta_1\xi_i)$, $i=1,\ldots,n$. Gleser (1987) examined (2.12) for both the structural and functional models under the assumption that λ is known and rescaled to one. He found a lower bound on the coverage probability as a function of γ and the so-called '**signal to noise**' ratio

$$\kappa=\begin{cases}\sigma^2/\sigma^2_\delta & \text{for the structural model}\\ \sum(\xi_i-\bar{\xi}_n)^2/n\sigma^2_\delta & \text{for the functional model.}\end{cases} \qquad (2.14)$$

He showed that if κ is at least one then (2.12) provides quite reasonable confidence interval for β.

Recall that $\kappa_\xi=\kappa/(\kappa+1)$ is called the *reliability ratio* (1.13). For $\kappa=1$, the reliability ratio is 0.5. The preceding arguments point out that one of the main practical ramifications of the Gleser–Hwang effect is that it is important to know or estimate the reliability ratio. See Hasabelnaby *et al.* (1989).

Though Gleser's results are based on knowledge of λ, it is likely that they can be extended to other assumptions; however, his method of proof no longer works. In applications, one would need to estimate κ and this introduces additional accuracy problems. One can use consistent estimators of σ^2 (or $\sum(\xi_i - \bar{\xi})^2/n$) and consistent estimators of σ_δ^2 to estimate κ.

Huwang (1996) improved the asymptotic confidence set in the structural model when σ_δ^2 is known. The usual asymptotic confidence set replaces the unknown parameters in the asymptotic covariance matrix by their estimates; however, Huwang used an approach based on an idea in Hwang (1995) and did not substitute an estimate for β_1 in the asymptotic covariance matrix. Because β_1 is difficult to estimate when σ^2 is small, leaving β_1 unestimated should reduce the error in this case. Huwang showed that the improved asymptotic confidence set has coverage probability converging to the nominal confidence level uniformly in all parameters; however, this confidence set will inevitably be unbounded because of the Gleser–Hwang effect. Theoretical and simulation results in Huwang (1996) show that the improved set outperforms the traditional set, especially when the signal to noise ratio is low.

2.4.3 Confidence sets using pivotal statistics: both error variances known

When both error variances are known, Brown (1957) proposed a confidence set for (β_0, β_1). The idea is to use a **pivotal quantity** (or **pivot**), that is, a function of the data and parameters whose distribution does not depend on any unknown parameters (see, for example, Casella and Berger, 1990; or Stuart and Ord, 1991, Section 31.21). To apply this idea to ME models, note that $y_i - \beta_0 - \beta_1 x_i$ is normal with mean 0 and variance $\sigma_\varepsilon^2 + \beta_1^2\sigma_\delta^2$ regardless of the assumptions on ξ_i, that is, no matter whether the model is ultrastructural, structural or functional (Brown only considered the structural relationship). When σ_ε^2 and σ_δ^2 are both known, rescale so that both are unity, and then

$$\sum(y_i - \beta_0 - \beta_1 x_i)^2/(1 + \beta_1^2)$$

is χ^2 with n degrees of freedom. For c_γ satisfying $P(\chi_n^2 \le c_\gamma) = 1 - \gamma$, the set

$$\left\{(\beta_0, \beta_1) \,\middle|\, \sum \frac{(y_i - \beta_0 - \beta_1 x_i)^2}{1 + \beta_1^2} \le c_\gamma\right\} \tag{2.15}$$

gives a $1 - \gamma$ confidence region for (β_0, β_1). Expression (2.15) is equivalent to (see Exercise 2.11)

$$\beta_0^2 + \left(s_{xx} - \frac{c_\gamma}{n}\right)\beta_1^2 - 2s_{xy}\beta_1 + \left(s_{xx} - \frac{c_\gamma}{n}\right) \le 0. \tag{2.16}$$

Equality in (2.16) defines a conic in the (β_0, β_1) plane. Brown's conic could be an ellipse, a hyperbola, degenerate or have no real values satisfying (2.16). The confidence region is bounded only when the conic is an ellipse.

One might think that Brown's result can be generalized to the λ known case because one can replace both error variances by their consistent estimates respectively. It was found by Okamoto (1983), however, that the (asymptotic) distribution of Brown's result is not χ^2 if the estimates are used. Okamoto (1983) made an adjustment for this situation.

Kendall and Stuart (1979, p. 416) made the comment that it is not efficient to use confidence limits for the mean of a normal distribution with known variance by using the χ^2 distribution of the sum of squares about the mean. Because the procedure proposed by Brown is based on an inefficient procedure, it is probably also inefficient.

One can modify Brown's approach for σ_ε^2 and σ_δ^2 known by substituting a normally distributed statistic for the χ^2 statistic because $(y_i - \beta_0 - \beta_1 x_i)/\sqrt{(1+\beta_1^2)}$ is standard normal (assuming both error variances are unity). This approach was proposed in Cheng and Van Ness (1994) and is summarized as follows. First, consider the special case when β_0 is zero. Then $\{n(1+\beta_1^2)\}^{-1/2} \sum (y_i - \beta_1 x_i)$ has a standard normal distribution, and the $1-\gamma$ confidence interval of β_1 can be obtained by solving

$$\left| n^{-1/2} \sum \frac{y_i - \beta_1 x_i}{(1+\beta_1^2)^{1/2}} \right| \leq z_{\gamma/2}$$

or

$$n(\bar{y} - \beta_1 \bar{x})^2 \leq z_{\gamma/2}^2 (1+\beta_1^2). \tag{2.17}$$

The resulting quadratic inequality is $a\beta_1^2 - 2b\beta_1 + c \leq 0$, where $a = \bar{x}^2 - z_{\gamma/2}^2/n$, $b = \bar{x}\bar{y}$, and $c = \bar{y}^2 - z_{\gamma/2}^2/n$. The roots of this quadratic are either (i) both real or (ii) both complex. In case (i), if $a > 0$, the $1-\gamma$ confidence interval is between the two roots, that is, the interval $[(b - \sqrt{b^2 - ac})/a, (b + \sqrt{b^2 - ac})/a]$, and if $a < 0$, it is the complement of this interval. In case (ii), the confidence interval is the real line because the quadratic function $a\beta_1^2 - 2b\beta_1 + c$ is either always negative or always positive and (2.17) is satisfied with strict inequality for $\beta_1 = \bar{y}/\bar{x}$.

One might be disappointed that, after all the mathematical labours, the confidence interval turns out to be the entire real line, but this is not a vacuous statement because it says that the data are such that they cannot specify β_1 any more precisely. This problem is similar to the Fieller problem (Fieller, 1954) of finding the confidence interval of the ratio of two normal means. Compare the Fieller problem for two independent samples with known variances. The estimate, $\hat{\beta}_1$, discussed is the ratio of two means. The difference is that, in the ME model, x and y are dependent, but the surprise is that the two solutions are identical (Cox and Hinkley, 1974, p. 233). We discuss this further below. For related 'improper confidence sets' having the property that with positive probability the confidence set gives a trivially true statement, see Scheffé (1970).

Now consider the intercept model. The $1-\gamma$ confidence region for (β_0, β_1) can be obtained by solving

$$\left| n^{-1/2} \sum \frac{y_i - \beta_0 - \beta_1 x_i}{(1+\beta_1^2)^{1/2}} \right| \leq z_{\gamma/2}.$$

Then with probability $1-\gamma$,

$$\left(\bar{x}^2 - \frac{z_{\gamma/2}^2}{n} \right) \beta_1^2 + 2\bar{x}\beta_1(\beta_0 - \bar{y}) + (\beta_0 - \bar{y})^2 \leq \frac{z_{\gamma/2}^2}{n}. \tag{2.18}$$

Equality in (2.18) defines a conic in the (β_0, β_1) plane. It can be shown that it always has real solutions and always defines a hyperbola. So the confidence region is the area between the two curves of the hyperbola and is unbounded.

The reason for the improper confidence sets is explained by the Gleser–Hwang effect and by this confidence region's relation to the Fieller problem just mentioned. The ME model (1.2) and (1.3), when $\beta_0 = 0$, can be made arbitrarily close to the model

$$x_i = \mu + \delta_i, \quad y_i = \beta_1 \mu + \varepsilon_i, \quad i = 1, \ldots, n.$$

This can happen, for example, when σ^2 is arbitrarily close to zero in the structural model. When both of the error variances are unity, then the (x_i, y_i) are independent and identically distributed with mean $(\mu, \beta_1 \mu)$ and identity covariance matrix. This is equivalent to the Fieller problem because β_1 is the ratio of these two means and the ratio is meaningless if $\mu = 0$. Similarly, the intercept model can be made arbitrarily close to the model (2.11), which is unidentifiable as mentioned earlier.

Because Brown's confidence region is sometimes bounded and (2.18) is always unbounded, it might appear that the former is superior in some sense. Cox and Hinkley (1974, p. 234) make the following comment on the Fieller problem:

> By using a less efficient test, we might, for a particular set of data, avoid effects of the type discussed above (the whole real line as the confidence region), but we would be ignoring some information about the parameter in question.... Difficulty only arises if we incorrectly attempt to deduce a probability statement about the parameter from the confidence region statement.

2.4.4 Confidence sets using pivotal statistics: λ known

Now consider the Creasy–Williams confidence set (Creasy, 1956; Williams, 1959; 1973). Creasy (1956) obtained a confidence interval for the slope of the structural relationship in the case where the ratio of the error variances λ is known. For simplicity assume that λ is unity. Creasy worked with the angles $\hat{\varphi} = \tan^{-1} \hat{\beta}_1$ and proved that

$$t = \frac{1}{2} \left[(n-2) \frac{(s_{xx} - s_{yy})^2 + 4s_{xy}^2}{s_{xx}s_{yy} - s_{xy}^2} \right]^{1/2} \sin 2(\hat{\varphi} - \varphi) \qquad (2.19)$$

is distributed as Student's t with $n - 2$ degrees of freedom. Some gaps in Creasy's proof were clarified by Schneeweiss (1982). A given value of $|t|$ in (2.19) could yield four values of φ, as a result of the periodicity of the sine function. Thus confidence sets consisting of the entire interval $[-\pi/2, \pi/2]$, two disjoint intervals or three disjoint intervals are possible. In order to avoid this difficulty, Creasy (1956) assumed that $P(|\hat{\varphi} - \varphi| > \pi/4)$ is negligible and an approximate confidence interval results. Anderson and Sawa (1982) derived the density function of $\hat{\beta}$ and showed that it is continuous and strictly positive on the real line. Consequently, $P(|\hat{\varphi} - \varphi| > \pi/4)$ is always positive though is difficult to calculate exactly. This casts some doubt on the assumption that it is negligible.

If $P(|\hat{\varphi} - \varphi| > \pi/4)$ can be assumed negligible, then (2.19) gives the following confidence interval for φ:

$$\hat{\varphi} \pm \frac{1}{2} \sin^{-1} \left[2t \left\{ \frac{s_{xx}s_{yy} - s_{xy}^2}{(n-2)(s_{xx} - s_{yy})^2 + 4s_{xy}^2} \right\} \right], \tag{2.20}$$

where t is the appropriate Student's t deviate for $n - 2$ degrees of freedom and the confidence level being used.

Creasy (1956) claimed the confidence interval (2.19) is also appropriate for the functional relationship based on the false statement that the distribution functions of $\hat{\beta}$ for both structural and functional relationships are identical. Actually, the ML estimates, not the distributions of the estimates, are the same for functional and structural relationships when the ratio of error variances is known. This was pointed out by Patefield (1981b), who further proved that (2.19) is suitable for the functional relationship when (1.48) holds.

Example 2.2
For the data of Example 1.11,

$$\hat{\varphi} = \tan^{-1} 1.99 = 1.105 = 0.352\pi.$$

The two-sided 95% confidence value for t_7 is $t = 2.365$. Substituting into (2.20) gives

$$1.105 \pm \frac{1}{2} \sin^{-1} \left[4.73 \left\{ \frac{(238 \times 906 - 451^2)^{1/2}}{\sqrt{7} \times 1122} \right\} \right] = 1.105 \pm 0.08856.$$

The resulting 95% confidence interval for β_1 is

$$[\tan(1.105 - 0.08856), \tan(1.105 + 0.08856)] = [1.61, 2.52].$$

The ML estimate, 1.99, is not the centre of this interval because of the transformation to φ used to obtain the interval. The resulting confidence interval is rather wide because of the small sample size.

Williams (1959; 1973) used the fact that $x_i + \beta_1 y_i$ and $\beta_1 x_i - y_i$ are jointly normal and uncorrelated. Let $r(\beta_1)$ be the sample correlation of $x_i + \beta_1 y_i$ and $\beta_1 x_i - y_i$; then

$$\tau = (n-2)^{\frac{1}{2}} \frac{r(\beta_1)}{(1 - r^2(\beta_1))^{1/2}} = (n-2)^{\frac{1}{2}} \frac{(\beta_1^2 - 1)\, s_{xy} + \beta_1(s_{xx} - s_{yy})}{(1 + \beta_1^2)(s_{xx}s_{yy} - s_{xy}^2)^{1/2}} \tag{2.21}$$

has a t distribution with $n - 2$ degrees of freedom. Schneeweiss (1982) made the remark that (2.19) and (2.21) are identical. In fact, these two statistics are identical up to a negative sign. Because the t distribution is symmetric, there is no difference in the distribution theory of (2.19) and (2.21). Thus the Williams $1 - \gamma$ confidence interval for β_1 is

$$\left\{ \beta_1 : \frac{(n-2)r^2(\beta_1)}{1 - r^2(\beta_1)} \leq F_{1,n-2,\gamma} \right\} \tag{2.22}$$

where $F_{1,n-2,\gamma}$ is the $100(1-\gamma)$ percentile of the F distribution. Casella and Berger (1990, p. 594) showed that the function describing the set (2.22) has two minima at which the function is zero (see Figure 2.1). The confidence set can therefore consist of two finite disjoint intervals, one finite and two infinite disjoint intervals, or the whole real line. Moreover, for every β_1, $-r(\beta_1) = r(-1/\beta_1)$; so if β_1 is in the confidence set, so is $-1/\beta_1$. In other words, β_1 and $-1/\beta_1$ cannot be distinguished and hence the sign of β_1 can never be determined from this confidence set. In many applications, however, the sign of β_1 is known a priori and the part of the confidence set that contains values of β_1 of the opposite sign can be ignored. Recall that $\hat{\beta}_1$ has the same sign as s_{xy} when λ is known and the errors are independent (see Section 1.3.1). Creasy's reparameterization from β_1 to φ, of course, guarantees bounded confidence intervals, but the Gleser–Hwang effect persists and any confidence set for φ with coverage probability greater than one-half will have positive probability of containing the whole parameter space.

Example 2.3
For the data of Example 1.11, the two-sided 95% confidence interval obtained from (2.22) is

$$\frac{7r^2(\beta_1)}{1-r^2(\beta_1)} \leq F_{1,7,0.95} = 5.59,$$

which implies

$$\begin{aligned} r^2(\beta_1) &= \frac{[(\beta_1^2-1)s_{xy} + \beta_1(s_{xx}-s_{yy})]^2}{(s_{xx}+2\beta_1 s_{xy}+\beta_1^2 s_{yy})(\beta_1^2 s_{xx} - 2\beta_1 s_{xy} + s_{yy})} \\ &= \frac{((\beta_1^2-1)451 + \beta_1(238-906))^2}{(238+2\beta_1 451+\beta_1^2 906)(\beta_1^2 238 - 2\beta_1 451 + 906)} = 0.444. \end{aligned}$$

A plot of r^2 is given in Figure 2.1.
Solving this yields

$$r(\beta_1) = \frac{(\beta_1^2-1)451 + \beta_1(238-906)}{(238+2\beta_1 451+\beta_1^2 906)^{1/2}(\beta_1^2 238 - 2\beta_1 451 + 906)^{1/2}} = \pm 0.66633.$$

Solving this in turn gives four solutions, $\beta_1 = -0.620$, $\beta_1 = -0.397$, $\beta_1 = 1.61$, and $\beta_1 = 2.52$. The resulting 95% confidence interval for β_1 is

$$[1.61, 2.52] \cup [-0.620, -0.397] = [1.61, 2.52] \cup [-1/1.61, -1/2.52].$$

If it is known a priori that β_1 is nonnegative, then the result is the same as in the previous example.

When one of the error variances is known, no pivot is available. This is why in these cases there is no interval estimation formula except the one obtained using the asymptotic confidence set. Such a set will be bounded and, therefore, must have confidence level zero for any finite n.

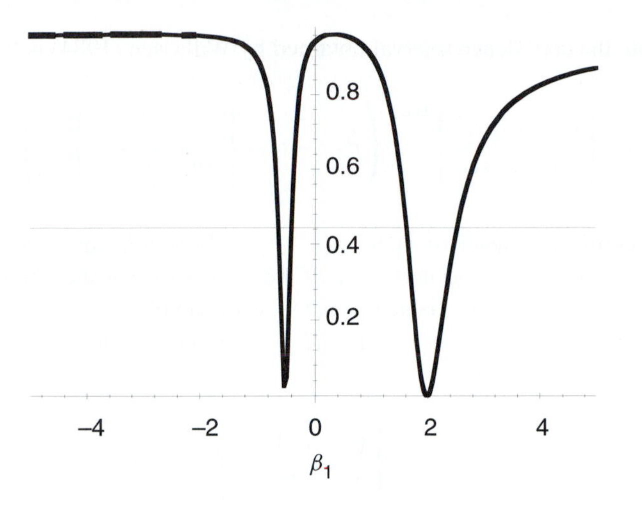

Figure 2.1: r^2 for the data of Example 1.11.

2.4.5 Confidence sets for unidentifiable models

Interestingly, one can find a confidence interval for β_1 even though the β_1 is not identifiable. An approach using (1.28) was suggested by Moran (1971, p. 237) and worked out in detail by Willassen (1984). Consider the structural model and assume that the true value of β_1 is positive. This is no restriction because one can always change the sign of y or x. Let $\beta_{1R} = \beta_1\sigma^2/(\sigma^2+\sigma_\delta^2)$ and $\beta_{1I} = \beta_1\sigma^2/(\beta_1^2\sigma^2+\sigma_\varepsilon^2)$ denote the true regression slopes of y on x and x on y respectively and let $\hat{\beta}_{1R} = s_{xy}/s_{xx}$ and $\hat{\beta}_{1I} = s_{xy}/s_{yy}$ be their respective estimates. Note that β_{1R} and β_{1I} are both positive because β_1 is. Because β_1 is contained in the interval $[\beta_{1R}, 1/\beta_{1I}]$, it suffices to construct a confidence interval for this interval. That is to say, one can find a $1-\gamma/2$ confidence interval (d_1, d_2) for β_{1R} and a $1-\gamma/2$ confidence interval (d_3, d_4) for $1/\beta_{1I}$ based on the two regressions, y on x and x on y, respectively. Then, by the Bonferroni inequality, one can take (d_1, d_4) as a $1-\gamma$ confidence interval for β_1. Note that this interval may be a bit too wide, it can even have infinite length, but it might still provide useful information even though the parameter is not identifiable.

Let

$$
\begin{aligned}
t_1 &= \frac{(\hat{\beta}_{1R} - \beta_{1R})\{(n-2)s_{xx}\}^{1/2}}{\{(1-r^2)s_{yy}\}^{1/2}}, \\
t_2 &= \frac{(\hat{\beta}_{1I} - \beta_{1I})\{(n-2)s_{yy}\}^{1/2}}{\{(1-r^2)s_{xx}\}^{1/2}},
\end{aligned}
$$

where $r^2 = s_{xy}^2/s_{xx}s_{yy}$ is the square of the sample correlation. Then t_1 and t_2 have t distributions with $n-2$ degrees of freedom. So a confidence interval for the interval $[\beta_{1R}, 1/\beta_{1\mathrm{I}}]$ is $(d_1, d_4) \equiv \{\beta_1 : t_1 \leq t_{\gamma/2}\} \cap \{\beta_1 : t_1 \leq t_{\gamma/2}\}$. After some algebraic manipulation, the confidence interval obtained by Willassen (1984) is

$$\left[\hat{\beta}_{1R} - t_{\gamma/2}\left\{\frac{(1-r^2)\,s_{yy}}{(n-2)\,s_{xx}}\right\}^{1/2}, \left(\hat{\beta}_{1I} - t_{\gamma/2}\left\{\frac{(1-r^2)\,s_{xx}}{(n-2)\,s_{yy}}\right\}^{1/2}\right)^{-1}\right]. \tag{2.23}$$

This has a confidence level of at least $1-\gamma$. Note that this interval tends to $[\beta_{1R}, 1/\beta_{1I}]$ as $n \to \infty$ and not to β_1. One cannot expect the confidence interval to tend to β_1 because there is no consistent estimate of β_1.

At first glance, it might appear that (2.23) violates the Gleser–Hwang effect; however, it must be true that

$$\hat{\beta}_{1I} - t_{\gamma/2}\left\{\frac{(1-r^2)\,s_{xx}}{(n-2)\,s_{yy}}\right\}^{1/2} > 0 \tag{2.24}$$

because otherwise the inequality reverses for the bound on $1/\beta_{1I}$. If this happens, then the confidence interval contains the entire positive real line, which is no restriction because β_1 was assumed positive. Furthermore, (2.24) is equivalent to

$$r^2 > \frac{t^2_{\gamma/2}}{n-2+t^2_{\gamma/2}}$$

whose right-hand side is strictly between 0 and 1 if $n > 2$. It is clear that

$$P\left\{r^2 > \frac{t^2_{\gamma/2}}{n-2+t^2_{\gamma/2}}\right\} < 1$$

because r has a continuous density that is strictly positive on $(-1, 1)$.

Even if this confidence interval is finite, intuitively one can expect trouble with this approach without employing the results of Gleser and Hwang. It is only known that $\beta_1 \in [\beta_{1R}, 1/\beta_{1I}]$. If $[\beta_{1R}, 1/\beta_{1I}]$ is too wide, the confidence interval will not be very informative even for large samples. Note that

$$[\beta_{1R}, 1/\beta_{1I}] = \left[\frac{\beta_1\kappa}{1+\kappa}, \beta_1 + \frac{1}{\beta_1\kappa_\varepsilon}\right] \tag{2.25}$$

where κ is the signal to noise ratio defined by (2.14), and $\kappa_\varepsilon \equiv \sigma^2/\sigma^2_\varepsilon$. To make the interval short one needs κ and κ_ε large. This is related to Gleser's assertion, mentioned earlier, that κ should be greater than 1 (see Section 2.4.2). Suppose, for example, that $\kappa = \kappa_\varepsilon = 1$; then if $\beta_1 = 1/100$, the interval is [0.005, 100.1]; whereas if $\beta_1 = 100$, the interval is [50, 100.1]. The problem persists if β_1 is either too large or too small. It seems reasonable to use (2.25) when it is short enough to give useful information. The real merit of Willassen's result is that one can make some inference even when the parameters are not uniquely determined.

Example 2.4
We apply (2.23) on the data of Example 1.11 to get a two-sided 95% confidence interval. First, $\hat{\beta}_{1R} = s_{xy}/s_{xx} = 451/238 = 1.895$, $\hat{\beta}_{1I} = s_{xy}/s_{yy} = 451/906 = 0.4978$, and $r^2 = s_{xy}^2/s_{xx}s_{yy} = 451^2/(238 \times 906) = 0.9433$. The interval (2.23) is

$$\left[1.895 - 2.365\left(\frac{0.0567 \times 906}{7 \times 238}\right)^{1/2}, \left(0.4978 - 2.365\left(\frac{0.0567 \times 238}{7 \times 906}\right)^{1/2}\right)^{-1}\right]$$
$$= [1.48, 2.57].$$

This interval is reasonably good because $\hat{\beta}_{1R} = 1.895$ is reasonably close to $1/\hat{\beta}_{1I} = 1/0.4978 = 2.009$. It is slightly larger than the intervals of Examples 2.1 and 2.2 but those intervals assumed that λ is known. If λ is not actually equal to one as assumed in Examples 2.1 and 2.2, then those intervals do not have a 95% confidence level, whereas the confidence interval given here is not affected by λ.

2.4.6 General comments

Theoretically, if the exact distribution of $\hat{\beta}_1$ were known, then there should not be any difficulty in deriving a confidence interval for β_1. But because of the complexity of the distributions derived thus far, calculating such a confidence interval analytically is not practical, though it might be possible numerically. Furthermore, all derived distributions for $\hat{\beta}_1$ are a function of other unknown parameters, in particular the signal to noise ratio (or the reliability ratio). For example, the Anderson and Sawa (1982) and Wong (1989) exact distributions for $\hat{\beta}_1$ both are functions of the signal to noise ratio. The signal to noise ratio is estimable and estimates could be substituted in the distributions in an attempt to obtain approximate confidence intervals. The signal to noise ratio has an accuracy problem of its own, however. Ideally, it would be better to have an independent experiment to estimate the signal to noise ratio. In any case, the signal to noise ratio (or reliability ratio) plays a key role here.

Finally, we make a point about the correct interpretation of the preceding discussion. The Gleser–Hwang effect is correct but this does not mean that for any given data set one will definitely have the zero confidence level effect. What it means is that one cannot take for granted, for example, that the asymptotic confidence of (2.12) will have the indicated coverage probability when the sample size is sufficiently large. Results in Gleser (1987) state that, if the signal is at least as large as the noise (in the sense of comparing their variances), then the coverage probability is close to the nominal one even if the sample size is small (as low as 25). But if the signal to noise ratio is less than one, there might be a problem. In practice, for a given data set, the structure is already fixed. One can then estimate the parameters when the model is identifiable and then use these estimates to compare the noise and signal to decide if one can obtain a good interval estimate. Of course, an independent estimate of the signal to noise ratio would be preferred.

Another possible way to construct a confidence interval that is not subject to the Gleser–Hwang effect is to use a sequential approach. Recently Hwang and Liu

(1992) showed that the Gleser–Hwang effect persists for finite staged sequential sampling but one can construct fully sequential nontrivial confidence sets for β_1 whose confidence level is at least $1 - \gamma$ for any given γ between 0 and 1. These are mainly existence results, however, and the resulting fully sequential confidence sets are probably too wide to be useful. Further investigation is necessary to provide practical implementation.

2.5 Prediction and Calibration under Measurement Error Models

This section discusses the situation in which the goal of the modelling is to predict rather than to estimate model parameters. It turns out that this goal leads to different methods than the parameter estimation methods discussed heretofore.

2.5.1 Prediction

There is an interesting but sometimes misleading statement regarding the prediction of y from x that asserts that the ordinary regression least-squares predictor should be used even when dealing with the ME model. This is true under certain circumstances but not in general (see Section 1.2.4).

Suppose that a data set (1.4) arises from a linear structural model and one is given a new value of the x variable, say x_0, and wants to predict the new y_0. If the data are independent and identically distributed normal random variables, as is the case with the normal structural model, a well-known property of the normal distribution implies that $E(y_0 \mid x_0) = \alpha_0 + \alpha_1 x_0$ and the coefficients α_0 and α_1 can be estimated by $\tilde{\alpha}_1 = s_{xy}/s_{xx}$, and $\tilde{\alpha}_0 = \bar{y} - \tilde{\alpha}_1 \bar{x}$, the ordinary least-squares estimators of α_1 and α_0 respectively. The parameters α_0 and α_1 are not the same as β_0 and β_1. The best mean squared error linear unbiased predictor of y_0 conditional on $(x_1, \ldots, x_n, x_0)$ is

$$\hat{y}_0 = \tilde{\alpha}_0 + \tilde{\alpha}_1 x_0 = \bar{y} + \tilde{\alpha}_1 (x_0 - \bar{x}). \tag{2.26}$$

Note that the estimator $\hat{y}_0$ is consistent (see Exercises 2.6 and 2.7).

If the joint distribution of (x_i, y_i) is normal, then both $E(y \mid x)$ and $E(y \mid \xi)$ are linear. If one starts with the linear structural model

$$y_i = \beta_0 + \beta_1 \xi_i + \varepsilon_i, \qquad x_i = \xi_i + \delta_i,$$

with normally distributed error $(\delta_i, \varepsilon_i)$, then $E(y \mid x)$ is linear if and only if ξ_i is normally distributed (see Lindley, 1947; Kendall and Stuart, 1979, p. 438). Note that one need not assume any additional identifiability assumption in order to use $E(y \mid x)$ for prediction because the βs are not involved. This, then, is the source of the folkloric notion that if the object is prediction, it is not necessary to adjust for measurement error. Note that if the underlying model is the normal ME model, then it is not required that δ_i and ε_i be uncorrelated.

On the other hand, one should realize that if one imposes the linear structural model, one has implied that there are 'true' variables (ξ_i, η_i) as well as the observed variables (x_i, y_i). One must now clarify what is meant by prediction. For example, one might wish to find $\hat{\eta}_0$ instead of $\hat{y}_0$. Note that $E(\eta \mid x) = E(y \mid x)$ only if $E(\varepsilon \mid x) = 0$. Thus one needs to assume that $E(\varepsilon \mid x) = 0$, which is true when δ_i is independent of ε_i, in addition to the usual assumption that $(\delta_i, \varepsilon_i)$ is independent of ξ_i if one wants the regression predictor to be the best predictor of η_0. Another possibility, which might be rare in practice, is that one is given the true value ξ_0 and wishes to predict the corresponding value η_0. In the no-equation-error model, the exact relation is

$$\eta_0 = \beta_0 + \beta_1 \xi_0.$$

In this case, the ME model estimators of β_0 and β_1, not least-squares estimators, should be used. That is,

$$\hat{\eta}_0 = \hat{\beta}_0 + \hat{\beta}_1 \xi_0,$$

where $\hat{\beta}_0$ and $\hat{\beta}_1$ are the (consistent) estimates of β_0 and β_1, respectively. But, in this case, one must have an additional identifiability assumption because otherwise the parameters β_0 and β_1 are not identifiable. For the structural equation error model, a similar conclusion can be made because

$$E(\eta_0 \mid \xi_0) = E(\beta_0 + \beta_1 \xi_0 + q_i \mid \xi_0) = \beta_0 + \beta_1 \xi_0.$$

If the data are from a functional relationship with $(\delta_i, \varepsilon_i)$ independent and identically normally distributed, the ordinary least-squares estimator will not yield unbiased predictors. First of all, the data (1.4) are no longer identically distributed. Suppose the value ξ_i is fixed; then the conditional distribution of y_i given x_i becomes (assuming $\sigma_{\delta\varepsilon} = 0$)

$$f(y_i \mid x_i, \xi_i) = \frac{1}{\sigma_\varepsilon \sqrt{2\pi}} \exp\left\{ -\frac{(y_i - \beta_0 - \beta_1 \xi_i)^2}{2\sigma_\varepsilon^2} \right\},$$

so that $E(y \mid x, \xi) = \beta_0 + \beta_1 \xi$, which involves both the incidental parameter and the parameters β_0 and β_1. Thus the least-squares estimator for prediction will no longer be unbiased. For the functional model, one can use the ME model predictor

$$\hat{\eta}_0 = \hat{\beta}_0 + \hat{\beta}_1 x_0,$$

for prediction given x_0.

If one is certain that the model is structural, one could still attempt to use measurement error estimates for prediction. If the model is normally distributed, then

$$E(y_0 \mid x_0) = E(\beta_0 + \beta_1 \xi_0 + \varepsilon_0 \mid x_0) = \beta_0 + \beta_1 E(\xi_0 \mid x_0),$$

if $\sigma_{\delta\varepsilon} = 0$. The conditional expectation of ξ_0 given x_0 is

$$E(\xi_0 \mid x_0) = \frac{\sigma_\delta^2 \mu + \sigma^2 x_0}{\sigma^2 + \sigma_\delta^2}.$$

It can be easily shown (see Exercise 2.7) that, if the parameters are identifiable, then this approach will yield a consistent predictor, which coincides with the least-squares predictor asymptotically. In other words, the seemingly more accurate (measurement error) model and the additional (identifiability) knowledge does not make the prediction more accurate. In fact, for all the side assumptions, (1.12) except (1.12e), this approach and the least-squares approach are identical for the finite-sample case (see Exercise 2.8). The preceding arguments apply to both the classical model and the equation error model.

2.5.2 Calibration

A problem closely related to prediction is that of calibration. Given n pairs of linearly related measurements (1.4) from a calibration experiment, the problem is to use these data, together with a new y_0 (or η_0) to compute an estimate of the corresponding unknown x_0 (or ξ_0). The calibration problem is sometimes considered the inverse of the prediction problem. If one uses an ordinary regression model for linear calibration,

$$y = \beta_0 + \beta_1 x + \varepsilon, \tag{2.27}$$

then the goal is to predict x from y. This leads to a rather unsatisfactory set of classical calibration techniques based on regression or inverse regression. If the model (2.27) is written as

$$x = -\frac{\beta_0}{\beta_1} + \frac{1}{\beta_1} y - \frac{\varepsilon}{\beta_1},$$

it is immediately seen that this is neither a standard regression model nor a standard inverse regression model. Some people simply write the calibration model as

$$x = \gamma_0 + \gamma_1 y + \delta, \tag{2.28}$$

in which case 'calibration' is nothing but standard regression of x on y and there is no difference between the calibration problem and the regression problem, and therefore no need to distinguish the two fields. Furthermore, (2.28) assumes that y is measured without error, which in many applications is unrealistic. An ME model seems much more natural for the calibration problem than either of the previous regression models. In this case, calibration becomes an ordinary ME regression or an ordinary ME prediction problem with quite satisfactory solutions immediately available.

Example 2.5

The calibration problem could arise as follows. You have an old, slow, expensive, relatively precise instrument, x, that measures some quantity and whose readings are well understood for the purposes at hand. A new instrument, y, is introduced, which has a much faster response. It is believed to be a linear function of the values being measured by the old instrument. You wish to convert the new instrument's reading to the old instrument's reading, that is, to calibrate the new instrument. If both instruments measure with some error then the ME model might well be used.

If the xs in the calibration are at fixed prechosen levels, the calibration is said to be controlled; but if the xs are random, the calibration is said to be random.

If the no-equation-error structural model with λ known is assumed, then the problem is symmetric in x and y and there is no real distinction between the two variables. In any case, if an ME model is used, the solution to the calibration problem is simply to use the ME prediction methods of the previous subsection.

It is clear, from the above discussion, that if one uses an ME model for the random calibration problem, then $E(\xi_0 \mid y_0)$ is the best unbiased estimator of ξ_0 when (x, y) is normally distributed because random calibration corresponds to the structural model. This then leads to the 'inverse regression' solution for prediction. But this is not true for controlled calibration, which corresponds to the functional model. Also, in some calibration problems one is primarily interested in the relationship between the two instruments being calibrated and therefore is primarily interested in β_0 and β_1 themselves. In these cases the ME calibration does not lead to either regression or inverse regression.

A discussion of the calibration problem in the ME model context can be found in Cheng and Tsai (1994) and Cheng and Van Ness (1997b). The former paper compares the three calibration estimating procedures, namely, regression, inverse regression and ME model estimation. The latter paper discusses robust calibration methods and compares the various calibration methods on Monte Carlo data (see Section 7.4.3).

2.6 Bibliographic Notes and Discussion

The large sample properties of ML estimators of the structural model in Section 1.3.1 are given by general ML theory. Due to their algebraic complexity, explicit formulae for the asymptotic covariance matrix of these estimators were not obtained until Robertson (1974). For convenience we first presented the asymptotic covariance formulae for the ultrastructural models and then the other models just become special cases. One must keep in mind that the behaviour of the ML method is quite different depending on whether or not there are incidental parameters.

The nonexistence of a finite first moment of the estimators described in Section 2.3 is somewhat surprising to those familiar with ordinary regression analysis. It implies that there is no comparing of ME model estimators to ordinary least-squares estimators in terms of unbiasedness and minimum mean squared error. One can instead use such things as $P(|\hat{\beta}_1 - \beta_1| > t)$ for measuring the quality of $\hat{\beta}_1$. Unfortunately, such probabilities are difficult to calculate for finite samples. Fuller (1987, pp. 163 ff.) modified the ME model estimators so that they possess finite moments, have finite-sample properties superior to the ML estimates, and retain the asymptotic optimality properties. A difficulty with these modified estimators is that they have very complicated distributions (Gleser, 1993, p. 114).

Carroll *et al.* (1995, p. 32) note that, if one ignores the measurement error and uses least squares, the resulting estimates are biased, and ME model estimates correct this bias, at least asymptotically, but result in increased variance. This occurs for all ME models, linear or nonlinear. Therefore, some trade-off between bias and variance seems to be called for. They suggest that the mean squared error, the sum of the variance and the square of bias, is a suitable criterion. Now the problem is that the ME model estimators do not posses finite first moments except for the reliability ratio known case. One must, therefore, take the asymptotic mean squared error as a comparison. Clearly, more studies are necessary for the finite-sample situation. If one can a priori bound β_1, then a consistent estimate with finite moments is possible, as mentioned in Section 2.3.

The zero confidence level effect serves a caution about the use of confidence regions in the ME model, especially asymptotic confidence regions; however, this effect does not preclude the use of confidence regions in ME models. The practical meaning of this effect is that we need to check the reliability (or the signal to noise ratio). It is intuitively clear that, if the noise is greater than the signal (in the sense of comparing their variances) then the resulting confidence regions might not be very useful.

The prediction problem was mentioned in Lindley (1947, p. 232) and Madansky (1959, p. 176). Basically, they noticed that the prediction from a true value of ξ_0 is quite different from that from an observed value x_0. Section 2.5 discusses this seemingly confusing situation.

2.7 Exercises

Problems 2.1–2.3 concern the nonexistence of the first moments of the estimators $\hat{\beta}_1$ in different cases. Assume the functional model with $(\delta_i, \varepsilon_i)$ normally distributed.

2.1 (*Fuller, 1987, p. 28, Problem 13*) *For the σ_δ^2 known case:*

- *(a) find $E(\hat{\beta}_1 \mid \delta_1, \ldots, \delta_n)$;*
- *(b) show that $E(\hat{\beta}_1)$ does not exist (in fact, both the positive part and the negative part of the integral (expectation) are infinite); and*
- *(c) if we drop the normality assumption on the errors, does this imply that the expectation of $\hat{\beta}_1$ exists? Why or why not?*

2.2 *For the σ_ε^2 known case, show that $E\hat{\beta}_1$ does not exist. (Hint: show that both positive and negative parts of $E(\hat{\beta}_1 \mid \varepsilon_1, \ldots, \varepsilon_n)$ are infinite. Is this true for other distributions besides the normal?)*

2.3 *When the intercept is known, show that $E\hat{\beta}_1$ does not exist. (Hint: see Exercise 2.2. Is this true for other distributions as well?)*

2.4 *In the normal structural model, when both error variances are known, show that the asymptotic variance of $\hat{\beta}_1$ has smallest variance if we use (1.29).*

2.5 *In the normal structural model, when the intercept is known (and assumed to be zero), there is a pivot, and we can use this pivot to construct a confidence interval* *for β_1. Note that $y_i - \beta x_i$ is normal with mean zero and variance $\sigma_\varepsilon^2 + \beta_1^2\sigma_\delta^2$, and that*

$$\frac{\sqrt{n-1}(\bar{y} - \beta_1\bar{x})}{\sqrt{\beta_1^2 s_{xx} - 2\beta_1 s_{xy} + s_{yy}}}$$

has a t distribution with $n-1$ degrees of freedom. Find the confidence interval of β_1 based on this pivot.

2.6 *Show that in the normal structural prediction problem, for a given x_0, the least-squares predictor (2.26) converges to $E(y_0 \mid x_0)$.*

2.7 *It is interesting that, in the normal structural prediction problem,*

$$E(y_0 \mid x_0) = E(\beta_0 + \beta_1\xi_0 + \varepsilon_0 \mid x_0) = \beta_0 + \beta_1 E(\xi_0 \mid x_0),$$

assuming δ_i and ε_i are independent. Under the normality assumption,

$$E(\xi_0 \mid x_0) = \frac{\sigma_\delta^2\mu + \sigma^2 x_0}{\sigma^2 + \sigma_\delta^2}.$$

Therefore, if the parameters are identifiable (see Section 1.3) then one can replace them by their consistent estimates. Show that $\hat{y}_0 = \hat{E}(y_0 \mid x_0)$ is a consistent predictor.

2.8 *For all the identifiability side conditions (1.12) except (1.12e), show that the ME model predictor in Exercise 2.7 coincides with the least-squares predictor.*

2.9 *Assume that the no-intercept ME model holds. Obtain Brown's confidence interval (Section 2.4.3) and discuss its properties.*

2.10 *Derive (2.16) from (2.15).*

2.11 *Construct a confidence region for (β_0, β_1) for the λ known case ($\lambda = 1$) using the Creasy–Williams idea. (Hint: $y_i - \beta_0 - \beta_1 x_i$ and $x_i + \beta_1 y_i$ are uncorrelated. What are the degrees of freedom? See also Exercise 4.6.)*

2.8 Research Problems

2.12 *Anderson and Sawa (1982) derived the exact distribution of $\hat{\beta}_1$ in the functional model (when the errors are normally distributed) when the ratio of error variances is known, whereas Wong (1989) derived the exact distribution of $\hat{\beta}_1$ in the structural model (under the normality assumption) when both error variances are known. No similar result for the remaining identifiability cases is known. Find the exact distribution of $\hat{\beta}_1$ in these cases.*

2.13 *Anderson (1976) showed that asymptotically, in the functional model under the normality assumption of the errors* $(\delta_i, \varepsilon_i)$,

$$P\{|\hat{\beta}_1 - \beta_1| \leq t\} \leq P\{|\hat{\beta}_{1R} - \beta_1| \leq t\}$$

for all $t > 0$, *where* $\hat{\beta}_1$ *is the ML estimator (1.29) when* $\lambda = 1$ *and* $\hat{\beta}_{1R}$ *is the ordinary least-squares estimator, except when the magnitude of the slope* β_1 *is small. This is a very useful result because it means that the ML estimator is better than the ordinary least-squares estimator because the former is closer to the true parameter. Derive similar results in other cases.*

3

Comparing Model Assumptions and Modifying Least Squares

There has been considerable progress in understanding the relationships among the various measurement error and related models as the ME model literature has developed. This continued research has exposed some former misunderstandings and some claims that were true only if additional restrictions were imposed. In this chapter, we take up some of these issues.

We first compare the identifiability assumptions introduced in Chapter 1. Then we introduce two new estimation methods not discussed heretofore: generalized least squares and modified least squares. An important alternative to the model assumptions discussed here is the use of instrumental variables, which will be taken up in Chapter 4. Chapter 4 also discusses estimation methods for the linear nonnormal ME model, which do not require either the additional assumptions discussed in this chapter or instrumental variables.

3.1 Issues Facing Users of ME Models

Usually the random quantities in the ME model are assumed to be normally distributed. If so, then the first requirement is that additional information be given in order to make the model identifiable and/or to make consistent estimation possible. These additional assumptions can sometimes present practical difficulties that could cause users, who would otherwise employ ME models, to revert to ordinary regression, which does not require any additional assumptions. If the error in measuring ξ is large, however, ordinary regression will be seriously biased and inconsistent. In this case, it would be much better to gather more information that would enable the user either to specify one of the additional assumptions or to use one of the methods of Chapter 4 and then proceed with the appropriate ME model.

Whether to use the functional, structural, or ultrastructural model is another important issue facing the ME model user. The ultrastructural model can be considered close to the functional model because both have incidental parameters whose number increases with the sample size. This causes their estimates to mimic each other's behaviour. The ultrastructural model also has a 'structural' version in the sense that one could allow the μ_i to be random and identically distributed. Most articles in the literature focus on just one model, but it is important to keep in mind that there are close relationships between these models and close relationships between the behaviours of their estimates. The next section presents a unified approach to the functional, structural, and ultrastructural models and discusses the various identifiability assumptions.

We will then show that least-squares methods can be adapted to ME models. Ordinary least squares does not work, but in Sections 3.4 and 3.5 we introduce two general modifications which work for the linear functional, structural, and ultrastructural models. It seems reasonable to attempt such modifications, because least squares gives such simple and elegant results for ordinary regression.

3.2 A Unified Approach to the Functional, Structural, and Ultrastructural Relationships

We now discuss the interrelationships between the various types of ME models. In the past there has been some confusion concerning the terms 'functional' and 'structural', but now the statistics community is converging on the definitions used here.

We have seen that the point estimates of the 'structural parameters', β_0 and β_1, can be taken to be the same in both the functional and structural models, but more detailed inspection reveals important differences between these two models. Some researchers (among them Nussbaum, 1977) have made use of the results of one type of ME model to infer results about another type of ME model. Gleser (1983) considered a unified approach to the two models with the following main results.

(a) Estimators of the structural parameters that are consistent in a functional relationship are also consistent in the corresponding structural relationship. The converse of this result is not true. This is illustrated by Example 3.1.
(b) If a parameter is not identifiable in the structural relationship, it is also not consistently estimable in the corresponding functional relationship. If a parameter is consistently estimable in the structural relationship, there is no guarantee that the same parameter is consistently estimable in the corresponding functional relationship.
(c) Under normality, if the ML estimators for the structural relationship are consistent and asymptotically normal for the functional relationship when the limits (1.48) exist, then these structural ML estimators are best asymptotic normal estimators (for the functional relationship) whose asymptotic covariance matrix depends on the sequences of the incidental parameters only through the limits (1.48).

Example 3.1 (*Gleser, 1983*)
Consider the simple no-intercept model

$$x_i = \xi_i + \delta_i, \qquad y_i = \beta_1 \xi_i + \varepsilon_i, \qquad i = 1, \ldots, n, \tag{3.1}$$

with the $(\delta_i, \varepsilon_i)$ independent and identically distributed $N\left(\mathbf{0}, \sigma_\delta^2 \mathbf{I}_2\right)$, where $\mathbf{I}_2$ is the 2×2 identity matrix. This corresponds to the case in which λ is known to be 1. The

ML estimate of β_1 for the functional relationship is (see (1.30))

$$\hat{\beta}_1 = \frac{\sum y_i^2 - \sum x_i^2 + \left\{\left(\sum y_i^2 - \sum x_i^2\right)^2 + 4\left(\sum x_i y_i\right)^2\right\}^{1/2}}{2\sum x_i y_i}$$

provided $\sum x_i y_i \neq 0$. Define

$$\tilde{\beta}_1 \equiv \hat{\beta}_1 + n^{-1}\sum(-1)^i x_i \,.$$

If the ξ_i are independent and identically distributed random variables, each with mean μ and variance σ^2, it can be shown (see Exercise 3.2) that

$$n^{-1}\sum(-1)^i x_i \to 0 \qquad \text{with probability 1 as } n \to \infty, \tag{3.2}$$

so that $\tilde{\beta}_1$ is strongly consistent under the structural version of model (3.1). On the other hand, consider the functional version of the same model with

$$\xi_i = \mu + (-1)^i \sigma, \qquad i = 1, \ldots, n;$$

then it is easily seen that $\tilde{\beta}_1 \to \beta_1 + \sigma$ as $n \to \infty$. This violates the converse of (a) above.

Note that in (a) and (b), there is no need to require that $(\delta_i, \varepsilon_i)$ (or ξ_i in the structural relationship) be multivariate normal.

Gleser's results suggest that, in a theoretical study of a new ME model, one might begin with the structural version. It is then much easier to check for the existence of consistent estimators because there is no general methodology for finding consistent estimators for the functional model. In functional relationships, the theory is complicated by the fact that the number of parameters increases with sample size. It is not difficult to obtain estimators, such as ML estimators or method of moments estimators, for structural relationships. Conversely, once the point estimates are obtained, an advantage of working with functional relationships is that, if one can prove consistency and asymptotic normality for a sequence of estimators in the functional relationship, then these results are proved for the corresponding structural relationship. For asymptotic efficiency one can again turn to structural relationships for which a well-established theory is available.

Example 3.2 (*Gleser, 1985*)
A demonstration that the parameters of the ultrastructural relationship are not all consistently estimable is provided by appealing to the corresponding 'structural' version of the ultrastructural model. More precisely, let $\mu_1, \ldots, \mu_n$ be independent and identically distributed $N\left(\mu, \sigma_\mu^2\right)$ random variables; then the (x_i, y_i) are independent and identically distributed $N(\boldsymbol{\mu}, \boldsymbol{\Omega})$ random variables, where

$$\boldsymbol{\mu} = \begin{pmatrix} \mu \\ \beta_0 + \beta_1\mu \end{pmatrix}, \qquad \boldsymbol{\Omega} = \begin{bmatrix} \sigma_\delta^2 + \tau & \beta_1\tau \\ \beta_1\tau & \lambda\sigma_\delta^2 + \beta_1^2\tau \end{bmatrix},$$

$\tau = \nu\sigma_\delta^2 + \sigma_\mu^2 = \sigma^2 + \sigma_\mu^2$, $\lambda = \sigma_\varepsilon^2/\sigma_\delta^2$, and $\nu = \sigma^2/\sigma_\delta^2$. This, then, is a structural model having covariance structure, $\mathbf{\Omega}$. Assume that λ is known. Note that both ν and σ_μ^2 are not identifiable parameters because they appear only through τ. That is, the distribution of the data is the same for any pairs $(\nu_1, \sigma_{\mu,1}^2)$ and $(\nu_2, \sigma_{\mu,2}^2)$ satisfying $\nu_1\sigma_\delta^2 + \sigma_{\mu,1}^2 = \nu_2\sigma_\delta^2 + \sigma_{\mu,2}^2$. Using (b) above, it follows that ν is not consistently estimable in the original ultrastructural model.

3.3 Identifiability Assumptions and the Equation Error Model

Chapter 1 presented several identifiability assumptions that are sufficient to allow estimation of the parameters of the normal structural model. In analysing these assumptions, it is convenient to note first that there are, in effect, really three different cases to discuss: the ratio of the error variances known, one error variances known and the intercept known. We say this for the following reasons. The assumption that the error variance ratio, λ, is known and the assumption that both error variances are known are essentially the same. There are some further restrictions in the latter overidentified case that ensure a global maximum for the ML equations in the structural relationship (see Section 1.3.1; and Birch, 1964). For the functional relationship, these restrictions are not necessary (Willassen, 1979). The assumption that σ_δ^2 is known is essentially the same as the assumption that σ_ε^2 is known. The reliability ratio known case is closely connected to the σ_δ^2 known case. The σ^2 known assumption is artificial. We will now discuss the advantages and disadvantages of the three basic assumptions.

First consider the assumptions from a theoretical viewpoint. A major advantage of the λ known case is that there is a pivot,

$$\frac{y_i - \beta_0 - \beta_1 x_i}{(\lambda + \beta_1^2)^{1/2}\sigma_\delta}. \tag{3.3}$$

A pivot was defined in Section 2.4.3 to be a function of the data and parameters whose distribution does not depend on any parameters. Pivots are useful in forming hypothesis tests and confidence intervals. Section 1.7 uses pivots to eliminate nuisance parameters for the purpose of ML estimation. The generalized least-squares method proposed by Lindley (1947) and Sprent (1966) uses pivots, as does a method proposed by Morton (1981). Brown's (1957) confidence region uses the pivot (3.3). When one error variance is known, there is no pivot and these methods cannot be used.

Consider the fitting of a regression line to data from a geometrical point of view. The least-squares approach, which minimizes the sum of squares of vertical distances from the data points to the regression line, does not work well for ME models. An alternative is to rescale so that $\lambda = 1$ and minimize the perpendicular distances from the data points to the line, that is, use orthogonal regression (recall Section 1.2.3 and Figure 1.2). This regression line has the nice geometric interpretation of minimizing the sum of squares of the shortest distances between the data points and the line. This idea was used by Adcock (1877; 1878) and Pearson (1901) and many other authors. The statistical model that corresponds to orthogonal regression is the linear

ME model whose covariance matrix for (δ, ε) is a constant times the identity matrix. For ML estimation in the functional relationship, Lindley (1947, p. 236) made what he called the 'most convenient' assumption of knowing the error variance ratio λ. Indeed, as discussed in Section 1.3.2, the other assumptions do not give satisfactory ML estimators for the functional relationship except for the case where both error variances are known. Perhaps this explains why many authors call the error variances ratio known case the classical approach.

In many applications, for example in many areas of econometrics (Johnston, 1972, p. 287), researchers do not find the λ known assumption very useful. Some prefer the equation error model introduced in Section 1.5:

$$x_i = \xi_i + \delta_i, \qquad y_i = \beta_0 + \beta_1 \xi_i + e_i \tag{3.4}$$

where e_i contains 'equation error', q_i, that is,

$$e_i = \varepsilon_i + q_i,$$

ε_i is the measurement error, and ξ_i, $(\delta_i, \varepsilon_i)$, and q_i are independent. Thus

$$x_i = \xi_i + \delta_i, \qquad y_i = \eta_i + \varepsilon_i, \qquad \eta_i = \beta_0 + \beta_1 \xi_i + q_i. \tag{3.5}$$

The equation error model has the effect of changing the type of side condition, used to get consistent estimates, that is realistic in applications. For example, in order to assume that the ratio of error variances is known, one needs to know $\lambda_t = (\sigma_\varepsilon^2 + \sigma_q^2)/\sigma_\delta^2$. An a priori estimate of λ_t is frequently not available in applications in which equation error is used because one cannot estimate σ_q^2. On the other hand, an a priori or independent estimate of σ_δ^2 is obtainable, as previously mentioned, if one has replications on the true variable.

The model (3.4)–(3.5) is called the equation error model of Type I (or the Type I model for short) if δ and ε are independent. The Type I model is analytically equivalent to the ordinary ME model except for the changes in the usefulness and the effectiveness of the identifiability assumptions. If σ_δ^2 is known or if β_0 is known or if the reliability ratio, κ_ξ, is known for the Type I model, then the problem reduces to the corresponding ordinary ME model with σ_δ^2 known or β_0 known or κ_ξ known, respectively. Of course, here σ_e^2 takes the role of σ_ε^2. The assumption that σ_ε^2 is known does not make the Type I model identifiable. If both σ_δ^2 and σ_ε^2 are known, then this is the same as the σ_δ^2 known case except for the minor advantage that now σ_q^2 can be estimated.

The equation error model accommodates correlated measurement errors, δ_i and ε_i, in (3.4)–(3.5). The typical identifiability assumption is that the measurement error covariance matrix, $\mathbf{\Omega}_{\delta\varepsilon}$, is known, while the equation error variance, σ_q^2, is unknown. This model is called the equation error model of Type II (or Type II model for short). If $\sigma_q^2 = 0$ (the degenerate case), then the Type II model reduces to the classical ME model with the measurement error covariance matrix known. The degenerate Type II model is an 'overidentified' model equivalent to the ME model with both error variances known and zero correlation between the measurement errors. Typically, the identifiability assumption, σ_δ^2 known, is used in the Type I model. Thus the degenerate case of the typical Type I model leads to an identified model while the degenerate case of the usual Type II model leads to an overidentified model. This is one reason for distinguishing the Type I and Type II models. For the typical Type I model, it does not matter whether it has equation error because this does not make any difference in the estimation of β_1. On the other hand, in the typical Type II model, the estimation of β_1 is different for the degenerate and nondegenerate models. Although it first appears that the nondegenerate case estimator

$$\hat{\beta}_1 = (s_{xy} - \sigma_{\delta\varepsilon})/(s_{xx} - \sigma_\delta^2) \tag{3.6}$$

works for both nondegenerate and degenerate Type II models, in the latter case, (3.6) is not the ML estimator for the normal model (see Section 1.3).

Another difference between the Type I model and the Type II model is that the typical Type II model with error covariance nonzero does not correspond to an ME model with no equation error because

$$\mathbf{\Omega}_{\delta e} = \begin{bmatrix} \sigma_\delta^2 & \sigma_{\delta\varepsilon} \\ \sigma_{\delta\varepsilon} & \sigma_\varepsilon^2 + \sigma_q^2 \end{bmatrix}$$

has all its elements known except the lower right, σ_e^2, which is only bounded below by the known σ_ε^2.

In practice, the intercept known assumption is frequently immediately obvious from the properties of the problem. For example, in some applications it is obvious that if $\xi = 0$ then $\eta = 0$. The intercept known assumption is fundamentally different from the other two assumptions because it is a 'first-order' assumption, that is, it is an assumption about means rather than variances. Furthermore, a pivot exists for the intercept known case (see Exercise 2.5).

An interesting point concerning the symmetry properties of ME models can also be made (recall Section 1.2.2). Many people consider symmetry between the independent and dependent variables an important property of certain ME models. Unlike regression models, certain ME models are symmetric in the sense that there is no need to distinguish independent and dependent variables, because the ME model can be written as (1.16). In other words, the roles of the independent and dependent variables are interchangeable. Of course, this symmetry is destroyed if the assumed side condition is not symmetric; for example, if the side condition is that σ_δ^2 is known. It is easy to see that such an interchange of variables is also not valid for nondegenerate equation error model.

3.4 Generalized Least Squares

Because the least-squares estimator minimizes the sum of the squares of the vertical distances, it only minimizes the errors in fitting the dependent variable y. It is intuitively clear that least squares might not be suitable for ME models because it does not take the errors in measuring x into account. It is also clear that the least-squares estimator is biased (see (1.21)) and inconsistent for the ME model because, from (1.10), the 'errors', ζ_i, are correlated with the 'independent variables', x_i, and hence the least-squares estimator is biased. Asymptotically, the least-squares estimator $\hat{\beta}_1 = s_{xy}/s_{xx}$ converges to $\beta_1/(1 + \sigma_\delta^2/w)$, where $w = \sigma^2$ for the structural model and $w = s_{\xi\xi}^*$ (see (1.48)) for the functional model.

Let us examine just how least squares breaks down when applied to the estimation of β_0 and β_1 in the ME model. Averaging (1.10) over the n pairs of observed values (x_i, y_i), $i = 1, \ldots, n$, yields

$$\bar{y} = \beta_0 + \beta_1 \bar{x} + \frac{1}{n} \sum (\varepsilon_i - \beta_1 \delta_i). \tag{3.7}$$

The last term on the right-hand side of (3.7) has a zero expectation, implying that the estimating equation

$$\bar{y} = \beta_0 + \beta_1 \bar{x} \tag{3.8}$$

is unbiased in the sense that both sides have the same expectation. Unbiased estimating equations are defined and discussed in Section 3.5. Substituting (3.8) into (1.10) gives

$$y_i - \bar{y} = \beta_1 (x_i - \bar{x}) + (\varepsilon_i - \beta_1 \delta_i).$$

Multiplying this by $x_i - \bar{x}$ and averaging, one obtains

$$\frac{1}{n} \sum (y_i - \bar{y})(x_i - \bar{x}) = \frac{\tilde{\beta}_1}{n} \sum (x_i - \bar{x})^2 + \frac{1}{n} \sum (x_i - \bar{x})(\varepsilon_i - \tilde{\beta}_1 \delta_i), \tag{3.9}$$

where $\tilde{\beta}_1$ is the estimator of β_1. If the last term of (3.9) vanished as $n \to \infty$, then the least-squares estimator, $\hat{\beta}_1 = s_{xy}/s_{xx}$, would asymptotically satisfy (3.9). This last term, however, tends to $-\beta_1 \sigma_\delta^2$ and the least-squares estimate does not asymptotically satisfy (3.9).

Adcock (1877; 1878) considered the appropriate estimator to be orthogonal regression, which has been rediscovered many times during the first half of the twentieth century. Lindley (1947, p. 243), however, considered a *weighted least squares* approach to the model (1.10). The weight was taken to be proportional to the reciprocal of the variance of the errors ζ_i, that is, $\sigma_\varepsilon^2 + \beta_1^2 \sigma_\delta^2$. Thus, one minimizes the weighted least squares,

$$Q(\beta_0, \beta_1) = \frac{1}{n(\sigma_\varepsilon^2 + \beta_1^2 \sigma_\delta^2)} \sum (y_i - \beta_0 - \beta_1 x_i)^2. \tag{3.10}$$

Weighted least squares has drawn much attention in the literature; see, for example, Sprent (1966), Nussbaum (1976), Höschel (1978a; 1978b) and Gleser (1981). Since Sprent (1966), the name has standardized to *generalized least squares*. A good way to envision generalized least squares is via the following geometrical interpretation. Classical least squares minimizes the sums of squares of vertical distance because only the ys are measured with errors. In ME models, both the measurement errors must be taken into account. Because the errors are independent, one minimizes

$$\sum \frac{(x_i - \xi_i)^2}{\sigma_\delta^2} + \sum \frac{(y_i - \beta_0 - \beta_1 \xi_i)^2}{\sigma_\varepsilon^2} = \sum \frac{\delta_i^2}{\sigma_\delta^2} + \sum \frac{\varepsilon_i^2}{\sigma_\varepsilon^2}. \tag{3.11}$$

If δ_i and ε_i have the same variance, this is the same as minimizing

$$\sum (x_i - \xi_i)^2 + \sum (y_i - \beta_0 - \beta_1 \xi_i)^2 = \sum \delta_i^2 + \sum \varepsilon_i^2.$$

By Pythagoras' theorem, this is the same as minimizing the sum of squares of perpendicular distances from the observed points (x_i, y_i) to the estimated line, that is, minimizing

$$\sum \frac{(y_i - \beta_0 - \beta_1 x_i)^2}{1 + \beta_1^2}.$$

This is classical orthogonal regression. Minimizing (3.11) is equivalent to maximizing the likelihood function of the normal functional relationship when λ is known (see also Section 1.5).

Although many authors consider generalized least-squares estimation only for the functional model, it also works for the structural and ultrastructural models, because the maximum likelihood estimators of the βs are the same as those of the functional model. The approach also generalizes to multivariate models, in which y and x are multidimensional, and to the mixed model, in which some of the predictor variables can be measured without error. One basic assumption remains unchanged: namely, the error covariance matrix has to be known up to a scalar multiple.

3.5 Modified Least Squares

The success of generalized least-squares might give people the impression that it is *the* least-squares method for the ME model. Madansky (1959, p. 177) commented that (generalized) least squares also works for the case of one error variance known, but this is incorrect. By direct calculation, it can be seen that the minimization of (3.10) does not lead to a solution if only σ_δ^2 or σ_ε^2 is known. In fact, generalized least-squares estimation only works for the no-equation-error model with the error covariance matrix known up to a scalar multiple.

Thus, a unified approach for modifying least squares to suit all the different assumptions on the error covariance structure is called for. *Modified least squares* is such an approach; it works for the structural, functional, and ultrastructural models simultaneously as well as for the equation error model. The normality assumption on the errors (and on the true variables for the structural and ultrastructural models) is not needed, only the existence of second moments.

First, consider the classical no-equation-error model in which the errors δ_i and ε_i are independent. Define the *residuals* to be

$$v_i = y_i - \beta_0 - \beta_1 x_i \ .$$

From (1.10), $v_i = \zeta_i = \varepsilon_i - \beta_1 \delta_i$ and it is clear that v_i are independent and identically distributed random variables with mean zero and variance $\sigma_\varepsilon^2 + \beta_1^2 \sigma_\delta^2$ regardless of whether the model is functional, structural or ultrastructural. Cheng and Tsai (1995) developed modified least-squares estimators for β_0 and β_1 by minimizing an unbiased and consistent estimator of the appropriate unknown error variance. The estimators are a function of the residuals. The details are as follows.

Case 1: σ_δ^2 known.

The unknown error variance is σ_ε^2 and

$$Q(\beta_0, \beta_1) = n^{-1} \sum \{(y_i - \beta_0 - \beta_1 x_i)^2 - \beta_1^2 \sigma_\delta^2\} \tag{3.12}$$

is an unbiased and consistent estimator of σ_ε^2. Minimizing Q with respect to β_0 and β_1, by taking partial derivatives of Q with respect to β_0 and β_1 and setting these partial derivatives equal to zero, yields

$$\hat{\beta}_0 = \bar{y} - \hat{\beta}_1 \bar{x}, \tag{3.13}$$

and

$$\hat{\beta}_1 = \frac{s_{xy}}{s_{xx} - \sigma_\delta^2}, \qquad \text{provided } s_{xx} - \sigma_\delta^2 > 0, \tag{3.14}$$

which are the same estimates as the ML estimates (1.35). Note that $s_{xx} - \sigma_\delta^2 > 0$ is required for this estimate though it is possible to have data for which this does not hold.

Case 2: σ_ε^2 known.

The unknown error variance is σ_δ^2 and

$$Q(\beta_0, \beta_1) = \frac{1}{n\beta_1^2} \sum \{(y_i - \beta_0 - \beta_1 x_i)^2 - \sigma_\varepsilon^2\}$$

is an appropriate estimator of σ_δ^2. After minimizing Q with respect to β_0 and β_1, $\hat{\beta}_0$ is the same as (3.13) and

$$\hat{\beta}_1 = \frac{s_{yy} - \sigma_\varepsilon^2}{s_{xy}}, \qquad \text{provided } s_{yy} - \sigma_\varepsilon^2 > 0 \text{ and } s_{xy} \neq 0,$$

which is the same as the ML estimate (1.38). Again, $s_{yy} - \sigma_\varepsilon^2 > 0$ is required in order to obtain the minimum.

Case 3: λ known.

Neither $\boldsymbol{\sigma_\varepsilon^2}$ nor $\boldsymbol{\sigma_\delta^2}$ is known. There are two possible error variance estimates that one can minimize. First, modified least-squares estimators can be obtained by minimizing the unknown error variance σ_δ^2. It is readily seen that

$$Q(\beta_0, \beta_1) = \frac{1}{n(\lambda + \beta_1^2)} \sum (y_i - \beta_0 - \beta_1 x_i)^2 \tag{3.15}$$

is an appropriate estimator of σ_δ^2. Minimizing Q with respect to β_0 and β_1 yields $\hat{\beta}_0$ the same as (3.13) and

$$\hat{\beta}_1 = \frac{s_{yy} - \lambda s_{xx} + \{(s_{yy} - \lambda s_{xx})^2 + 4\lambda s_{xy}^2\}^{1/2}}{2 s_{xy}}, \qquad \text{provided } s_{xy} \neq 0, \tag{3.16}$$

which is the same as the ML estimate (1.30). Second, if the estimating function is obtained by minimizing σ_ε^2, then

$$Q'(\beta_0, \beta_1) = \frac{1}{n(1+\beta_1^2/\lambda)} \sum (y_i - \beta_0 - \beta_1 x_i)^2$$

is an estimator of σ_ε^2. It can easily be shown that the minimization of Q' leads to the same estimators, (3.13) and (3.16). In fact, it is intuitively clear that minimization of either σ_δ^2 or σ_ε^2 will result in the same estimator because their ratio is a known constant.

3.5.1 Discussion

Modified least squares leads to the same estimates as maximum likelihood, but without requiring the normality assumption.

A natural restriction in modified least squares is that $Q(\hat{\beta}_0, \hat{\beta}_1) \geq 0$ because $Q(\hat{\beta}_0, \hat{\beta}_1)$ is an estimate of variance. This constraint is automatically satisfied for case 3, but $s_{yy} \geq s_{xy}^2/(s_{xx} - \sigma_\delta^2)$ where $s_{xx} > \sigma_\delta^2$, and $s_{xx} \geq s_{xy}^2/(s_{yy} - \sigma_\varepsilon^2)$ where $s_{yy} - \sigma_\varepsilon^2 > 0$, are required for cases 1 and 2 respectively. These restrictions for the modified least-squares estimators are the same as those needed for the ML estimators for the normal structural model under the corresponding error variance assumptions.

The modified least-squares approach can be extended to more complicated models. The first extension is to allow the measurement errors, δ and ε, to be correlated. Let $\mathbf{\Omega}$ be the measurement error covariance matrix with elements σ_δ^2, $\sigma_{\delta\varepsilon}$, and σ_ε^2; and assume that the error covariance matrix is known up to a scalar multiple. In other words, assume (1.59). The modified least-squares estimators obtained in the correlated error model are the same as the generalized least-squares estimators given in Sprent (1966) and Gleser (1981). This was shown in Cheng and Tsai (1995). The concepts behind these two estimators are different, however. Generalized least squares minimizes a scalar function of the normalized error (see Gleser, 1981, p. 25), but modified least squares minimizes the unknown error variance. Thus, modified least squares is simpler to adapt and to generalize to other models than is generalized least squares.

A second extension is to equation error models. For the Type I model, (3.4)–(3.5), with both σ_ε^2 and σ_q^2 unknown, the modified least-squares estimators of β_0 and β_1 are obtained by minimizing (3.12), which estimates the sum of the two unknown error variances σ_ε^2 and σ_q^2. Alternatively, one can view $e_i = \varepsilon_i + q_i$ as simply one error. The modified least-squares approach is then to minimize the unknown error variance $\sigma_e^2 = \sigma_\varepsilon^2 + \sigma_q^2$. For the Type II model, (3.5), the unknown error variance is σ_q^2, and, from the identity

$$v_i = y_i - \beta_0 - \beta_1 x_i = \varepsilon_i + q_i - \beta_1 \delta_i,$$

it follows that

$$\sigma_v^2 = \sigma_\varepsilon^2 + \sigma_q^2 - 2\beta_1 \sigma_{\delta\varepsilon} + \beta_1^2 \sigma_\delta^2.$$

The modified least-squares estimator is obtained by minimizing

$$Q(\beta_0, \beta_1) = n^{-1} \sum \{(y_i - \beta_0 - \beta_1 x_i)^2 - (\sigma_\varepsilon^2 - 2\beta_1 \sigma_{\delta\varepsilon} + \beta_1^2 \sigma_\delta^2)\}. \tag{3.17}$$

Minimization of (3.17) leads to (3.13) and to (3.14) and

$$\hat{\beta}_1 = \frac{s_{xy} - \sigma_{\delta\varepsilon}}{s_{xx} - \sigma_\delta^2}, \qquad \text{provided } s_{xx} - \sigma_\delta^2 > 0.$$

The requirement that $Q(\hat{\beta}_0, \hat{\beta}_1) \geq 0$ implies

$$s_{yy} - \sigma_\varepsilon^2 \geq \frac{\left(s_{xy} - \sigma_{\delta\varepsilon}\right)^2}{s_{xx} - \sigma_\delta^2}.$$

Note that these constraints coincide with those of the ML estimate in the normal structural model; see Section 1.4.1.

The large-sample behaviour of modified least-squares estimators can be derived by standard techniques. One can also use the results of M-estimation theory (see, for example, Hampel *et al.*, 1986; Huber, 1981). M-estimation theory was developed for robust statistics (see Chapter 7), but it is suitable in many situations; for example, maximum likelihood estimators are M-estimators. In fact, the 'M' in M-estimator stands for 'maximum likelihood type'. The idea is to look for unbiased estimating equations for the unknown parameters of interest. This is the first step in obtaining a consistent estimator.

Definition 3.1 *Let* $\mathbf{z}_1, \mathbf{z}_2, \ldots, \mathbf{z}_n$ *be independent random vectors,* $\boldsymbol{\theta}$ *be a p-dimensional unknown parameter vector, and* $\boldsymbol{\psi}$ *be a p-dimensional score function. Then the estimating equation*

$$\sum_{i=1}^{n} \boldsymbol{\psi}(\mathbf{z}_i, \boldsymbol{\theta}) = 0 \tag{3.18}$$

is said to be unbiased if the true parameter, $\boldsymbol{\theta}$*, satisfies*

$$E_{\boldsymbol{\theta}}\{\boldsymbol{\psi}(\mathbf{z}_i, \boldsymbol{\theta})\} = \mathbf{0}, \qquad i = 1, \ldots, n. \tag{3.19}$$

A solution, $\tilde{\boldsymbol{\theta}}$, of (3.18) over the possible values of the parameter is called an **M-estimator** of $\boldsymbol{\theta}$ (Huber, 1964). Under certain regularity conditions, Huber (1967) showed that $\tilde{\boldsymbol{\theta}}$ is consistent and $n^{1/2}(\tilde{\boldsymbol{\theta}} - \boldsymbol{\theta})$ converges in distribution to a normal distribution with mean $\mathbf{0}$ and covariance matrix $\boldsymbol{\Gamma} = \mathbf{A}^{-1}\mathbf{B}(\mathbf{A}^{-1})'$, where

$$\begin{aligned}
\mathbf{A} &= \lim_{n\to\infty} n^{-1} \sum E\left[\left\{\frac{\partial \boldsymbol{\psi}(\mathbf{z}_i, \mathbf{u})}{\partial \mathbf{u}^t}\right\}_{\mathbf{u}=\boldsymbol{\theta}}\right], \\
\mathbf{B} &= \lim_{n\to\infty} n^{-1} \sum E\{\boldsymbol{\psi}(\mathbf{z}_i, \boldsymbol{\theta})\boldsymbol{\psi}'(\mathbf{z}_i, \boldsymbol{\theta})\}.
\end{aligned}$$

In the structural case,

$$\mathbf{A} = E\left[\left\{\frac{\partial \boldsymbol{\psi}(\mathbf{z}, \mathbf{u})}{\partial \mathbf{u}^t}\right\}_{\mathbf{u}=\boldsymbol{\theta}}\right], \qquad \text{and} \qquad \mathbf{B} = E\{\boldsymbol{\psi}(\mathbf{z}, \boldsymbol{\theta})\boldsymbol{\psi}^t(\mathbf{z}, \boldsymbol{\theta})\}.$$

For a summary of unbiased estimating equations, see Carroll *et al.* (1995, pp. 261 ff.).

It is straightforward to check that modified least squares yields unbiased estimating equations for β_0 and β_1 and hence the M-estimation theory is applicable (see Exercise 3.8).

Modified least squares can be used to estimate the unknown parameters in linear ME models under a variety of error assumptions. These estimators are the same as the ML estimators for structural normal models. They can be viewed as method of moments estimators for the functional model and the ultrastructural model for which, under the normality assumption, the solutions to the ML equations fail to provide reasonable estimators. This happens, for example, in the case of one error variance known. We emphasize again that the modified least-squares estimator coincides with the generalized least-squares estimator when the error covariance matrix is known up to a scalar multiple even though the two estimation philosophies are different. The generalized least-squares approach only works for the no-equation-error model with the error covariance matrix known, at least up to a scalar multiple. It does not work for the equation error model, nor for the no-equation-error model when one error variance is known.

One drawback of the modified least-squares method is that it does not apply to the so-called 'overidentified' case when both error variances are known because there is no unknown error variance to be minimized. In this case one can use generalized least squares, which leads to the same estimates as in the case of known ratio of error variances.

3.6 Bibliographic Notes and Discussion

Traditionally, the functional and structural models have been treated separately. Although some authors have made use of the connections between these two models in areas such as point estimation, the connection was not bridged formally until Gleser (1983). As we have seen in this chapter, the connections are important and useful in working with these models.

There is disagreement on the relative importance in applications of the various functional, structural, and ultrastructural models, but this is not the issue. What is important is that the proper model choice should be made for a given application, and this requires careful consideration by the practitioner. Understanding the ramifications of selecting the various models should help one understand the finer points of model selection.

Carroll *et al.* (1995, p. 6) suggest the use of the terms *functional modelling* and *structural modelling*, which do not correspond directly to the usage in this book. They use the term 'functional modelling' to refer to the situation in which the true variable ξ is either fixed or random, but if it is random, then no, or at least minimal, distributional assumptions are imposed on ξ. On the other hand, 'structural modelling' refers to the situation in which ξ is random, and distributional assumptions, usually parametric, are made on the true variable ξ. For example, if one assumes that the ξ_i are independent and identically distributed but endowed with no specific distribution, then this is a functional modelling, whereas if some specific parametric family of distributions is assumed (e.g., normal) then it is a structural modelling.

Section 3.4 gave the background for least-squares methods in ME models. It is not clear when the term *orthogonal regression* first appeared, but the term is now well understood in the context of ME models. The term *generalized least squares* was first used by Sprent (1966). Gleser (1981) gave a general approach to generalized least squares in which the minimization is taken with respect to any orthogonally invariant norm in the multivariate model (where both $\mathbf{x}$ and $\mathbf{y}$ are vectors) and then proved the associated large sample results. Many people have investigated this approach, for example, Höschel (1978a; 1978b) and Nussbaum (1976). Again, Höschel and Nussbaum focus on the functional model. The term 'generalized least squares' has a different meaning in regression analysis (Carroll and Ruppert, 1988) and covariance structure analysis.

A unified approach via *modified least squares* was introduced by Cheng and Tsai (1995). It provides least-squares methods for all kinds of ME models: functional, structural or ultrastructural, and with or without equation error. The beauty of modified least squares is that it is a unified approach; however, it does not introduce any new estimators that were not already available from maximum likelihood or the method of moments. It does clarify the use of least squares in ME models. Moreover, modified least squares is useful in other aspects of model analysis, such as diagnostics (Cheng and Tsai, 1993), because it gives simple unbiased estimating equations for the parameters.

The popular generalized least-squares approach, in its simplest version, is just orthogonal regression, which seems to dominate the ME model literature. This is partly due to its long history and partly because of its interesting geometrical interpretation. We have to bear in mind, however, that orthogonal regression and its simple geometry is restricted to the no-equation-error model with the error covariance matrix known at least up to an unknown multiplicative constant. It is not the solution for any other identifiability side assumption, (1.12). Importantly, it also does not work for the equation error model. Modified least squares, on the other hand, does not have any simple geometrical interpretation, except when it coincides with generalized least squares. It has been our experience that some people are disappointed with methods that lack a geometrical interpretation and are, therefore, less interested in such methods. Geometrical interpretations are very helpful but are not necessarily associated with any meaningful statistical properties. For example, the estimator that minimizes the triangular areas in Figure 1.3 has a nice geometrical interpretation, but it is a saddle point solution without any special statistical properties (see Section 1.6.2). See also Carroll and Ruppert (1996) for a relevant discussion.

3.7 Exercises

3.1 *For each model – the structural, the functional, and the ultrastructural – give some practical examples for which that model might be appropriate.*

3.2 *Use the strong law of large numbers to show (3.2).*

3.3 (*Monte Carlo computer experiment*) *In order to get a feel for the behaviour of the estimators for the structural model, perform m replications of the following experiment.*

1. *For some choice of parameter values, n an integer,* $\sigma^2 > 0$, $\sigma_\delta^2 > 0$, $\sigma_\varepsilon^2 > 0$, *and* β_1, *use a random number generator to generate independent random values,*

$$\xi_1, \ldots, \xi_n \sim N(0, \sigma^2), \tag{3.20}$$

$$\delta_1, \ldots, \delta_n \sim N(0, \sigma_\delta^2), \qquad \varepsilon_1, \ldots, \varepsilon_n \sim N\left(0, \sigma_\varepsilon^2\right).$$

2. *Form the data*

$$x_i = \xi_i + \delta_i, \qquad y_i = \beta_1 \xi_i + \varepsilon_i, \qquad i = 1, \ldots, n.$$

3. *Calculate the estimates,* $\hat{\beta}_0$ *and* $\hat{\beta}_1$, *given in Chapter 1 for the* λ *known case. Use the true* λ *in your estimates.*

Let m be as large as can be conveniently handled by your computer (say, $m = 100$*). Try, for example,* $n = 20$ *and* 100; $\sigma^2 = 9$ *and* 36; $\sigma_\delta^2 = \sigma_\varepsilon^2 = 1$ *and* 9; *and* $\beta_1 = 1$ *and* 4. *From each replication you will obtain estimates,* $\hat{\beta}_0$ *and* $\hat{\beta}_1$, *giving m independent values of each estimate:*

$$\hat{\beta}_{0,1}, \ldots, \hat{\beta}_{0,m}, \qquad \hat{\beta}_{1,1}, \ldots, \hat{\beta}_{1,m}.$$

Measure the accuracy of the estimation procedure by finding the sample mean absolute errors (MAEs):

$$\widehat{MAE}(\hat{\beta}_0) = \frac{1}{m}\sum_{j=1}^{m} |\hat{\beta}_{0,1} - \beta_0|,$$

$$\widehat{MAE}(\hat{\beta}_1) = \frac{1}{m}\sum_{j=1}^{m} |\hat{\beta}_{1,1} - \beta_1|.$$

Note that these sample MAEs are not estimates of the true MAEs, which are ∞, *they are just measures of the accuracy of the estimates. Compare and discuss the effects of the parameter values on the estimates.*

3.4 *Do Exercise 3.3 for the functional model. Instead of using (3.20), generate the true values by, say,*

$$\xi_i = \frac{10i}{n}, \qquad i = 1, \ldots, n.$$

Compare and discuss the effects of the parameter values on the estimates and compare with Exercise 3.3.

3.5 *Do Exercise 3.3 for the ultrastructural model. Instead of using (3.20), generate the true values by, say,*

$$\xi_i \sim N\left(\frac{10i}{n}, \sigma^2\right), \qquad i = 1, \ldots, n.$$

Compare and discuss the effects of the parameter values on the estimates and compare with Exercises 3.3 and 3.4.

3.6 *Do a simulation study of the effect of choosing λ wrongly, which could easily happen in practice. Repeat Exercise 3.3, but now instead of using the true λ in the estimates, try using wrong choices for λ. For example, if you used $\sigma_\delta^2 = \sigma_\varepsilon^2 \Rightarrow \lambda = 1$, then find the estimates, $\hat{\beta}_0$ and $\hat{\beta}_1$, assuming λ known but chosen to be, say, $\lambda = 1.1$ and 2. Discuss the effect of a mischoice of λ. This issue is discussed in Lakshminarayanan and Gunst (1984).*

3.7 *For the σ_δ^2 known ME model, derive the unbiased estimating equations from (3.12) by differentiating Q. Find the asymptotic variance of $\hat{\beta}_1$ for the nonnormal structural model.*

3.8 *For the λ known ME model, derive the unbiased estimating equations from (3.15) by differentiating Q. Find the asymptotic variance of $\hat{\beta}_1$ for the nonnormal structural model. Suppose now that both σ_δ^2 and σ_ε^2 are known. Comparing the asymptotic variance here with that in the previous exercise, conclude that one estimator does not dominate the other as is true for the normal model.*

4

Alternative Approaches to the Measurement Error Model

In this chapter alternative methods for estimating the parameters of an ME model are presented. Section 4.2 discusses the use of instrumental variables in place of side conditions to make the model identifiable. Section 4.3 discusses grouping methods, which, under suitable assumptions, are special cases of instrumental variable methods. Section 4.4 discusses the use of ranking methods, which are another modification of instrumental variable methods. Section 4.5 gives an overview of the higher-order moment methods and cumulant methods, which can be used on nonnormal models.

4.1 Introduction and Overview

The previous chapters discussed the normal model and the use of side conditions (1.12) to make the parameters identifiable and/or consistently estimable. Some users object to the use of these side conditions and prefer other methods of approaching the ME model. This is common, for example, in the econometrics literature.

One of the most popular alternative methods of supplying supplementary information to make the parameters consistently estimable is to use *instrumental variables*. Such variables do allow for consistent estimation, but they can be difficult to obtain in practice, as we shall see. This limits their usefulness. In an effort to provide more practically realizable instrumental variables, a special case of instrumental variables, called *grouping variables*, has been widely studied. It turns out that grouping variables are also hard to come by in practice, and therefore, some authors have suggested a modification of grouping variables that violates the instrumental variable assumptions. This modification involves grouping on the observable independent variable, x. It will be seen that this modified grouping variable approach can lead to very poor estimates.

If the model is not normal, there are several alternative estimation methods based on higher-order moments or higher-order cumulants that do not need any side conditions. Note that if a nonnormal model is close to a normal model, then presumably these methods become unstable. This will be apparent in the formulations of the estimates.

4.2 Instrumental Variable Estimators

We motivate the general instrumental variable approach for ME model estimation, by first discussing instrumental variables for ordinary regression.

4.2.1 Definition of instrumental variables

Consider the ordinary regression model with one explanatory variable x,

$$y_i = \beta_0 + \beta_1 x_i + e_i, \tag{4.1}$$

with $\operatorname{cov}\left((e_1, \ldots, e_n)'\right) = \sigma_e^2 \mathbf{I}$ and the x_i independent and identically distributed. The least- squares estimator $\hat{\beta}_{1R} = s_{xy}/s_{xx}$ is unbiased and consistent only if x and e are uncorrelated (see Exercise 4.1). If x is correlated with e, then the least-squares estimator is biased and inconsistent. Suppose, in this situation, that a random sample $(z_1, \ldots, z_n)$ is available such that

(i) z is uncorrelated with e, and
(ii) z is correlated with x.

Observe that

$$\begin{aligned}
\frac{1}{n}\sum(z_i - \bar{z})y_i &= \frac{1}{n}\sum(z_i - \bar{z})\beta_0 + \frac{1}{n}\sum(z_i - \bar{z})x_i\beta_1 + \frac{1}{n}\sum(z_i - \bar{z})e_i \\
&= \frac{1}{n}\sum(z_i - \bar{z})x_i\beta_1 + \frac{1}{n}\sum(z_i - \bar{z})e_i \\
&\to \frac{1}{n}\sum(z_i - \bar{z})x_i\beta_1, \qquad \text{as } n \to \infty,
\end{aligned}$$

by the properties of z. This suggests the following estimator of β_1:

$$\hat{\beta}_{1IV} = \frac{\sum(z_i - \bar{z})y_i}{\sum(z_i - \bar{z})x_i}, \tag{4.2}$$

and the intercept β_0 is estimated by $\hat{\beta}_{0IV} = \bar{y} - \hat{\beta}_{1IV}\,\bar{x}$. The supplementary variable z is called an **instrumental variable**, because it is used merely as an instrument in the estimation of the parameters. The estimator (4.2) is called the instrumental variable estimator of β_1. Property (ii) of z says that the denominator of β_1 in (4.2) will not go to zero when the sample size is large. There is no normality assumption made on any variable. The instrumental variable can even be discrete (see Sections 4.3 and 4.4). The only assumptions on the instrumental variable are the properties (i) and (ii) above. Note that if x were uncorrelated with e and one took $z = x$, then one would get the standard least-squares estimate for β_1.

It can be shown that the instrumental variable estimator (4.2) is consistent and asymptotically normal with variance (see Exercise 4.2) $\sigma_e^2\sigma_z^2/\sigma_{zx}^2$, where σ_z^2 is the variance of z and σ_{zx} is the covariance between z and x.

Now consider the structural ME model

$$x_i = \xi_i + \delta_i, \qquad y_i = \beta_0 + \beta_1 \xi_i + \varepsilon_i,$$

which can be written as (4.1) with $e_i = \varepsilon_i - \beta_1 \delta_i$. Clearly the error e_i is correlated with x_i. Therefore, if a third variable, z, is available with the properties

(i′) z is uncorrelated with (δ, ε) and

(ii′) z is correlated with ξ,

then the instrumental variable estimator of β_1 is just (4.2), which is consistent and asymptotically normal with variance

$$\frac{\sigma_e^2 \sigma_z^2}{\sigma_{z\xi}^2}, \tag{4.3}$$

where $\sigma_{z\xi}$ is the covariance between z and ξ and $\sigma_e^2 = \sigma_\delta^2 - 2\beta_1 \sigma_{\delta\varepsilon} + \beta_1^2 \sigma_\varepsilon^2$. Note that $\sigma_{z\xi} = \sigma_{zx}$ and that the errors, δ and ε, can be correlated. From (4.3), it is seen that a new variable z is needed that is uncorrelated with the error (δ, ε) but is strongly correlated with ξ because one wants the (asymptotic) variance of the instrumental variable estimator as small as possible. We will return to this later.

Example 4.1

We now generate an artificial set of data to demonstrate the instrumental variable method. We performed the following experiment:

1. Generate 36 independent values; $\xi_1, \ldots, \xi_{36}$; of $\xi \sim N(0, \sigma^2)$.
2. Generate 36 independent values; $\delta_1, \ldots, \delta_{36}$; of $\delta \sim N(0, \sigma_\delta^2)$.
3. Generate 36 independent values; $\varepsilon_1, \ldots, \varepsilon_{36}$; of $\varepsilon \sim N(0, \sigma_\varepsilon^2)$.
4. Generate 36 independent values; $\gamma_1, \ldots, \gamma_{36}$; of $\varepsilon \sim N(0, \sigma_\gamma^2)$.
5. Form the data

$$\begin{pmatrix} x_i \\ y_i \\ z_i \end{pmatrix} = \begin{pmatrix} \xi_i + \delta_i \\ \beta_1 \xi_i + \varepsilon_i \\ \xi_i + \gamma_i \end{pmatrix}, \qquad i = 1, \ldots, 36.$$

Thus z is an instrumental variable, correlated with ξ, but not correlated with the errors. For $\beta_0 = 0$, $\beta_1 = 1$, $\sigma^2 = 9$, $\sigma_\delta^2 = \sigma_\varepsilon^2 = 1$, and $\sigma_\gamma^2 = 9$ the data are given in Table 4.1.

Table 4.1 Thirty-six ME model data with instrumental variable: $\beta_0 = 0,\ \beta_1 = 1,\ \sigma^2 = 9,$ and $\sigma_\delta^2 = \sigma_\varepsilon^2 = \sigma_\gamma^2 = 1$

	x	y	z
1	0.62	1.30	2.72
2	−1.07	−0.89	−1.90
3	−0.35	−1.03	−1.80
4	−2.88	−0.82	−3.68
5	3.79	3.70	−0.36
6	−1.90	−0.46	−1.37
7	−6.10	−6.56	−3.27
8	0.53	−1.11	2.56
9	1.12	−0.79	−2.28
10	−3.43	−2.14	−2.09
11	2.24	2.87	3.89
12	−1.07	−1.59	−1.97
13	−0.74	−1.23	−3.58
14	7.14	7.11	6.31
15	4.74	2.41	5.64
16	3.09	4.02	5.40
17	1.23	2.53	7.84
18	2.14	1.91	0.62

	x	y	z
19	3.91	3.72	4.37
20	−8.94	−9.37	−9.58
21	−0.81	−1.14	5.09
22	−3.16	−3.56	−9.22
23	−3.47	−3.18	−5.62
24	6.62	5.05	4.44
25	−2.75	−3.07	−6.30
26	1.96	3.83	6.43
27	−0.03	−1.02	−0.45
28	−2.14	−1.08	2.79
29	0.08	−1.47	−5.15
30	−1.37	−0.23	2.94
31	1.95	1.49	1.91
32	5.08	2.94	5.03
33	−0.03	−0.06	1.65
34	3.46	0.98	3.31
35	−3.47	−3.39	−4.55
36	3.36	2.08	4.87

Note that $\text{corr}(\xi, z) = \sigma^2/(\sigma\sqrt{\sigma^2 + \sigma_\gamma^2}) = 9/(3\sqrt{9+9}) = 0.707$ for the model generating Table 4.1. This is reasonable but not extremely high. Even so, evaluating (4.2) gives $\hat{\beta}_{1IV} = 1.003$, which is quite accurate. Furthermore $\hat{\beta}_{0IV} = \bar{y} - \hat{\beta}_{1IV}\bar{x} = -0.212$ for these data. Compare Example 4.8.

The instrumental variable approach can be extended to the functional model by requiring that the instrumental variable, z, satisfy

(i″) $\lim_{n\to\infty} n^{-1} \sum (z_i - \bar{z})(\delta_i, \varepsilon_i) = (0, 0)$ and
(ii″) $\lim_{n\to\infty} n^{-1} \sum (z_i - \bar{z})\xi_i \neq 0$.

For the no-intercept model ($\beta_0 = 0$), the term $\bar{z}$ is dropped in both conditions. It is interesting to note that, in the no-intercept model, the instrumental variable can be constant, that is, $z_i \equiv 1$ is a legitimate instrumental variable provided $\lim_{n\to\infty} n^{-1} \sum \xi_i \neq 0$. In this case, the instrumental variable estimator for β_1 is just $\bar{y}/\bar{x}$, which is the estimate obtained earlier for the intercept known model, (1.41), with $\beta_0 = 0$.

The major practical difficulty in using instrumental variables is finding a variable known to be correlated with ξ and known not to be correlated with δ or ε. Carroll *et al.* (1995, p. 107) noted that:

> One possible source of an instrumental variable is a second measurement on **X** (ξ) obtained by an independent method. This second measurement need not be unbiased for **X** (ξ). Thus the assumption that a variable is an instrument is weaker than the assumption that it is a replicate measurement.

A second difficulty is that the justification for the use of instrumental variables is based on consistency considerations only, and this does not guarantee good performance in applications involving large sample variances and moderate sample sizes, in which the consistency properties cannot be brought to bear. Furthermore, the choice of an instrumental variable might not be unique and there is the possibility of very large variations in the estimators resulting from different instrumental variables. A possible remedy is to investigate the efficiency of the instrumental variable estimators being considered (see Exercise 4.3).

4.2.2 Maximum likelihood estimation via instrumental variables

The previous subsection made no distributional assumptions on the true variable, ξ, the error, (δ, ε), or the instrumental variable, z. Now consider ML estimators under a normality assumption. We follow the treatment of Leamer (1978) (see also Carter and Fuller, 1980) and assume that the trivariate observed random variables (x_i, y_i, z_i) satisfy

$$\begin{aligned} x_i &= \xi_i + \delta_i, &(4.4)\\ y_i &= \beta_0 + \beta_1 \xi_i + \varepsilon_i, &(4.5)\\ z_i &= \rho_0 + \rho_1 \xi_i + u_i, &(4.6) \end{aligned}$$

where $(\xi_i, \delta_i, \varepsilon_i, u_i)$ are independent identically distributed jointly normal random variables. The errors $\delta_i, \varepsilon_i, u_i$ have mean zero and variances σ_δ^2, σ_ε^2 and σ_u^2 respectively and are independent of each other. Moreover, the error (δ, ε, u) is independent of the true variable ξ, which has mean μ and variance σ^2. Equations (4.4) and (4.5) constitute the usual structural model, and equation (4.6) defines an instrumental variable, z, having the properties (i′) and (ii′) above.

The vector (x_i, y_i, z_i) is normally distributed with mean $(\mu, \beta_0 + \beta_1\mu, \rho_0 + \rho_1\mu)$ and covariance matrix

$$\mathbf{K} = \begin{bmatrix} \sigma^2 + \sigma_\delta^2 & \beta_1\sigma^2 & \rho_1\sigma^2 \\ \beta_1\sigma^2 & \beta_1^2\sigma^2 + \sigma_\varepsilon^2 & \rho_1\beta_1\sigma^2 \\ \rho_1\sigma^2 & \rho_1\beta_1\sigma^2 & \rho_1^2\sigma^2 + \sigma_u^2 \end{bmatrix}. \tag{4.7}$$

The following result is due to Leamer (1978).

Theorem 4.1 *Maximization of the likelihood function for the above normal model, subject to the constraint that the variances σ^2, σ_δ^2, σ_ε^2 and σ_u^2 are nonnegative, gives the following ML estimator of β_1:*

(a) if $\hat\beta_{1R}$ and $\hat\beta_{1IV}$ have the same sign, then $\hat\beta_1 = \text{median}(\hat\beta_{1R}, 1/\hat\beta_{1I}, \hat\beta_{1IV})$*; or*

(b) if $\hat{\beta}_{1R}$ and $\hat{\beta}_{1IV}$ are of opposite sign, then

$$\hat{\beta}_1 = \begin{cases} 1/\hat{\beta}_{1I} & \text{if } |\rho_{xz}| < |\rho_{yz}|, |\rho_{xy}| \\ \hat{\beta}_{1R} & \text{if } |\rho_{yz}| < |\rho_{xz}|, |\rho_{xy}| \\ \hat{\beta}_{1IV} & \text{if } |\rho_{xy}| < |\rho_{xz}|, |\rho_{yz}| \end{cases}$$

where $\hat{\beta}_{1I} = s_{xy}/s_{yy}$ is the inverse regression slope estimate, $\hat{\beta}_{1R} = s_{xy}/s_{xx}$ is the ordinary regression slope estimate, and $\hat{\beta}_{1IV} = s_{yz}/s_{xz}$ is the instrumental variable ME model regression slope estimate.

Proof (Leamer 1978). Maximum likelihood gives

$$(\hat{\mu}, \hat{\beta}_0 + \hat{\beta}_1\hat{\mu}, \hat{\rho}_0 + \hat{\rho}_1\hat{\mu}) = (\bar{x}, \bar{y}, \bar{z}) .$$

Define

$$\mathbf{S} \equiv \begin{bmatrix} s_{xx} & s_{xy} & s_{xz} \\ s_{xy} & s_{yy} & s_{yz} \\ s_{xz} & s_{yz} & s_{zz} \end{bmatrix} .$$

If the constraints are not binding, then the ML estimates are obtained by solving

$$\hat{\mathbf{K}} = \mathbf{S} \tag{4.8}$$

where $\hat{\mathbf{K}}$ is the matrix (4.7) with the parameters replaced by their estimates. This happens if and only if all the variances obtained by solving (4.8) are nonnegative.

There are four variance terms in $\hat{\mathbf{K}}$: $\hat{\sigma}^2$, $\hat{\sigma}^2_\delta$, $\hat{\sigma}^2_\varepsilon$, and $\hat{\sigma}^2_u$. If one or more of these estimates is negative, then the maximum of the likelihood occurs on the boundary of the constraint set. It cannot be in the interior because the likelihood function does not have more than one local maximum. The constrained maximum can be found by imposing the constraints one at a time and then selecting the constraint that yields the highest likelihood value provided that no other constraints are violated once one of the constraints is imposed.

Assume, without loss of generality, that $s_{xy} > 0$. This holds if and only if $\hat{\beta}_{1R} > 0$ and $1/\hat{\beta}_{1I} > 0$. The constraint $0 \leq \hat{\sigma}^2 = s_{xy}s_{xz}/s_{yz}$ holds if and only if $\hat{\beta}_{1R}$ and $\hat{\beta}_{1IV}$ have the same sign. The constraints $\hat{\sigma}^2_\delta \geq 0, \hat{\sigma}^2_\varepsilon \geq 0$, and $\hat{\sigma}^2_u \geq 0$ imply respectively that

$$1/\hat{\beta}_{1I} = s_{yy}/s_{xy} \geq s_{yz}/s_{xz} = \hat{\beta}_{1IV}, \tag{4.9}$$

$$\hat{\beta}_{1IV} = s_{yz}/s_{xz} \geq s_{xy}/s_{xx} = \hat{\beta}_{1R}, \tag{4.10}$$

and

$$s_{xy} \geq s_{yz}s_{xz}/s_{zz}. \tag{4.11}$$

Thus $0 \leq \hat{\beta}_{1R} \leq \hat{\beta}_{1IV} \leq 1/\hat{\beta}_{1I}$.

First consider maximizing the likelihood subject to the constraint $\sigma_\varepsilon^2 = 0$. In this case the model can be written

$$\begin{aligned} x_i &= (y_i - \beta_0)/\beta_1 + \delta_i, \\ y_i &= \beta_0 + \beta_1 \xi_i, \\ z_i &= \rho_0 + \rho_1 (y_i - \beta_0)/\beta_1 + u_i. \end{aligned}$$

The likelihood function can be written

$$\begin{aligned} & L\left(\mu, \beta_0, \beta_1, \rho_0, \rho_1, \sigma^2, \sigma_\delta^2, \sigma_u^2\right) \\ & \propto \sigma_\delta^{-n} \exp\left[-\frac{1}{2\sigma_\delta^2} \sum \left(x_i - (y_i - \beta_0)/\beta_1\right)^2\right] \\ & \quad \times \sigma_u^{-n} \exp\left[-\frac{1}{2\sigma_u^2} \sum \left(z_i - \rho_0 - (y_i - \beta_0)\rho_1/\beta_1\right)^2\right] \\ & \quad \times (\beta_1 \sigma)^{-n} \exp\left[-\frac{1}{2\beta_1^2 \sigma^2} \sum (y_i - \beta_0)^2\right]. \end{aligned}$$

The ML estimators are the estimators obtained by calculating standard regressions of x on y and z on y. The results are

$$\begin{aligned} \hat{\beta}_1^{(\varepsilon)} &= s_{yy}/s_{xy}, & \hat{\rho}_1^{(\varepsilon)}/\hat{\beta}_1^{(\varepsilon)} &= s_{xz}/s_{yz}, \\ \left(\hat{\sigma}_\delta^{(\varepsilon)}\right)^2 &= s_{xx} - s_{xy}^2/s_{yy}, & \left(\hat{\sigma}_u^{(\varepsilon)}\right)^2 &= s_{zz} - s_{yz}^2/s_{yy}, \\ \left(\hat{\beta}_1^{(\varepsilon)} \hat{\sigma}^{(\varepsilon)}\right)^2 &= s_{yy}. \end{aligned}$$

Note that the variance estimates are all nonnegative. The value of the likelihood at these estimates is

$$L_\varepsilon = c\left[s_{yy}\left(s_{xx} - s_{xy}^2/s_{yy}\right)\left(s_{zz} - s_{yz}^2/s_{yy}\right)\right]^{-n/2},$$

where c is the appropriate constant. Similarly, given $\sigma_\delta^2 = 0$, we obtain $\hat{\beta}_1^{(\delta)} = s_{xy}/s_{xx}$ and a likelihood of

$$L_\delta = c\left[s_{xx}\left(s_{yy} - s_{xy}^2/s_{xx}\right)\left(s_{zz} - s_{xz}^2/s_{xx}\right)\right]^{-n/2};$$

and, given $\sigma_u^2 = 0$, we obtain $\hat{\beta}_1^{(u)} = s_{yz}/s_{xz}$ and a likelihood of

$$L_u = c\left[s_{zz}\left(s_{yy} - s_{yz}^2/s_{zz}\right)\left(s_{zz} - s_{xz}^2/s_{zz}\right)\right]^{-n/2}.$$

The assumption that $\sigma^2 = 0$ gives a likelihood of

$$L_\sigma = c\left[s_{xx} s_{yy} s_{zz}\right]^{-n/2},$$

which is clearly smaller than the other likelihoods, so this constraint cannot maximize the likelihood.

To compare the three likelihoods, neglect the exponent, $n/2$, and the constant, c, and multiply each by $s_{xx}s_{yy}s_{zz}$. In terms of correlations, the resulting three values are respectively $[(1-r_{xy}^2)(1-r_{yz}^2)t]^{-1}$, $[(1-r_{xy}^2)(1-r_{xz}^2)]^{-1}$, and $[(1-r_{yz}^2)(1-r_{xz}^2)]^{-1}$. The first of these is the largest if r_{xz}^2 is the smallest, the second is largest if r_{yz}^2 is the smallest, and the third is largest if r_{xy}^2 is the smallest. This proves the second part of the theorem.

It remains to show part (a). If $\hat{\beta}_{1R}$ and $\hat{\beta}_{1IV}$ have the same sign, then if (4.11) is violated $0 < r_{xy} < r_{xz}r_{yz}$. Because $r_{xz} < 1$ and $r_{yz} < 1$, r_{xy} is the smallest and (a) is true. If $\hat{\beta}_{1IV} < \hat{\beta}_{1R}$, then $0 < r_{yz}/r_{xz} < r_{xy}$, which implies $r_{yz}^2 < r_{xy}^2 r_{xz}^2$ and r_{yz}^2 is the smallest and again (a) is true. Finally, if $1/\hat{\beta}_{1I} < \hat{\beta}_{1IV}$, then $0 < r_{xy}^{-1} < r_{yz}/r_{xz}$, which implies $r_{xz}^2 < r_{xy}^2 r_{yz}^2$. □

Let us investigate this result. First consider the case when $\hat{\beta}_{1R}$ and $\hat{\beta}_{1IV}$ are of the same sign. This means that all three estimates, $\hat{\beta}_{1R}$, $1/\hat{\beta}_{1I}$, and $\hat{\beta}_{1IV}$, are of the same sign. If $\hat{\beta}_{1IV}$ is between the other two estimates then the ML estimate is just $\hat{\beta}_{1IV}$. This is true when all the estimates of the variances σ^2, σ_δ^2, σ_ε^2 and σ_u^2 are nonnegative. In this case, the instrumental variable estimates satisfy inequalities similar to those for the ML estimate based on the likelihood of (x, y) alone (see (1.27)).

The other situations might appear confusing. If $\hat{\beta}_{1IV}$ is less in absolute value than $\hat{\beta}_{1R}$, then the ML estimate becomes $\hat{\beta}_{1R}$, the ordinary least-squares estimate; however, as was pointed out, this estimate is biased and inconsistent for the ME model and is not recommended. Maximum likelihood is saying that the additional information from the instrumental variable is not needed and that $\hat{\sigma}_\delta^2 = 0$, which means there is no measurement error in the x variable. Looked at this way, the result makes perfect sense because this situation is just the standard regression of y on x and the ordinary least-squares estimate works well. If this situation occurs, one has modelled the data with an ME model, but the data are saying that σ_δ^2 is zero. Generally speaking, the reason for using maximum likelihood depends on its large sample properties: consistency, asymptotic normality and asymptotic efficiency. Thus, if the sample is large, the constraints should be satisfied automatically if the model is true, which implies that $\hat{\beta}_{1IV}$ should be the ML estimate. For finite samples, the constraints can be violated; in which case, using least squares is theoretically correct but one simply does not have any sound justification for doing this. When this happens, one could either gather more data in the hope that this problem goes away or consider a different model. In taking the former action, one believes the model is correct and the violation of the constraint is caused by the small sample. The latter action means that the ME model might not be correct and one believes the data, which indicates that the measurement error in the independent variable is very small and can be ignored (see Section 1.3).

If $\hat{\beta}_{1IV}$ is greater in absolute value than $1/\hat{\beta}_{1I}$, then the ML estimate is just $1/\hat{\beta}_{1I}$. This really means that maximum likelihood and the data indicate that $\hat{\sigma}_\varepsilon^2 = 0$ and the only error in the model is the error in x. Therefore a standard regression model

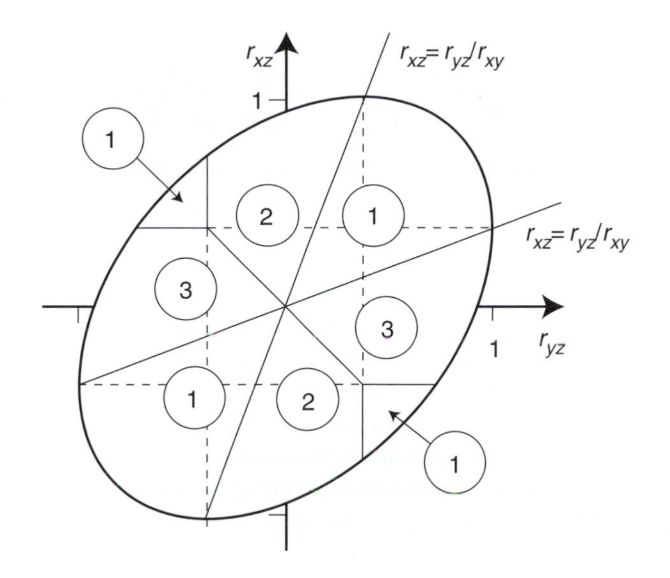

Figure 4.1: Regions for the alternative ML estimators of β_1

regressing x on y is indicated. Similar comments to those in the previous paragraph can be made.

The case when $\hat{\beta}_{1R}$ and $\hat{\beta}_{1IV}$ have opposite sign suggests that the data offer inconclusive evidence about the sign of β_1. Leamer (1978, p. 967) suggested that this might indicate a bimodal likelihood function requiring a more complex model of the data.

Finally, Theorem 4.1 implies that a high correlation between x and z can shift the ML estimate of β_1 from the instrumental variable estimate to the least-squares estimate (Leamer 1978, p. 966). One might also consider this evidence that y should be modelled as a function of z instead of ξ. If $|r_{xz}|$ is large, this means that σ_δ^2 and σ_u^2 must both be small and the least-squares estimate should be reasonable. Leamer has a discussion of this and gives the diagram reproduced as Figure 4.1. The elliptical area is the area det $(S) \geq 0$; the areas labelled '1' are those in which $\hat{\beta}_{1IV}$ is the ML estimate, the areas labelled '2' are those in which $\hat{\beta}_{1R}$ is the ML estimate, and the areas labelled '3' are those in which $1/\hat{\beta}_{1I}$ is the ML estimate.

4.2.3 Confidence intervals with instrumental variables

An instrumental variable not only enables one to estimate β_1 consistently but also allows one to obtain a confidence region for (β_0, β_1).

Durbin (1954) developed the following argument. The random variable $y - \beta_0 - \beta_1 x = \varepsilon - \beta_1 \delta$ is known to have mean 0 and is uncorrelated with z because both δ and ε are uncorrelated with z. Assuming that (δ, ε) and z are normally distributed, let

$$r = \frac{\sum(z_i - \bar{z})(y_i - \beta_0 - \beta_1 x_i)}{\left[\sum(z_i - \bar{z})^2 \sum(y_i - \beta_0 - \beta_1 x_i)^2\right]^{1/2}}$$

be the observed correlation between z and $y - \beta_0 - \beta_1 x$; then

$$t = \left[\frac{(n-1)r^2}{1-r^2}\right]^{1/2} \tag{4.12}$$

has a t distribution with $n - 1$ degrees of freedom. One might expect, from the usual results on sample covariances (see, for example, Kendall and Stuart, 1979, Section 26.23), the degrees of freedom to be $n - 2$, but this is not true because the mean of $y - \beta_0 - \beta_1 x$ is known (see Anderson, 1958, Section 4.2). Because $y - \beta_0 - \beta_1 x = \varepsilon - \beta_1 \delta$ no matter whether the model is structural, functional or even ultrastructural, the confidence region (interval) obtained in this subsection is good for all three types of ME model. If $t^2_{n-1,1-\gamma}$ satisfies

$$P\left(t^2 \leq t^2_{n-1,1-\gamma}\right) = 1 - \gamma,$$

then, because $r^2 = t^2/\{t^2 + (n-1)\}$ is monotone increasing in t^2,

$$P\left(\frac{\left[\sum(z_i - \bar{z})(y_i - \beta_0 - \beta_1 x_i)\right]^2}{\sum(z_i - \bar{z})^2 \sum(y_i - \beta_0 - \beta_1 x_i)^2} \leq r^2_{1-\gamma}\right) = 1 - \gamma$$

or

$$P\left(\frac{\left[\sum(z_i - \bar{z})y_i\right]^2 - 2\beta_1 \sum(z_i - \bar{z})y_i \sum(z_i - \bar{z})x_i + \beta_1^2\left[\sum(z_i - \bar{z})x_i\right]^2}{\sum(z_i - \bar{z})^2 \left(\sum y_i^2 + n\beta_0^2 - 2\beta_1 \sum x_i y_i + \beta_1^2 \sum x_i^2 - 2n\beta_0 \bar{y} + 2n\beta_0\beta_1\bar{x}\right)} \leq r^2_{1-\gamma}\right) = 1 - \gamma, \tag{4.13}$$

where $r^2_{1-\gamma} = t^2_{n-1,1-\gamma}/(t^2_{n-1,1-\gamma} + n - 1)$. Equation (4.13) depends only on β_0 and β_1, apart from the observables x, y, and z. It defines a quadratic confidence region in the (β_0, β_1) plane, with confidence level $1 - \gamma$.

A confidence interval for β_1 can be obtained in like manner. Because z is uncorrelated with $y - \beta_1 x = \beta_0 + \varepsilon - \beta_1 \delta$, let r^* be the observed correlation,

$$r^* = \frac{\sum(z_i - \bar{z})[y_i - \beta_1 x_i - (\bar{y} - \beta_1 \bar{x})]}{\left[\sum(z_i - \bar{z})^2 \sum(y_i - \bar{y} - \beta_1(x_i - \bar{x}))^2\right]^{1/2}};$$

then

$$t^* = \left\{\frac{(n-2)r^{*2}}{1 - r^{*2}}\right\}^{1/2}$$

has a t distribution with $n-2$ degrees of freedom (see Anderson, 1958, Section 4.2). The resulting confidence interval for β_1 is

$$P\left(\frac{\left[\sum(z_i-\bar{z})(y_i-\bar{y}-\beta_1(x_i-\bar{x}))\right]^2}{\sum(z_i-\bar{z})^2\sum(y_i-\bar{y}-\beta_1(x_i-\bar{x}))^2}\leq r_{1-\gamma}^{*2}\right)=1-\gamma \tag{4.14}$$

or

$$P\left(\frac{\left[\sum(z_i-\bar{z})y_i\right]^2-2\beta_1\sum(z_i-\bar{z})y_i\sum(z_i-\bar{z})x_i+\beta_1^2\left[\sum(z_i-\bar{z})x_i\right]^2}{\sum(z_i-\bar{z})^2\left[\sum(y_i-\bar{y})^2-2\beta_1\sum(x_i-\bar{x})y_i+\beta_1^2\sum(x_i-\bar{x})^2\right]}\right.$$

$$\left.\leq r_{1-\gamma}^{*2}\right)=1-\gamma,$$

where $r_{1-\gamma}^{*2}=t_{1-\gamma}^2/\left(t_{1-\gamma}^2+n-2\right)$.

Fuller (1987, p. 55 (1.4.18)) obtained a confidence interval for β_1 using a similar approach. It can be shown that this confidence interval is equivalent to (4.14).

The confidence intervals (regions) given above are, of course, also subject to the Gleser–Hwang effect. For example, (4.14) might be an interval, the complement of an interval or the entire real line. For details, see Konijn (1981).

4.3 Grouping Methods

We have mentioned the practical difficulty in finding instrumental variables that are known to be correlated with ξ and known not to be correlated with the errors. Because of this, special discrete instrumental variables, called **grouping variables**, have been proposed with the idea that they are easier to obtain in practice. We discuss grouping variables because they are sometimes found in the literature, particularly in the econometrics literature. Regrettably, there are serious reservations about the grouping method, which will be pointed out as we go along.

A grouping variable is a discrete instrumental variable; that is, it classifies an observation according to which of a finite number of groups that observation falls into. Usually a unique integer is associated with each group and the grouping variable assigns the integer associated with a group to each observation in that group. It is assumed that the user knows, from the nature of the problem, to which group each observation belongs, that this assignment depends only on the value of ξ, and has no connection with the observational errors, ε and δ. In reality, though, it appears that this special instrumental variable is not much more likely to occur in practice than a general continuous instrumental variable.

4.3.1 Two groups

Consider the case when the grouping variable, z, is Bernoulli, taking on only two possible values.

Example 4.2
Recall Example 1.1 concerning Boyle's law, which governs the pressure–volume relationship of a gas. Taking logarithms gives

$$\log P = \log C - \gamma \log V,$$

which is linear with $\beta_0 = \log C$ and $\beta_1 = \gamma$. Suppose now that the determinations of volume have been made sometimes by one method and sometimes by another; and suppose it is known that method 1 produces slightly different results from method 2. The method classification will then be correlated with the volume determination. The errors in this determination, and certainly those in the pressure determination (which is supposed to be made in the same way for all observations), might be uncorrelated with method classification. Thus, there is an instrumental variable of a special kind – a two-group grouping variable.

Suppose, for simplicity, that n, the number of observations, is even, and that the observations are divided into two equal groups of $m = n/2$ observations each. Let $\bar{x}_1$, $\bar{x}_2$, $\bar{y}_1$, and $\bar{y}_2$ be the respective means of the two groups. Then β_1 can be estimated by setting an instrumental variable z equal to $+1$ for each observation in the first group and -1 for each observation in the second group. The instrumental variable estimator (4.2) becomes

$$\hat{\beta}_{1G} \equiv \frac{\bar{y}_2 - \bar{y}_1}{\bar{x}_2 - \bar{x}_1}, \tag{4.15}$$

and the estimator of the intercept is

$$\hat{\beta}_{0G} \equiv \bar{y} - \hat{\beta}_{1G}\,\bar{x} = \frac{1}{2}\{(\bar{y}_2 + \bar{y}_1) - \hat{\beta}_{1G}(\bar{x}_2 + \bar{x}_1)\},$$

where $\bar{y}$ and $\bar{x}$ are the respective means of y and x for the entire sample.

Geometrically, this means that, in the (x, y) plane, the points are divided into equal groups according to some criterion (related only to ξ) and then the centre of gravity is determined for each group. The estimator is the slope of the line connecting these two centres, namely, $(\bar{x}_1, \bar{y}_1)$ and $(\bar{x}_2, \bar{y}_2)$. The problem is how to divide the data into the two groups. Unfortunately, it is hard to imagine a practical situation in which one could base the grouping criterion solely on ξ because ξ is not observable. Many authors instead base the division on x, therefore, and make various heuristic arguments as to why this is allowable. As we shall see, the estimates for groupings based on x do not generally behave well.

Wald (1940), to whom grouping estimators are due, showed that $\hat{\beta}_{1G}$ in (4.15) is consistent if the grouping is independent of the errors, (ε, δ), and if the true values ξ_i satisfy

$$\lim_{n\to\infty} \inf |\bar{\xi}_2 - \bar{\xi}_1| > 0, \tag{4.16}$$

which corresponds to (ii′) and (ii″) of Section 4.2.1, because here $\mathrm{cov}(z, \xi) = \bar{\xi}_2 - \bar{\xi}_1$. Note that (4.16) clearly will not be satisfied if the observations are randomly allocated into two groups because then $\lim_{n\to\infty} |\bar{\xi}_2 - \bar{\xi}_1| = 0$. Nor is it satisfactory to allocate the m smallest observed xs to one group and the m largest to the other because this allocation is not independent of the errors. Neyman and Scott (1951) show that, for this latter grouping method, the estimator, (4.15), will not be consistent (see also Theorem 4.2 below). If one can guarantee that the grouping according to the x values is the same as the one according to ξ values, then the resulting estimator will be consistent. (Note that such a guarantee would mean that the errors, δ, are smaller than the spacings between the ξs and, therefore, cannot be made for the normal model.) This means that in order to use Wald's method on the normal model, one still needs additional prior information independent of the errors.

Equation (4.15) can be used to obtain estimators of the error variances. By (1.22),

$$\begin{aligned} \sigma_\delta^2 &= \sigma_x^2 - \sigma_{xy}/\beta_1, \\ \sigma_\varepsilon^2 &= \sigma_y^2 - \beta_1 \sigma_{xy}, \end{aligned} \tag{4.17}$$

and substituting the consistent estimators, s_{xx}, s_{yy}, and s_{xy} (multiplying by $n/(n-1)$ to remove the bias) for the variances and covariances and (4.15) for β_1 yields the estimators

$$\begin{aligned} \hat{\sigma}_{\delta G}^2 &= \frac{n}{n-1}\left(s_x^2 - \frac{s_{xy}}{\hat{\beta}_{1G}}\right), \\ \hat{\sigma}_{\varepsilon G}^2 &= \frac{n}{n-1}(s_y^2 - \hat{\beta}_{1G} s_{xy}). \end{aligned} \tag{4.18}$$

As we shall see in the following examples, these estimators are not very reliable for small or moderate sample sizes. In fact, they can all too frequently give negative values for the estimated variances. Obviously, estimates that dominate (4.18) would be

$$\begin{aligned} \tilde{\sigma}_{\delta G}^2 &= \max\left[0, \frac{n}{n-1}\left(s_x^2 - \frac{s_{xy}}{\hat{\beta}_{1G}}\right)\right], \\ \tilde{\sigma}_{\varepsilon G}^2 &= \max\left[0, \frac{n}{n-1}\left(s_y^2 - \hat{\beta}_{1G} s_{xy}\right)\right]. \end{aligned}$$

Example 4.3 (*Kendall and Stuart, 1979, p. 425*)
Once again, consider the data of Example 1.11. These data are not grouped, but for illustrative purposes we violate the instrumental variable assumptions and divide the data into two groups using the x values. It is not uncommon in practice to find the grouping done in this way even though this means that the instrumental variable theory no longer holds.

Because there are an odd number of observations, we do not include the (9.3, 24.9) observation in either group:

$$\begin{aligned} G_1 &= \left\{ \begin{pmatrix} 1.8 \\ 6.9 \end{pmatrix}, \begin{pmatrix} 4.1 \\ 12.5 \end{pmatrix}, \begin{pmatrix} 5.8 \\ 20.0 \end{pmatrix}, \begin{pmatrix} 7.5 \\ 15.7 \end{pmatrix} \right\}, \\ G_2 &= \left\{ \begin{pmatrix} 10.6 \\ 23.4 \end{pmatrix}, \begin{pmatrix} 13.4 \\ 30.2 \end{pmatrix}, \begin{pmatrix} 14.7 \\ 35.6 \end{pmatrix}, \begin{pmatrix} 18.9 \\ 39.1 \end{pmatrix} \right\}. \end{aligned}$$

Then

$$\bar{x}_1 = 4.80, \qquad \bar{x}_2 = 14.4,$$
$$\bar{y}_1 = 13.775, \qquad \bar{y}_2 = 32.075,$$

so that

$$\hat{\beta}_{1G} = \frac{32.075 - 13.775}{14.40 - 4.80} = 1.91,$$

which is reasonably close to the true value, 2. For the eight observations used,

$$s_{xx} = 29.735, \qquad s_{yy} = 112.709, \qquad s_{xy} = 56.764.$$

Substituting into (4.18) gives

$$\hat{\sigma}^2_{\delta G} = -0.054 \qquad \text{and} \qquad \hat{\sigma}^2_{\varepsilon G} = 5.16,$$

which are obviously very bad estimators of the true values, 1.

We now do a simulation in which we divide the data into two equal-sized groups with the grouping related to ξ and independent of the errors. It is hard to imagine a situation such as the following occurring in practice, but this Example is at least an illustration of the two-group method on data that strictly satisfies all the instrumental variable assumptions.

Example 4.4
We performed 500 replications of the following experiment:

1. Generate n independent values; $\xi_1, \ldots, \xi_n$; of $\xi \sim N(0, \sigma^2)$.
2. Generate n independent values; $\delta_1, \ldots, \delta_n$; of $\delta \sim N(0, \sigma^2_\delta)$.
3. Generate n independent values; $\varepsilon_1, \ldots, \varepsilon_n$; of $\varepsilon \sim N(0, \sigma^2_\varepsilon)$.
4. Form the ME model observations using

$$\begin{pmatrix} x_i \\ y_i \end{pmatrix} = \begin{pmatrix} \xi_i + \delta_i \\ \beta_1 \xi_i + \varepsilon_i \end{pmatrix}, \qquad i = 1, \ldots, n.$$

5. Form the grouping variable observations using

$$z_i = \frac{\xi_i - \text{median}(\xi_i)}{|\xi_i - \text{median}(\xi_i)|}, \qquad i = 1, \ldots, n.$$

Thus, the observations corresponding to ξ_is greater than the median of the ξ_is are assigned grouping variable value 1, and those below the median are assigned grouping variable value -1. The results are summarized in the following tables. In order to get some feel for the accuracy of the estimates, the entries in the tables give the sample means of the estimates followed in parentheses by their sample mean absolute errors over the 500 replications. Recall that the true means and mean absolute errors are not finite (see Exercise 4.8), so that these sample values are not estimating any true values, but rather giving some idea about how well the estimators did on these 500 replications. Thus, for example, for $\sigma^2 = 4,\ \sigma_\delta^2 = \sigma_\varepsilon^2 = 1$, Table 4.2 gives

$$\frac{1}{500}\sum_{j=1}^{500}\hat{\beta}_{1G,j} = 1.03 \qquad \text{and} \qquad \frac{1}{500}\sum_{j=1}^{500}|\hat{\beta}_{1G,j} - 1| = 0.18,$$

where $\hat{\beta}_{1G,j}$ is the value of (4.15) for the jth replication.

Table 4.2 Estimate means and absolute errors, two groups based on ξ: 500 replications, $\beta_0 = 0,\ \beta_1 = 1,\ n = 20$

σ^2	σ_δ^2	σ_ε^2	$\hat{\beta}_{0G}$	$\hat{\beta}_{1G}$	$\hat{\sigma}_{\delta G}^2$	$\hat{\sigma}_{\varepsilon G}^2$
4	1	1	−0.01 (0.28)	1.03 (0.18)	1.01 (0.51)	0.96 (0.51)
		4	0.00 (0.42)	1.02 (0.27)	1.63 (1.90)	3.96 (1.41)
	4	1	0.05 (0.52)	1.15 (0.39)	3.96 (1.38)	0.91 (0.92)
		4	−0.09 (0.63)	1.00 (0.57)	5.51 (3.41)	3.97 (1.69)
16	1	1	−0.03 (0.26)	1.01 (0.09)	0.99 (0.89)	0.97 (0.87)
		4	−0.01 (0.42)	1.01 (0.13)	0.71 (1.43)	3.98 (1.75)
	4	1	0.02 (0.43)	1.03 (0.14)	4.08 (1.78)	0.89 (1.43)
		4	0.01 (0.54)	1.04 (0.18)	4.14 (2.22)	3.74 (2.17)

The estimates of β_0 and β_1 are quite reasonable for this sample size; however, the estimates for the error variances, while reasonably centred near their true values, have far too high variation. Thus, (4.18) is frequently giving poor estimates for the error variances. Many times these variance estimates were negative. Note also that when the signal to noise ratio gets low ($\sigma^2 = \sigma_\delta^2 = \sigma_\varepsilon^2 = 4$), $\hat{\beta}_{1G}$ begins to have high variation.

Table 4.3 Estimate means and absolute errors, two groups based on ξ: 500 replications, $\beta_0 = 0,\ \beta_1 = 1,\ n = 100$

σ^2	σ_δ^2	σ_ε^2	$\hat{\beta}_{0G}$	$\hat{\beta}_{1G}$	$\hat{\sigma}_{\delta G}^2$	$\hat{\sigma}_{\varepsilon G}^2$
4	1	1	0.00 (0.11)	1.00 (0.10)	0.96 (0.28)	0.99 (0.28)
		4	−0.01 (0.18)	0.99 (0.15)	0.94 (0.47)	4.08 (0.78)
	4	1	0.00 (0.18)	1.00 (0.15)	4.01 (0.74)	1.01 (0.44)
		4	−0.01 (0.14)	1.03 (0.18)	4.09 (0.88)	3.99 (0.74)
16	1	1	−0.02 (0.12)	1.01 (0.05)	1.10 (0.51)	0.90 (0.50)
		4	−0.01 (0.19)	0.99 (0.07)	0.91 (0.72)	4.07 (0.92)
	4	1	−0.01 (0.18)	1.00 (0.07)	3.81 (0.86)	1.04 (0.67)
		4	−0.01 (0.24)	1.01 (0.09)	4.10 (01.13)	3.96 (0.99)

If the sample size is increased to 100, better results are, of course, obtained (Table 4.3). The standard deviations of the error variance estimates are still somewhat large and negative estimates of the error variances still commonly occur.

If β_1 is increased to 2, matters grow worse. The estimate, $\hat{\beta}_{1G}$, does reasonably well but the estimates, $\hat{\beta}_{0G}$ and (4.18), deteriorate noticeably.

For comparison, let us consider the same data as in the previous example, but now violate the instrumental variable assumptions and separate the groups at the median of x instead of the median of ξ. Remember that, if one groups on x, this model is not identifiable so examples which group on x are for illustration purposes only.

Example 4.5

Consider the same simulation study as in Example 4.4, except that, instead of step (5), we form the grouping variable observations using

$$z_i = \frac{x_i - \text{median}(x_i)}{|x_i - \text{median}(x_i)|}, \qquad i = 1, \ldots, n.$$

Table 4.4 gives the means and sample mean absolute errors of the estimates as explained in Example 4.4.

Table 4.4 Estimate means and absolute errors, two groups based on x: 500 replications, $\beta_0 = 0,\ \beta_1 = 1,\ n = 100$

σ^2	σ_δ^2	σ_ε^2	$\hat{\beta}_{0G}$	$\hat{\beta}_{1G}$	$\hat{\sigma}_{\delta G}^2$	$\hat{\sigma}_{\varepsilon G}^2$
4	1	1	0.00 (0.10)	0.80 (0.20)	-0.04 (1.04)	1.83 (0.84)
		4	0.01 (0.17)	0.80 (0.21)	-0.04 (1.04)	4.81 (0.92)
	4	1	−0.01 (0.14)	0.51 (0.49)	-0.06 (4.06)	3.01 (2.01)
		4	−0.01 (0.19)	0.51 (0.49)	-0.15 (4.15)	6.05 (2.05)
16	1	1	−0.02 (0.12)	0.94 (0.06)	-0.01 (1.02)	1.95 (0.96)
		4	0.02 (0.19)	0.94 (0.07)	-0.02 (1.08)	4.94 (1.09)
	4	1	−0.01 (0.16)	0.80 (0.20)	0.02 (3.99)	4.16 (3.16)
		4	−0.01 (0.23)	0.80 (0.20)	0.02 (3.98)	7.15 (3.15)

Comparing Tables 4.3 and 4.4, one sees that there has been considerable deterioration in the performance of all the estimates except $\hat{\beta}_{0G}$, particularly the estimates of the error variances.

The results of Example 4.5 indicate that grouping based on x can give very poor results.

Inaccuracies in the estimates, (4.18), are quite likely, as can be easily seen. Look, for example, at the estimate

$$\hat{\sigma}_{\delta G}^2 = \frac{ns_{xx}}{n-1}(1 - \frac{s_{xy}}{s_{xx}}\frac{1}{\hat{\beta}_{1G}}) = \frac{ns_{xx}}{n-1}(1 - \frac{\hat{\beta}_{1R}}{\hat{\beta}_{1G}}).$$

If $\hat{\beta}_{1R}$ and $\hat{\beta}_{1G}$ are close together then small errors in $\hat{\beta}_{1G}$ can cause large errors in $\hat{\sigma}_{\delta G}^2$. Note that it is no longer guaranteed that $|\hat{\beta}_{1R}| \le |\hat{\beta}_{1G}|$.

The results obtained in Example 4.5 can also be explained theoretically. Theorem 4.2 states that, under the model of Example 4.5, $\hat{\beta}_{1G}$ converges to the expected value of the least-squares estimate,

$$p \lim_{n\to\infty} \hat{\beta}_{1G} = \kappa_\xi \beta = \frac{\sigma^2}{\sigma^2 + \sigma_\delta^2}\beta. \tag{4.19}$$

Note that the results of Example 4.5 are quite consistent with (4.19). The average estimates in the $\hat{\beta}_{1G}$ column of Table 4.4 are almost exactly the value given by (4.19).

Confidence regions based on two groups

A confidence interval for β_1 was obtained by Wald (1940). Using the two groups, G_1 and G_2, define the pooled estimators

$$\begin{aligned} s_{xx}^{(p)} &\equiv \frac{1}{n-2}\left\{\sum_{G_1}(x_i-\bar{x}_1)^2+\sum_{G_2}(x_i-\bar{x}_2)^2\right\}, \\ s_{yy}^{(p)} &\equiv \frac{1}{n-2}\left\{\sum_{G_1}(y_i-\bar{y}_1)^2+\sum_{G_2}(y_i-\bar{y}_2)^2\right\}, \\ s_{xy}^{(p)} &\equiv \frac{1}{n-2}\left\{\sum_{G_1}(x_i-\bar{x}_1)(y_i-\bar{y}_1)+\sum_{G_2}(x_i-\bar{x}_2)(y_i-\bar{y}_2)\right\}. \end{aligned} \tag{4.20}$$

Under the normality assumption, $s_{xx}^{(p)}$, $s_{yy}^{(p)}$, and $s_{xy}^{(p)}$ are distributed independently of $\bar{x}_1, \bar{x}_2, \bar{y}_1$, and $\bar{y}_2$ and, therefore of $\hat{\beta}_{1G}$. Substituting into (4.17) gives

$$\begin{aligned} s_{\delta\delta}^{(p)} &= s_{xx}^{(p)} - s_{xy}^{(p)}/\beta_1 \\ s_{\varepsilon\varepsilon}^{(p)} &= s_{yy}^{(p)} - \beta_1 s_{xy}^{(p)}. \end{aligned} \tag{4.21}$$

Now define

$$\begin{aligned} S^2 &\equiv s_{\varepsilon\varepsilon}^{(p)} + \beta_1^2 s_{\delta\delta}^{(p)} = s_{yy}^{(p)} + \beta_1^2 s_{xx}^{(p)} - 2\beta_1 s_{xy}^{(p)} \\ &= \frac{1}{n-2}\left[\sum_{G_1}\{(y_i-\beta_0-\beta_1x_i)-(\bar{y}_1-\beta_0-\beta_1\bar{x}_1)\}^2\right. \\ &\qquad \left.+\sum_{G_2}\{(y_i-\beta_0-\beta_1x_i)-(\bar{y}_2-\beta_0-\beta_1\bar{x}_2)\}^2\right]; \end{aligned}$$

then $(n-2)S^2$ is the sum of two sums of squares. The terms $(y_i-\beta_0-\beta_1x_i)$ are independent $N(0, \sigma_\varepsilon^2+\beta_1^2\sigma_\delta^2)$ random variables so that

$$\frac{(n-2)S^2}{\sigma_\varepsilon^2+\beta_1^2\sigma_\delta^2} \sim \chi_{n-2}^2.$$

Note that

$$\begin{aligned} u &\equiv \frac{1}{2}(\bar{x}_2-\bar{x}_1)(\hat{\beta}_{1G}-\beta_1) = \frac{1}{2}\{(\bar{y}_2-\bar{y}_1)-\beta_1(\bar{x}_2-\bar{x}_1)\} \\ &= \frac{1}{2}\{(\bar{\varepsilon}_2-\beta_1\bar{\delta}_2)-(\bar{\varepsilon}_1-\beta_1\bar{\delta}_1)\} \sim N\left(0, \frac{\sigma_\varepsilon^2+\beta_1^2\sigma_\delta^2}{n}\right). \end{aligned}$$

Because u is a function only of $\bar{x}_1$, $\bar{x}_2$, $\bar{y}_1$, and $\bar{y}_2$, it is independent of S^2, and

$$T \equiv \frac{n^{\frac{1}{2}}u}{S} = \frac{n^{1/2}(\bar{x}_2 - \bar{x}_1)(\hat{\beta}_{1G} - \beta_1)}{2(s_{yy}^{(p)} + \beta_1^2 s_{xx}^{(p)} - 2\beta_1 s_{xy}^{(p)})^{1/2}} \sim t_{n-2}.$$

For confidence level $1 - \gamma$ let $P(T^2 \le t_{n-2,1-\gamma}) = 1 - \gamma$. Solving for β_1 requires solving

$$(\bar{x}_2 - \bar{x}_1)^2(\hat{\beta}_{1G} - \beta_1)^2 = \frac{4t_{n-2,1-\gamma}^2}{n}(s_{yy}^{(p)} + \beta_1^2 s_{xx}^{(p)} - 2\beta_1 s_{xy}^{(p)}) \tag{4.22}$$

or

$$\begin{aligned} 0 \;=\; & \beta_1^2\left\{\frac{4t_{n-2,1-\gamma}^2}{n}s_{xx}^{(p)} - (\bar{x}_2 - \bar{x}_1)^2\right\} \\ & +2\beta_1\left\{\hat{\beta}_{1G}(\bar{x}_2 - \bar{x}_1)^2 - \frac{4t_{n-2,1-\gamma}^2}{n}s_{xy}^{(p)}\right\} \\ & +\left\{\frac{4t_{n-2,1-\gamma}^2}{n}s_{yy}^{(p)} - \hat{\beta}_{1G}(\bar{x}_2 - \bar{x}_1)^2\right\} = a\beta_1^2 + b\beta_1 + c. \end{aligned} \tag{4.23}$$

The discriminant of this quadratic is

$$4\left(\frac{4t_{n-2,1-\gamma}^2}{n}\right)^2(s_{xy}^{(p)} - s_{xx}^{(p)}s_{yy}^{(p)}) + \frac{16t_{n-2,1-\gamma}^2}{n}(\bar{x}_2 - \bar{x}_1)^2(\hat{\beta}_{1G}^2 s_{xx}^{(p)} - 2\hat{\beta}_{1G}s_{xy}^{(p)} + s_{yy}^{(p)}). \tag{4.24}$$

The first term of this discriminant is negative by the Cauchy–Schwarz inequality and is of order $\frac{1}{n^2}$, and the second term is positive because the term in parentheses is $\frac{1}{n-2}\sum_i(y_i - \hat{\beta}_{1G}x_i)^2$, and of order $\frac{1}{n}$. If n is large enough, the discriminant will be positive and the two real roots of (4.22) are the confidence limits for β_1.

Example 4.6

Consider again the model of Example 4.4. Using the same 500 replications, we calculated the nominal 90% confidence intervals from (4.22). We took $\beta_0 = 0$, $\beta_1 = 1$, and separated the groups based on ξ. We then gave the actual coverage probability for the 500 replications and the number of replications for which the discriminant is negative. The empirical coverage probability is then (the number of intervals that contained 1) $\div$ (the number of replications that gave a positive discriminant). Because we are claiming a 90% confidence interval, the number in the 'coverage' column of Table 4.5 should be close to 0.90.

Table 4.5 Two-group confidence interval coverage: group on ξ, $\beta_1 = 1$

			$n = 20$		$n = 100$	
σ^2	σ_δ^2	σ_ε^2	Coverage	No. complex	Coverage	No. complex
4	1	1	0.898	0	0.890	0
		4	0.902	0	0.926	0
	4	1	0.860	0	0.896	0
		4	0.879	5	0.896	0
16	1	1	0.902	0	0.904	0
		4	0.894	0	0.902	0
	4	1	0.914	0	0.910	0
		4	0.904	0	0.894	0

Table 4.5 shows that for this particular model, the confidence interval, based on two groups separated based on ξ, is very accurate even for a sample of size 20. Only for the low signal to noise ratio and sample size 20 case did the discriminant, (4.24), occasionally come out negative.

If β_1 is increased to 2, the results deteriorate somewhat for $n = 20$ but are still good for $n = 100$ (Table 4.6).

Table 4.6 Two-group confidence interval coverage: group on ξ, $\beta_1 = 2$

			$n = 20$		$n = 100$	
σ^2	σ_δ^2	σ_ε^2	Coverage	No. complex	Coverage	No. complex
4	1	1	0.896	0	0.910	0
		4	0.880	0	0.876	0
	4	1	0.765	1	0.918	0
		4	0.782	0	0.894	0
16	1	1	0.998	0	0.874	0
		4	0.998	0	0.862	0
	4	1	0.976	0	0.876	0
		4	0.976	0	0.906	0

Example 4.7

Consider now the model of Example 4.5. Using the same data as in the previous example we now group based on x instead of ξ. For $\beta_1 = 1$, the results are given in Table 4.7.

Table 4.7 Two-group confidence interval coverage: group on x, $\beta_1 = 2$

			$n = 20$		$n = 100$	
σ^2	σ_δ^2	σ_ε^2	Coverage	No. complex	Coverage	No. complex
4	1	1	0.672	0	0.180	0
		4	0.838	0	0.480	0
	4	1	0.224	0	0.000	0
		4	0.466	0	0.008	0
16	1	1	0.868	0	0.604	0
		4	0.858	0	0.768	0
	4	1	0.606	0	0.044	0
		4	0.726	0	0.182	0

It is obvious that the confidence interval (4.22) does not work on this model when the groups are separated based on x. This is easily explained by the bias of the estimate of β_1 given by Theorem 4.2 below. Note that, for the larger sample size, this bias does more harm. The results are better for larger signal to noise ratios ($\sigma^2 = 16$) because there the group separation based on x is closer to that based on ξ.

A joint confidence region can be obtained for (β_0, β_1) (see Kendall and Stuart, 1979, Section 29.39).

4.3.2 Three groups

It was pointed out by Nair and Shrivastava (1942) and by Bartlett (1949) that the efficiency of the grouping method might be increased by using three groups instead of two. We shall investigate this using the approach of Pakes (1982).

The idea is to group the n observations into three groups, G_1, G_2, and G_3, with G_1 containing np_1 observations, G_2 containing np_2 observations, and G_3 containing np_3 observations, where $0 < p_1, p_2, p_3$ and $p_1 + p_2 + p_3 = 1$. If one could make the division into groups based on the data ordered according to the values of ξ and independently of the errors, then the instrumental variable is

$$z_i = \begin{cases} -1 & \text{if } i\text{th observation in the first group,} \\ 0 & \text{if } i\text{th observation in the second group,} \\ 1 & \text{if } i\text{th observation in the third group.} \end{cases}$$

The estimator of β_1 is defined as the slope of the line connecting $(\bar{x}_1, \bar{y}_1)$ and $(\bar{x}_3, \bar{y}_3)$,

$$\hat{\beta}_{1T} = \frac{\bar{y}_3 - \bar{y}_1}{\bar{x}_3 - \bar{x}_1}, \tag{4.25}$$

where $\bar{x}_3$ and $\bar{y}_3$ are the sample means over G_3 and $\bar{x}_1$ and $\bar{y}_1$ are the sample means over G_1. For example, $\bar{x}_1 = \frac{1}{\|G_1\|} \sum_{G_1} x_i$ where $\|G_1\|$ denotes the number of points in G_1. Thus, in estimating β_1, the data in the second group are ignored. The estimator of the intercept is still

$$\hat{\beta}_{0T} \equiv \bar{y} - \hat{\beta}_{1T}\,\bar{x},$$

where $\bar{y}$ and $\bar{x}$ are the respective means of y and x for the entire sample. These estimators are consistent under the same conditions as before.

Nair and Shrivastava (1942) and Bartlett (1949) studied the case $p_1 = p_2 = p_3 = 1/3$. In this case (4.20) becomes

$$s_{xx}^{(p)} \equiv \frac{1}{n-3}\left\{\sum_{G_1}(x_i - \bar{x}_1)^2 + \sum_{G_2}(x_i - \bar{x}_2)^2 + \sum_{G_3}(x_i - \bar{x}_3)^2\right\}, \quad (4.26)$$

$$s_{yy}^{(p)} \equiv \frac{1}{n-3}\left\{\sum_{G_1}(y_i - \bar{y}_1)^2 + \sum_{G_2}(y_i - \bar{y}_2)^2 + \sum_{G_3}(y_i - \bar{y}_3)^2\right\}, \quad (4.27)$$

$$s_{xy}^{(p)} \equiv \frac{1}{n-3}\left\{\sum_{G_1}(x_i - \bar{x}_1)(y_i - \bar{y}_1) + \sum_{G_2}(x_i - \bar{x}_2)(y_i - \bar{y}_2) + \sum_{G_3}(x_i - \bar{x}_3)(y_i - \bar{y}_3)\right\}. \quad (4.28)$$

Furthermore, (4.21) still defines $s_{\delta\delta}^{(p)}$ and $s_{\varepsilon\varepsilon}^{(p)}$ and $S^2 \equiv s_{\varepsilon\varepsilon}^{(p)} + \beta_1^2 s_{\delta\delta}^{(p)}$. Then

$$\frac{(n-3)S^2}{\sigma_\varepsilon^2 + \beta_1^2\sigma_\delta^2} \sim \chi_{n-3}^2,$$

$$u \equiv \left(\frac{1}{6}\right)^{1/2}(\bar{x}_3 - \bar{x}_1)(\hat{\beta}_{1T} - \beta_1) \sim N\left(0, \frac{\sigma_\varepsilon^2 + \beta_1^2\sigma_\delta^2}{n}\right)$$

independently of S^2, and

$$T \equiv \left(\frac{1}{6}\right)^{1/2}\frac{(\bar{x}_3 - \bar{x}_1)(\hat{\beta}_{1T} - \beta_1)}{S} \sim t_{n-3},$$

which can be used to obtain confidence intervals as before.

Table 4.8 Thirty-six ME model data with ξs: $\beta_0 = 0,\ \beta_1 = 1,\ \sigma^2 = 9,$ and $\sigma_\delta^2 = \sigma_\varepsilon^2 = 1$

	x (ξ)	y		x (ξ)	y
1	0.62 (2.20)	1.30	19	3.91 (3.96)	3.72
2	−1.07 (−0.38)	−0.89	20	−8.94 (−8.69)	−9.37
3	−0.35 (−1.16)	−1.03	21	−0.81 (−0.94)	−1.14
4	−2.88 (−2.07)	−0.82	22	−3.16 (−3.84)	−3.56
5	3.79 (4.84)	3.70	23	−3.47 (−4.02)	−3.18
6	−1.90 (−0.66)	−0.46	24	6.62 (5.77)	5.05
7	−6.10 (−6.26)	−6.56	25	−2.75 (−3.14)	−3.07
8	0.53 (1.88)	−1.11	26	1.96 (2.30)	3.83
9	1.12 (0.42)	−0.79	27	−0.03 (−0.97)	−1.02
10	−3.43 (−3.71)	−2.14	28	−2.14 (−0.70)	−1.08
11	2.24 (3.05)	2.87	29	0.08 (−0.99)	−1.47
12	−1.07 (−1.99)	−1.59	30	−1.37 (−0.02)	−0.23
13	−0.74 (−0.79)	−1.23	31	1.95 (1.27)	1.49
14	7.14 (7.53)	7.11	32	5.08 (3.68)	2.94
15	4.74 (3.39)	2.41	33	−0.03 (0.04)	−0.06
16	3.09 (3.17)	4.02	34	3.46 (1.71)	0.98
17	1.23 (2.11)	2.53	35	−3.47 (−2.69)	−3.39
18	2.14 (1.80)	1.91	36	3.36 (1.53)	2.08

Example 4.8

The 36 observations in Table 4.8 are the same as those for Example 4.1 except that the instrumental variable, z, is omitted and the unobservable ξs are listed so that the reader can follow the grouping procedure. The numbers in parentheses are the unobservable ξ values corresponding to the respective x values. Imagine that somehow the observations can be separated into three groups with $p_1 = p_3 = 1/3$ based on the ξ values. The grouped data are given in Table 4.9.

The data in Table 4.9 give

$$\bar{x}_1 = -2.97, \qquad \bar{x}_2 = 0.33, \qquad \bar{x}_3 = 3.41,$$
$$\bar{y}_1 = -3.10, \qquad \bar{y}_2 = 0.05, \qquad \bar{y}_3 = 3.20,$$

and by (4.25)

$$\hat{\beta}_{1T} = \frac{3.20 + 3.10}{3.41 + 2.97} = 0.987,$$

which is quite accurate. Next

$$\hat{\beta}_{0T} = 0.048 - 0.987\,(0.260) = -0.208.$$

Table 4.9 Data from Table 4.8 grouped on ξ

	Group 1	
	x	y
1	−0.35	−1.03
2	−2.88	−0.82
3	−6.11	−6.57
4	−3.43	−2.14
5	−1.07	−1.59
6	−8.94	−9.37
7	−3.16	−3.56
8	−3.47	−3.18
9	−2.75	−3.07
10	−0.03	−1.02
11	0.08	−1.47
12	−3.47	−3.39

	Group 2	
	x	y
1	−1.07	−0.89
2	−1.90	−0.46
3	1.12	−0.79
4	−0.74	−1.23
5	2.14	1.91
6	−0.81	−1.14
7	−2.14	−1.08
8	−1.37	−0.23
9	1.96	1.48
10	−0.03	−0.06
11	3.46	0.98
12	3.36	2.08

	Group 3	
	x	y
1	0.62	1.30
2	3.79	3.70
3	0.54	−1.11
4	2.24	2.87
5	7.14	7.11
6	4.74	2.41
7	3.09	4.02
8	1.23	2.53
9	3.91	3.72
10	6.62	5.05
11	1.96	3.83
12	5.08	2.94

The two-group method (4.15) gives $\hat{\beta}_{1G} = 0.893$. $\hat{\beta}_{1T}$ is not as good as the instrumental variable estimate ($\hat{\beta}_{1IV} = 1.003$) from Example 4.1, but in that example we were given a continuous instrumental variable, whereas here we are only given a legitimate three-grouping of the data.

The 90% confidence interval is calculated as follows. From (4.26)–(4.28) we obtain

$$s_{xx}^{(p)} = 5.21, \qquad s_{yy}^{(p)} = 3.99, \qquad s_{xy}^{(p)} = 3.88.$$

Substituting these and $t_{33,0.95} = 1.692$ into (4.23) yields the coefficients

$$a = -38.2102, \qquad b = 76.6581, \qquad c = -37.7645.$$

The solution of the quadratic gives the confidence interval (0.87, 1.14).

Example 4.9

This is a Monte Carlo study of the accuracy of the confidence interval (4.22). We perform the same simulation as in Example 4.6, except we now estimate the parameters based on three groups with $p_1 = p_3 = 1/3$. We again perform 500 replications and calculate the nominal 90% confidence intervals from (4.22). When we take $\beta_0 = 0$, $\beta_1 = 1$, and separate the three groups based on ξ, the results are very similar to those of Table 4.5. There is no obvious improvement in using three groups instead of two groups. Again the values in the 'coverage' column should be close to 0.9 (see Table 4.10).

Table 4.10 Three-group confidence interval coverage: group on ξ, $\beta_1 = 1$

			$n = 20$		$n = 100$	
σ^2	σ_δ^2	σ_ε^2	Coverage	No. complex	Coverage	No. complex
4	1	1	0.916	0	0.896	0
		4	0.930	0	0.862	0
	4	1	0.884	0	0.898	0
		4	0.871	3	0.904	0
16	1	1	0.918	0	0.908	0
		4	0.896	0	0.868	0
	4	1	0.914	0	0.872	0
		4	0.902	0	0.880	0

Three groups based on the x values

Again, it is common in practice to base the grouping on the ordering of the xs. So, for the remainder of this section, the grouping criterion will be based on the quantiles of the x variable and, hence, the instrumental variable requirements are violated. We follow the treatment of Pakes (1982).

Suppose that $A \leq B$; define the groups, $G_1 = \{x : x \leq A\}$ and $G_2 = \{x : x > B\}$. First some notation: for any variable, w, occurring with x as a pair (x, w), let $\bar{w}_A$ and $\bar{w}_B$ be the average values of w for $x < A$ and $x > B$, respectively; then the estimator is (4.25), where $\bar{x}_1 = \bar{x}_A$, $\bar{x}_3 = \bar{x}_B$, $\bar{y}_1 = \bar{y}_A$ and $\bar{y}_3 = \bar{y}_B$. Note that the middle group of the data, associated with the x values between A and B, is discarded. The special case when $A = B$ is the two-group method.

If the model is structural and ξ, δ, and ε are mutually independent variables with finite variances, then

$$p\lim \bar{y}_A = E[y \mid x - A < 0] = \beta_0 + \beta_1 E[\xi \mid x - A < 0] = \beta_0 + \beta_1 \bar{\xi}_A,$$
$$p\lim \bar{y}_B = E[y \mid x - B > 0] = \beta_0 + \beta_1 E[\xi \mid x - B > 0] = \beta_0 + \beta_1 \bar{\xi}_B,$$

and

$$p\lim \bar{x}_A = \bar{\xi}_A + E[\delta \mid x - A < 0],$$
$$p\lim \bar{x}_B = \bar{\xi}_B + E[\delta \mid x - B > 0].$$

If ξ could be observed, then the slope of the line joining $(\bar{y}_A, \bar{\xi}_A)$ to $(\bar{y}_B, \bar{\xi}_B)$ would provide a consistent estimator of β_1. Unfortunately, when x is used, a problem arises when values of ξ greater than A associated with sufficiently large negative values of δ give values of x less than A, while values of ξ smaller than A associated with sufficiently large positive values of δ give values of x greater than A, so that $E[\delta \mid x - A < 0] \leq 0$ and $E[\delta \mid x - B > 0] \geq 0$. Thus, if the distance between A and B is small enough compared to the range of δ (which it frequently is in applications), then there is a positive probability that the grouping on the basis of the observed values, x, will be different from the grouping on the basis of the true values, ξ. One then gets strict inequality in the previous expectations and in most cases of practical interest

$$|p\lim \hat{\beta}_{1T}| = |\beta_1| \left| \frac{\bar{\xi}_B - \bar{\xi}_A}{\bar{\xi}_B - \bar{\xi}_A - E[\delta \mid x - A < 0] + E[\delta \mid x - B > 0]} \right| < |\beta_1|.$$

In practice, the cut-off points A and B are often constructed by using estimates of pth quantiles of the x variable. For example, in the two-group case, one could take the cut-off point to be the median of x, as was done in Example 4.5. In the three-group case, work by Bartlett (1949), Theil and van Yzeren (1956), and Gibson and Jowett (1957) indicates that one should take the first group to be those observations corresponding to xs less than or equal to $n\hat{q}_{1/3}$ and the second group to be those observations corresponding to xs greater than or equal to $n\hat{q}_{2/3}$, where n is the sample size and $\hat{q}_p$ is the sample pth quantile of the xs.

It is clear that for $(\xi, \delta, \varepsilon)$ normal, the grouping method does not provide a consistent estimator because Reiersol (1950) showed that β_1 is unidentifiable in this case. It is interesting to compare the grouping estimator with the least-squares estimator. Pakes (1982) showed that the asymptotic bias of $\hat{\beta}_{1T}$ is precisely the same as that of the least-squares estimator $\hat{\beta}_{1R}$:

Theorem 4.2 *Assume that $(\xi, \delta, \varepsilon)$ is normal and that the grouping is based on x; then, regardless of the cut-off points chosen,*

$$p\lim \hat{\beta}_{1T} = \kappa_\xi \beta_1.$$

Proof. Write the grouping estimator as

$$\hat{\beta}_{1T} = \beta_1 - \beta_1 \left(\frac{\bar{\delta}_B - \bar{\delta}_A}{\bar{x}_B - \bar{x}_A} \right) + \left(\frac{\bar{\varepsilon}_B - \bar{\varepsilon}_A}{\bar{x}_B - \bar{x}_A} \right).$$

Because the numerator of the last term of the equation above is independent of x, and therefore, of the selection criterion,

$$p\lim \hat{\beta}_{1T} = \beta_1 - \beta_1 p\lim \left(\frac{\bar{\delta}_B - \bar{\delta}_A}{\bar{x}_B - \bar{x}_A} \right).$$

Note that

$$\begin{aligned} p\lim \bar{\delta}_B &= E[\delta \mid x - B > 0] = (1 - \kappa_\xi) E[x \mid x - B > 0], \\ p\lim \bar{\delta}_A &= E[\delta \mid x - A < 0] = (1 - \kappa_\xi) E[x \mid x - A < 0], \\ p\lim \bar{x}_B &= E[x \mid x - B > 0], \qquad p\lim \bar{x}_A = E[x \mid x - A < 0], \end{aligned}$$

where κ_ξ is the reliability ratio defined in (1.13). The result follows. □

Pakes's theorem states that, under normality, there is no asymptotic advantage to grouping on x over simply using the ordinary least-squares estimator. If the ME model is not normal, then it is identifiable and the relative merits of the grouping method are not clear.

If the grouping is based on the values of y, reasoning similar to that given previously shows that the resultant estimator has the same limiting behaviour of $1/\hat{\beta}_{1I}$, where $\hat{\beta}_{1I}$ is the least-squares estimator of x regressing on y.

Example 4.10

Taking the same data as in Example 4.9 and grouping based on x instead of ξ, one obtains results almost identical to those in Table 4.7. Thus, the confidence intervals are still quite poor when one uses three groups based on x.

4.4 Methods Based on Ranks

Theil (1950) proposed another method called the **method of ranks** that can be viewed as related to the instrumental variable approach. The data are ordered according to either the x variable or the y variable. For simplicity, throughout this section, assume that the ordering is based on the x variable; that is, the xs are ordered to obtain the order statistics, $x_{(1)}, \ldots, x_{(n)}$. Then the 'instrumental variable', which is not a true instrumental variable because it is based on x, could be taken as $z_i = j$, where x_i corresponds to the jth order statistic, $x_{(j)}$. The corresponding ordering of the data is

$$\begin{pmatrix} x_{(1)} \\ y_{(1)} \end{pmatrix}, \ldots, \begin{pmatrix} x_{(n)} \\ y_{(n)} \end{pmatrix}. \tag{4.29}$$

Note that the original pairs x_i, y_i are kept together and the xs and the ys have not been ordered separately and then recombined to form new data pairs.

Now the strong assumption is made that the values of ξ are so spread out compared with σ_δ^2 that the series of observed xs is in the same order as the unobserved ξs. This is equivalent to the assumption that the ordering is based on ξ. That is,

$$x_i \leq x_j \Leftrightarrow \xi_i \leq \xi_j. \tag{4.30}$$

Note that, with high probability, this assumption would not be satisfied for the normal structural model if n is moderately large and σ_δ^2 is not very close to zero. Note that (4.30) is actually a new identifiability side condition, different from those in (1.12). Except for very small sample sizes, this assumption would tend to hold for the structural model when either ξ is not normal or when σ_δ^2 is very small. In the former case one would have a nonnormal structural model and could then use the methods of the next section. In the latter case, one is saying that the reliability ratio, κ_ξ, is known to be close to one and one could use least squares. These alternatives might be more efficient than the ranking methods about to be described.

The standard assumptions, (1.7), are made on the error structure and it is assumed that n is even, $n = 2m$, and that $x_i \neq x_j$ for all $i \neq j$.

Now, from the ordered data, (4.29), form either of two common sets of statistics: the lagged slopes,

$$b_i = \frac{y_{(m+i)} - y_{(i)}}{x_{(m+i)} - x_{(i)}}, \qquad i = 1, \ldots, m, \tag{4.31}$$

or the paired slopes,

$$b_{i,j} = \frac{y_{(j)} - y_{(i)}}{x_{(j)} - x_{(i)}}, \qquad i = 1, \ldots, j-1; \; j = 2, \ldots, n.$$

There are $n(n-1)/2$ of the paired slopes. Furthermore, from either the lagged slopes or the paired slopes one can form estimates of β_1 using either their mean or median:

$$\begin{aligned}
\hat{\beta}_{1,lm} &= \frac{1}{m}\sum b_i, \\
\hat{\beta}_{1,lmed} &= \text{median}(b_i), \\
\hat{\beta}_{1,pm} &= \frac{2}{n(n-1)}\sum_{j=2}^{n}\sum_{1\le i<j} b_{i,j}, \\
\hat{\beta}_{1,pmed} &= \text{median}(b_{i,j}).
\end{aligned}$$

The median estimates have the advantage that they will tend to be more robust and, as we see below, one can easily obtain confidence intervals for them.

Example 4.11

Ordering the data in Example 4.8 based on the x values and forming the lagged differences (4.31), one obtains

$$\begin{gathered}
0.88,\ 0.82,\ 1.15,\ 0.52,\ 1.00,\ 0.99,\ 0.96,\ 1.02,\ 0.90, \\
0.90,\ 0.49,\ 0.57,\ 0.94.\ 1.03,\ 0.67,\ 0.73,\ 0.77,\ 1.13\,.
\end{gathered}$$

These give

$$\begin{aligned}
\hat{\beta}_{1,lm} &= \frac{1}{18}\sum b_i = 0.86, \\
\hat{\beta}_{1,lmed} &= \text{median}(b_i) = 0.90.
\end{aligned}$$

Using all the paired slopes yields

$$\begin{aligned}
\hat{\beta}_{1,pm} &= \frac{2}{1260}\sum_{j=2}^{36}\sum_{1\le i<j} b_{i,j} = 4.44, \\
\hat{\beta}_{1,pmed} &= \text{median}(b_{i,j}) = 0.86.
\end{aligned}$$

Note the poor performance of $\hat{\beta}_{1,pm}$. This is undoubtedly because the average is sensitive to outliers in the b_{ij}. The other three estimators give values more or less in line with Theorem 4.2 because, for this example,

$$\kappa_\xi = \frac{\sigma^2}{\sigma^2 + \sigma_\delta^2} = 0.9.$$

4.4.1 Confidence intervals for median ranking methods

Confidence intervals can be obtained using an even milder condition than the normality of the errors. It is only required that the terms $(\varepsilon_i - \beta_1\delta_i)$ be independent and identically distributed with some continuous distribution. It then follows that for all $i \neq j$

$$P(\varepsilon_i - \beta_1\delta_i > \varepsilon_j - \beta_1\delta_j) = 1/2.$$

By (4.31)

$$\begin{aligned} b_i - \beta_1 &= \frac{(y_{(m+i)} - \beta_1 x_{(m+i)}) - (y_{t(i)} - \beta_1 x_{(i)})}{x_{(m+i)} - x_{(i)}} \\ &= \frac{(\varepsilon_{(m+i)} - \beta_1\delta_{(m+i)}) - (\varepsilon_{(i)} - \beta_1\delta_{(i)})}{x_{(m+i)} - x_{(i)}}. \end{aligned}$$

The denominator is not zero; so $P\left(b_i - \beta_1 > 0\right) = 1/2$. From the properties of the errors, the $b_i - \beta_1$ are independent Bernoulli(1/2) random variables. Thus the probability that exactly j of the $b_i - \beta_1$ exceed zero, that is $\beta_1 < b_i$, is given by $\binom{m}{j}\frac{1}{2^m}$. Thus, if $b_{(1)}, \ldots, b_{(m)}$ are the order statistics of the bs, then

$$P(b_{(r)} < \beta_1 < b_{(m-r+1)}) = 1 - 2\sum_{j=0}^{r}\binom{m}{j}\frac{1}{2^m}, \tag{4.32}$$

which provides a $1 - 2\sum_{j=0}^{r}\binom{m}{j}\frac{1}{2^m}$ confidence interval for β_1.

If it is assumed that δ and ε have zero medians, then

$$P(y_i - \beta_1 x_i > \beta_0) = 1/2.$$

Given any β_1, the quantities $c_i \equiv y_i - \beta_1 x_i$ can be ordered as $c_{(1)}, \ldots, c_{(m)}$ and this again gives

$$P(c_{(r)} < \beta_0 < c_{(m-r+1)}) = 1 - 2\sum_{j=0}^{r}\binom{m}{j}\frac{1}{2^m}.$$

It does not appear that this method will yield a joint confidence region for both β_0 and β_1 except for an upper confidence bound for the confidence coefficient; see Exercise 4.5.

The use of the paired slopes is more complicated because the distributions are no longer binomial.

These methods can be generalized to ME models in k variables. See, for example, Kendall and Stuart (1979, Section 29.45).

4.5 Methods of Higher-order Moments and Product Cumulants

So far, we have essentially been considering situations in which identifiability is ensured by some form of additional knowledge such as a side condition or an instrumental variable. In Appendix A, we describe the identifiability characteristics (of the slope) in the structural model with normal error. Reiersol (1950) showed that β_1 is identifiable if there exists a nonzero, finite or infinite, cumulant $\kappa(r, s)$ (defined in subsection 4.5.1) of the joint distribution function of x and y, with $r = 1$ and $s > 1$ or with $r > 1$ and $s = 1$. This implies that β_1 is identifiable if ξ is not normally distributed.

In this section, assume that ξ is not normally distributed and that all the suitable moments used below exist. We follow the treatment of Pal (1980) and van Montfort *et al.* (1987). The model considered is the structural relationship

$$x_i = \xi_i + \delta_i, \qquad y_i = \beta_0 + \beta_1 \xi_i + e_i, \qquad i = 1, \ldots, n, \tag{4.33}$$

with all the ξ_i, δ_i and ε_i independent of each other. Possible equation error is allowed in the above model by permitting the error to be of the form $e = \varepsilon + q$, where ε is the measurement error and q is the equation error. We introduce the following notation:

$$c_{rs} = E(x - \mu_x)^r (y - \mu_y)^s, \qquad v_{\xi t} = E(\xi - \mu_\xi)^t,$$
$$v_{\delta t} = E\delta^t, \qquad v_{\varepsilon t} = E\varepsilon^t.$$

It is easily seen that

$$\begin{array}{ll} \mu_y = \beta_0 + \beta_1 \mu_x & c_{20} = \sigma^2 + \sigma_\delta^2 \\ c_{02} = \beta_1^2 \sigma^2 + \sigma_\varepsilon^2 & c_{11} = \beta_1 \sigma^2 \\ c_{30} = v_{\xi 3} + v_{\delta 3} & c_{21} = \beta_1 v_{\xi 3} \\ c_{12} = \beta_1^2 v_{\xi 3} & c_{03} = \beta_1^3 v_{\xi 3} + v_{\varepsilon 3}. \end{array} \tag{4.34}$$

If $v_{\xi 3} \neq 0$, $c_{12} \neq 0$, and $c_{21} \neq 0$, then (4.34) provides eight equations in eight unknowns with a unique solution for the parameters, which yield the estimates

$$\begin{array}{ll} \hat{\beta}_0 = \bar{y} - \bar{x}\hat{c}_{12}/\hat{c}_{21}, & \hat{\beta}_1 = \hat{c}_{12}/\hat{c}_{21}, \\ \hat{\sigma}^2 = (\hat{c}_{11}\hat{c}_{21})/\hat{c}_{12}, & \hat{v}_{\xi 3} = \hat{c}_{21}^2/\hat{c}_{12}, \\ \hat{\sigma}_\delta^2 = \hat{c}_{20} - (\hat{c}_{11}\hat{c}_{21})/\hat{c}_{12}, & \hat{v}_{\delta 3} = \hat{c}_{30} - \hat{c}_{21}^2/\hat{c}_{12}, \\ \hat{\sigma}_\varepsilon^2 = \hat{c}_{02} - (\hat{c}_{11}\hat{c}_{12})/\hat{c}_{21}, & \hat{v}_{\varepsilon 3} = \hat{c}_{03} - \hat{c}_{12}^2/\hat{c}_{21}, \end{array} \tag{4.35}$$

where, for example,

$$\hat{c}_{rs} = \frac{1}{n} \sum (x_i - \bar{x})^r (y_i - \bar{y})^s \quad \text{and} \quad \hat{v}_{\xi t} = \frac{1}{n} \sum (\xi_i - \bar{\xi})^t. \tag{4.36}$$

Obviously the resulting estimators are consistent because the sample moments are consistent estimators of the true moments. It should be noted that the resulting estimates of σ^2, σ_δ^2, and σ_ε^2 are no longer guaranteed to be nonnegative for finite samples. Furthermore, if c_{12} and/or c_{21} are close to zero, these estimates will probably be very unstable unless the sample size is very large. If, for example, ε and δ are symmetrically distributed about zero and ξ is symmetrically distributed about its mean, then $c_{12} = c_{21} = 0$ and the estimates (4.35) cannot be used. One can then turn to fourth-order moments (see Section 4.5.1).

An interesting special case is when either δ or ε is symmetrically distributed (and ξ is not symmetrically distributed about its mean) as commonly occurs in error models. Durbin (1954) pointed out that this leads to multiple estimates. For if δ is symmetric (or simply has zero third moment) then $v_{\delta 3} = 0$, and

$$\beta_1 = c_{12}/c_{21} = c_{21}/c_{30}.$$

On the other hand, if ε is symmetric (or simply has zero third moment), then $v_{\varepsilon 3} = 0$, and

$$\beta_1 = c_{12}/c_{21} = c_{03}/c_{12}.$$

If both errors are symmetrically distributed, then

$$\begin{aligned} \beta_1 = c_{03}/c_{12} = c_{12}/c_{21} &= c_{21}/c_{30}, \\ \beta_1^3 &= c_{03}/c_{30}, \end{aligned} \tag{4.37}$$

and

$$\beta_1^2 = c_{03}/c_{21} = c_{12}/c_{30}. \tag{4.38}$$

The sign of β_1 in (4.38) should be the same as that of any of the other estimates. In fact, one can find many or even infinitely many consistent estimators of β_1 based on moments up to order 3. Pal (1980) noted that there are infinitely many consistent estimates by forming weighted arithmetic or geometric means of the three estimators defined by (4.37).

As is generally the case with method of moments estimators, even if neither of the errors is symmetric, one can find infinitely many consistent estimators of β_1 by using the higher (than third) moments. Using the relation (4.33), one can equate fourth, fifth, etc., moments (assuming they exist) and solve the resulting equations to obtain higher-order moment estimators. It would seem that the obvious choice is the lowest-order moments, that is, the third-order moments estimator, because the estimates of lower-order moments generally have lower variations. For nonsymmetric errors, van Montfort *et al.* (1987) showed that the estimator $\hat{\beta}_1 = \hat{c}_{12}/\hat{c}_{21}$ is optimal in the sense of minimal asymptotic variance in the class of all consistent estimators that are functions of the moments up to order 3.

If both errors are symmetric, the optimal estimator is the weighted mean

$$\hat{\beta}_1^{\text{opt}} = \frac{\mathbf{u}'\mathbf{V}^{-1}\mathbf{t}}{\mathbf{u}'\mathbf{V}^{-1}\mathbf{u}},$$

where $\mathbf{t}' = (\hat{\beta}_{11}, \hat{\beta}_{12}, \hat{\beta}_{13})$, $\mathbf{u}' = (1, 1, 1)$, $\mathbf{V}$ is the asymptotic covariance matrix of $\mathbf{t}$ and

$$\hat{\beta}_{11} = \hat{c}_{03}/\hat{c}_{12}, \qquad \hat{\beta}_{12} = \hat{c}_{12}/\hat{c}_{21}, \qquad \hat{\beta}_{13} = \hat{c}_{21}/\hat{c}_{30}.$$

Note that usually $\mathbf{V}$ is unknown, so one needs to replace $\mathbf{V}$ by a consistent estimator. The resulting estimator is no longer a function of moments up to order 3 only, but it is asymptotically equivalent to $\hat{\beta}_1^{\text{opt}}$. It should be pointed out that the asymptotic covariance matrix $\mathbf{V}$ has a complicated form.

Pal (1980) stated that, under mild conditions, all the moment estimators obtained so far have asymptotic normal distributions. He also gave the asymptotic variances (but not the covariances) of these estimators when (δ, ε) are normally distributed. On the other hand, van Montfort *et al.* (1987) showed that the asymptotic variance of $\hat{\beta}_1^{\text{opt}}$ has the form $(\mathbf{u}'\mathbf{V}^{-1}\mathbf{u})^{-1}$. They also gave the asymptotic covariance matrix $\mathbf{V}$ of $\mathbf{t}$ and a consistent estimate of $\mathbf{V}$. For more details, see van Montfort (1988, Chapter 1).

The estimators obtained via moments are also consistent in the functional relationship, provided certain limits of the true variables exist. This follows from the fact that they are method of moments estimators for the functional model.

Example 4.12

This example uses an asymmetric nonnormal model for the ξs. We generated 36 observations from the experiment in Example 4.8 except that step (1) was replaced by:

1.′ generate 36 independent values; $\xi_1, \ldots, \xi_{36}$; from a chi-square distribution with four degrees of freedom.

The data are given in Table 4.11.

Table 4.11 Thirty-six ME model data with $\xi \sim \chi_4^2$: $\beta_0 = 0,\ \beta_1 = 1,$ and $\sigma_\delta^2 = \sigma_\varepsilon^2 = 1$

	x	y		x	y		x	y
1	9.41	9.23	13	5.67	5.31	25	0.11	0.66
2	1.66	4.29	14	1.71	3.52	26	0.83	2.17
3	7.52	7.31	15	6.72	8.49	27	7.36	6.90
4	3.57	5.82	16	5.93	5.15	28	8.95	7.53
5	6.35	6.51	17	5.50	4.18	29	5.43	5.64
6	−0.06	0.85	18	1.76	1.66	30	2.46	1.24
7	1.39	2.14	19	2.13	0.79	31	12.01	12.33
8	1.57	3.27	20	0.85	1.06	32	3.79	0.66
9	3.57	3.61	21	14.58	12.70	33	2.17	1.87
10	4.11	5.55	22	2.77	2.16	34	1.44	3.19
11	2.93	0.30	23	0.19	1.13	35	9.23	10.61
12	1.97	2.01	24	0.19	3.00	36	0.76	3.36

Then

$\bar{x} = 4.07,$ $\bar{y} = 4.34,$

$\hat{c}_{12} = 38.94,$ $\hat{c}_{21} = 43.59,$

$\hat{\beta}_1 = \hat{c}_{12}/\hat{c}_{21} = 38.94/43.59 = 0.893,$ $\hat{\beta}_0 = 4.34 - 0.893 \times 4.07 = 0.705.$

The two-group method (4.15), separating on x, gives $\hat{\beta}_{1G} = 0.805$ and $\hat{\beta}_{0G} = 1.06$.

Because the errors are symmetrically distributed, (4.37) can be used to estimate the parameters, giving:

$\hat{c}_{03} = 33.02$

$\hat{\beta}_1' = \hat{c}_{03}/\hat{c}_{12} = 33.02/38.94 = 0.848$ $\hat{\beta}_0' = 4.34 - 0.848 \times 4.07 = 0.889$

and

$$\hat{c}_{30} = 47.62$$
$$\hat{\beta}_1'' = \hat{c}_{21}/\hat{c}_{30} = 43.59/47.62 = 0.915 \qquad \hat{\beta}_0'' = 4.34 - 0.915 \times 4.07 = 0.616.$$

The method of moments estimates in Example 4.12 are reasonable for a sample of size 36. It has been the authors' experience that the method of moments methods and Geary's method, discussed next, require moderately large sample sizes (see Madansky, 1959). This is easily explained by looking at the variances of the moment estimates. For example, suppose that $\xi \sim N(0, 9)$, $\delta \sim N(0, 1)$, $\varepsilon \sim N(0, 1)$, $\beta_0 = 0$, and $\beta_1 = 1$; then

$$\operatorname{var}\left(\frac{1}{n}\sum_{i=1}^{n} y_i^2\right) = \frac{1}{n}E(\xi+\varepsilon)^4 = \frac{1}{n}E(\xi^4+2\xi^2\varepsilon^2+\varepsilon^4) = \frac{1}{n}(243+18+3) = \frac{264}{n}$$

whereas

$$\begin{aligned}
\operatorname{var}\left(\frac{1}{n}\sum_{i=1}^{n} x_i y_i^2\right) &= \frac{1}{n}E(\xi+\delta)^2(\xi+\varepsilon)^4 \\
&= \frac{1}{n}E(\xi^2+2\delta\xi+\varepsilon^2)(\xi^4+2\xi^2\varepsilon^2+\varepsilon^4) \\
&= \frac{1}{n}E(\xi^6+3\xi^4\varepsilon^2+3\xi^2\varepsilon^4+\varepsilon^6) \\
&= \frac{1}{n}(9^3\times 15+3\times 243+9+15) \\
&\doteq \frac{11\,688}{n}.
\end{aligned}$$

Thus the estimates of third-order moments have much larger variances than the estimates of second order moments, in this case by a factor of over 44.

4.5.1 Geary's method using product cumulants

An approach closely related to the method using moments is a method based on product cumulants proposed by Geary (1942; 1943). In fact, the method of cumulants leads to the same estimators as those obtained via higher-order moments. The method of cumulants yields a particularly simple formula for a series of estimators of β_1. Bivariate cumulants are discussed, for example, in Stuart and Ord (1994, Section 3.27). Briefly, if X and Y are jointly distributed random variables, then, provided the expansions are permissible, one can write the natural logarithm of the joint characteristic function as

$$\psi(t_1, t_2) = \log\phi(t_1, t_2) = \log[E(\mathrm{e}^{\mathrm{i}t_1X+\mathrm{i}t_2Y})] = \sum_{r,s=0}^{\infty}\kappa(r, s)\frac{(\mathrm{i}t_1)^r}{r!}\frac{(\mathrm{i}t_2)^s}{s!}. \qquad (4.39)$$

ψ is called the joint cumulant generating function and if $r \neq 0$ and $s \neq 0$, then $\kappa\,(r, s)$ is called the r, s product cumulant of X and Y. These cumulants can be related to the moments (Stuart and Ord, 1994, Chapter 3).

Consider the classical no-equation-error structural model

$$x_i = \xi_i + \delta_i, \quad y_i = \eta_i + \varepsilon_i, \quad \eta_i = \beta_0 + \beta_1 \xi_i, \qquad i = 1, \ldots n.$$

First measure (ξ_i, η_i) from their true means so that the intercept is absorbed into the relationship and write the structural relationship in the symmetric form

$$\alpha_1 \xi_i + \alpha_2 \eta_i = 0. \tag{4.40}$$

The following important properties of bivariate cumulants are used (see, for example, Stuart and Ord, 1994, Chapter 12).

(a) The cumulant of a sum of independent random variables is the sum of the cumulants of the variables.
(b) The bivariate cumulant of any order p_1, p_2, where both p_1 and p_2 are positive, of independent random variables is zero.
(c) Cumulants are invariant under the change of origin, except for the first cumulant.

The joint cumulant generating function, ψ, of (x, y) is the sum of the joint cumulant generating functions of (ξ, η) and (δ, ε) by property (a). Moreover, the bivariate cumulants of any order p_1, p_2, where both p_1 and p_2 are positive, of (δ, ε) are zero by the second property. Finally, centring on the mean does not affect the estimation by property (c). Write $\kappa_{(x,y)}$ for the cumulants of (x, y), $\kappa_{(\xi,\eta)}$ for the cumulants of (ξ, η); then

$$\kappa_{(x,y)}(p_1, p_2) = \kappa_{(\xi,\eta)}(p_1, p_2), \tag{4.41}$$

provided both p_1 and p_2 are positive. Thus the product cumulants of (ξ, η) can be estimated by estimating those of (x, y) and from now on the index (x, y) or (ξ, η) will be dropped.

The joint characteristic function of (ξ, η) is

$$\phi(t_1, t_2) = E\{\exp(\theta_1 \xi + \theta_2 \eta)\}, \tag{4.42}$$

where $\theta_j = \mathrm{i}t_j$ for $j = 1, 2$. Equations (4.42) and (4.40) yield

$$\alpha_1 \frac{\partial \phi}{\partial \theta_1} + \alpha_2 \frac{\partial \phi}{\partial \theta_2} = E\{(\alpha_1 \xi + \alpha_2 \eta)\ \exp(\theta_1 \xi + \theta_2 \eta)\} = 0 \tag{4.43}$$

and, for the cumulant generating function $\psi = \log \phi$,

$$\alpha_1 \frac{\partial \psi}{\partial \theta_1} + \alpha_2 \frac{\partial \psi}{\partial \theta_2} = \frac{1}{\psi} \left(\alpha_1 \frac{\partial \phi}{\partial \theta_1} + \alpha_2 \frac{\partial \phi}{\partial \theta_2} \right) = 0. \tag{4.44}$$

By (4.39) and (4.44), for all nonnegative p_1 and p_2,

$$\alpha_1 \kappa(p_1 + 1, p_2) + \alpha_2 \kappa(p_1, p_2 + 1) = 0. \tag{4.45}$$

This relation is also true for the product cumulants of the observed (x, y) by (4.41), provided both p_1 and p_2 are both positive.

The same arguments hold for the functional relationship in the following sense. The true values, $\xi_1, \ldots, \xi_n$ are now fixed, but if ξ is regarded as sampling from the finite population $\{\xi_1, \ldots, \xi_n\}$, which is exhaustively sampled without replacement, the argument remains intact.

Before proceeding, note that the arguments leading to (4.45) also hold for the equation error model. This model was not discussed in Geary (1942; 1943); however, the method is easily adapted to this model. To see this, recall that the true variables η and ξ satisfy the linear relation

$$\eta_i = \beta_0 + \beta_1 \xi_i + q_i.$$

Define

$$\eta_i' = \beta_0 + \beta_1 \xi_i$$

so that

$$y_i = \eta_i' + q_i + \varepsilon_i = \eta_i' + e_i.$$

The preceding arguments are true with η_i replaced by η_i' and ε_i replaced by e_i. Thus, the cumulant estimators for ME models are the same with or without equation error.

In the original model (4.33), $\alpha_1 = \beta_1, \alpha_2 = -1$ because one is measuring ξ and η from their means. Therefore (4.45) becomes

$$\beta_1 \kappa(p_1 + 1, p_2) - \kappa(p_1, p_2 + 1) = 0$$

or, if $\kappa(p_1 + 1, p_2) \neq 0$,

$$\beta_1 = \frac{\kappa(p_1, p_2 + 1)}{\kappa(p_1 + 1, p_2)}. \tag{4.46}$$

This holds for any $p_1, p_2 > 0$. Consistent estimators of β_1 are therefore obtained by replacing the cumulants $\kappa(p_1, p_2)$ by their sample versions, $K(p_1, p_2)$, that is, the sample cumulant of order (p_1, p_2) of the empirical distribution, F_n, of the (x_i, y_i), where

$$F_n(x, y) \equiv \frac{1}{n} \sum_{i=1}^{n} I_{(-\infty, x]}(x_i) I_{(-\infty, y]}(y_i)$$

and I_A is the indicator function of the set A. The sample cumulants can be obtained by simply substituting the moment estimates into the formulae that express the cumulants in terms of the moments (see, for example, Stuart and Ord, 1994, Section 3.29). For example,

$$K(3, 1) = \hat{c}_{31} - 3\hat{c}_{20}\hat{c}_{11} = \hat{c}_{31} - 3 s_{xx} s_{xy},$$

where $\hat{c}_{ij}$ is defined by (4.36). In general, the only estimator based on cumulants of order 3 is

$$\hat{\beta}_1 = K(1, 2) / K(2, 1). \tag{4.47}$$

If δ and ε are symmetrically distributed about 0, then there are two additional estimators, namely, $K(0, 3) / K(1, 2)$ and $K(2, 1) / K(3, 0)$. Note that $\kappa(r, s) = c_{rs}$ for

$r + s = 3$, where c_{rs} is the central moment defined earlier (4.36). Therefore, these estimators are the same as those just obtained via the third-order moments.

If the marginal distributions of (x, y) are symmetric about 0, then all odd-order product central moments and hence all odd-order product cumulants will be zero. For example,

$$c_{21} = Ex^2y = Ex^2[\beta_1(x - \delta) + \varepsilon] = \beta_1 Ex^3 - \beta_1 Ex^2\delta + Ex^2\varepsilon = 0.$$

In this case (4.47) cannot be used and the lowest-order cumulant estimators are

$$\begin{array}{llll} \text{(a)} & p_1 = 1, & p_2 = 2: & \hat{\beta}_1 = K(1,3)/K(2,2), \\ \text{(b)} & p_1 = 2, & p_2 = 1: & \hat{\beta}_1 = K(2,2)/K(3,1). \end{array} \tag{4.48}$$

Unfortunately, this cumulant method for estimating the slope β_1 does not work if (x, y) is jointly normally distributed. All cumulants of order greater than or equal to 3 are zero in normal systems and (4.46) is useless in this case. This is not surprising, because the normal model is not identifiable, and no further assumption has been made to render the model identifiable. Presumably, as a nonnormal model gets closer and closer to a normal model, the cumulant estimates become more and more unstable.

It is clear, from the previous discussion, that the method of cumulants results in the same estimators as the method of higher-order moments. The method of cumulants, on the other hand, does allow a simple explicit formula, (4.46), for the estimators of β_1 of any order provided the denominator of (4.46) is nonzero.

Example 4.13

For the data of Example 4.12, Geary's method using fourth-order cumulants gives

$$\begin{aligned} &K(1,3) = 35.18, \\ &\hat{\beta}_1 = K(1,3)/K(2,2) = 35.18/59.73 = 0.589, \\ &K(2,2) = 59.73, \\ &\hat{\beta}_0 = 4.34 - 0.589 \times 4.07 = 1.943, \end{aligned}$$

and

$$\begin{aligned} &K(3,1) = 82.88, \\ &\hat{\beta}_1' = K(2,2)/K(3,1) = 59.73/82.88 = 0.721, \\ &\hat{\beta}_0' = 4.34 - 0.721 \times 4.07 = 1.406. \end{aligned}$$

These results are not as good as the method of moments, as one would expect because the method of moments uses lower-order moment estimates.

Example 4.14

We thought it would be interesting to compare the sample mean absolute errors on symmetric nonnormal data (see Example 4.4) for the three methods: method of moments using third-order moments; Geary's method using fourth-order cumulants;

and the grouping method with two groups based on x. We performed 500 replications of the following experiment:

1. Generate 100 independent values $\xi_1, \ldots, \xi_{100}$ of $\xi \sim$ uniform on $[-5, 5]$.
2. Generate 100 independent values $\delta_1, \ldots, \delta_{100}$ of $\delta \sim N(0, 1)$.
3. Generate 100 independent values $\varepsilon_1, \ldots, \varepsilon_{100}$ of $\varepsilon \sim N(0, 1)$.
4. Form the ME model data.

$$\begin{pmatrix} x_i \\ y_i \end{pmatrix} = \begin{pmatrix} \xi_i + \delta_i \\ \beta_1 \xi_i + \varepsilon_i \end{pmatrix}.$$

The results are given in Table 4.12 for $\beta_1 = 1$ and in Table 4.13 for $\beta_1 = 2$.

Table 4.12 Sample mean absolute errors for the $\hat{\beta}$s, comparing three estimation methods on symmetric data: 500 replications, $\beta_0 = 0,\ \beta_1 = 1,\ n = 100$

Method	Mean $\hat{\beta}_0$	MAE $\hat{\beta}_0$	Mean $\hat{\beta}_1$	MAE $\hat{\beta}_1$
Two groups	−0.010	0.114	0.919	0.084
Method of moments	−0.440	0.747	−2.796	4.860
Geary, (4.48a)	−0.010	0.117	0.997	0.055
Geary, (4.48b)	−0.010	0.118	1.001	0.056

Table 4.13 Sample mean absolute errors for the $\hat{\beta}$s, comparing three estimation methods on symmetric data: 500 replications, $\beta_0 = 0,\ \beta_1 = 2,\ n = 100$

Method	Mean $\hat{\beta}_0$	MAE $\hat{\beta}_0$	Mean $\hat{\beta}_1$	MAE $\hat{\beta}_1$
Two groups	-0.002	0.165	1.845	0.157
Method of moments	-0.211	0.541	2.053	2.200
Geary, (4.48a)	-0.002	0.172	2.002	0.082
Geary, (4.48b)	-0.002	0.173	2.003	0.090

The method of moments using third-order moments does not work well, as expected, because the data are symmetrically distributed. Geary's method using fourth-order cumulants does slightly better that the grouping method with two groups based on x.

The method of higher-order moments or the product cumulant method, while free from additional assumptions, is vulnerable in a rather unexpected way. It always estimates β_1 by a ratio of cumulants, and if the denominator cumulant is near zero, one would expect large fluctuations in the estimator. This is not a phenomenon that disappears as sample size increases – indeed it might get worse. The key assumption in these methods is that ξ is nonnormal while the major known merit of these methods is the consistency of the resulting estimators. In practice, one should be aware that the nonnormality needs to be ensured. Such estimators are not robust in the sense that the closer to normal the data are, the more unstable the estimates become.

Some authors, for example Pal(1980), commented that the method of moment (or cumulant) estimators can be regarded as a special kind of instrumental variable estimator with instruments being taken from x and/or y itself. For example, the estimator $\hat{\beta}_{11} = c_{03}/c_{12}$ is with 'instrumental variable' $z_{1i} = (y_i - \bar{y})^2$, the estimator $\hat{\beta}_{12} = c_{12}/c_{21}$ is with 'instrumental variable' $z_{2i} = (x_i - \bar{x})(y_i - \bar{y})$ and $\hat{\beta}_{13} = c_{21}/c_{30}$ is with 'instrumental variable' $z_{3i} = (x_i - \bar{x})^2$. This is not quite true; these variables are not really instrumental because they are clearly correlated with the errors δ or ε. This violates the first assumption of the instrumental variables; though it satisfies the second assumption that the 'instrumental variables' are correlated with the true explanatory variable, ξ.

4.6 Bibliographic Notes and Discussion

We have discussed some alternative approaches to the estimation of the parameters of ME models in this chapter. We now review and comment on these alternatives.

The term 'instrumental variable' was coined by Reiersol (1941); however, instrumental variable estimation methods were used as early as the 1920s; see Goldberger (1972) for a historical review. As was pointed out earlier, instrumental variable estimation is associated with ME models; however, it suits all kinds of regression with random regressors for which the explanatory variables are correlated with the errors. Therefore, instrumental variables are used in the estimation of a variety of models, and the ME model is a special case. For example, in econometrics, an application of instrumental variables is the two-stage least-squares estimator for a single equation in a simultaneous system of equations. Schneeweiss (1985) examined the relationship between the two approaches, using the side condition of knowing the error variance and using an instrumental variable. Instrumental variables can be used to obtain estimates of the error variances, which, in turn can be used to construct estimates of the βs. For more details of general treatment of instrumental variables, see White (1984) and Bowden and Turkington (1984).

Perhaps the greatest drawback to the use of instrumental variables in ME models is the practical difficulty in finding such variables. As was seen in the discussion of the maximum likelihood approach in Section 4.2, instrumental variables can also present contradictory information about the sign of the slope. In our opinion, instrumental variables provide just another form of additional information that can be used to obtain consistent estimates. They are not automatically available in applications.

Wald's grouping method and the instrumental variable method are popular in econometrics. They are considered to be the way out of the underidentification problem in ME models. Since the first appearance of the grouping method in Wald's (1940) paper, there have been many discussions on improving the efficiency of the grouping method, such as groups of unequal size and more than two groups. See, for example, Nair and Banerjee (1942), Bartlett (1949), Dorff and Gurland (1961), and Ware (1972). In particular, Richardson and Wu (1970) compare the grouping method generally with the least-squares estimator. Their overall recommendation is that it is safest to use at least 15 groups to avoid the possibility of a large increase in mean squared error. Grouping variables were introduced partly because they were supposedly easier to obtain in practice. Fifteen groups seems too many for a grouping method to have this advantage. As Pakes (1982) points out, the real problem with the grouping method is consistency. The two conditions (see Section 4.3) required for consistency of grouping estimators are rarely satisfied in practice.

The relation between the grouping method and the instrumental method is also clarified in Section 4.3. Unless the allocation of groups does not depend on the error δ, the grouping method is not a special case of an instrumental variable method.

Nevertheless, the grouping method is still important in practice, and there are times when it is appropriate. For example, Anderson (1951) and Cox (1976) studied the replicated model with the observations $\{(x_{ij}, y_{ij});\ i = 1, \ldots, k,\ j = 1, \ldots n_i\}$, where

$$x_{ij} = \xi_{ij} + \delta_{ij}, \quad y_{ij} = \beta_0 + \beta_1 \xi_{ij} + \varepsilon_{ij}.$$

The true independent variables ξ_{ij} are assumed to have independent normal distribution with means μ_i ($\mu_i \neq \mu_j,\ i \neq j$) and a common variance σ^2, whereas the errors δ_{ij} have independent normal distribution with mean zero and variance σ_δ^2, and the errors ε_{ij} have independent normal distribution with mean zero and variance σ_ε^2. Moreover, the errors δ_{ij} and ε_{ij} are assumed to be independent of each other and of the ξ_{ij}. When $k = 2$, the ML estimate of the slope β_1 (under some restrictions) is just the grouping estimate, namely, the slope of the line connecting the two group means, where group 1 is the first set of replications and group 2 is the second set. This is not completely surprising because, when $k = 2$, the data given have already been allocated into two distinct groups. This satisfies the normally difficult problem of finding a proper grouping, namely, a group allocation that is not dependent on the errors. Secondly, the distinction of the group means μ_1 and μ_2 fulfils the condition (4.16). Interestingly, this model is related to change-point (in the mean of ξ) structural models. When $k = 2$, this model is a structural model with exactly one change point, and the grouping estimator is also the ML estimator under the normality assumption. Some related results can be found in Chang and Huang (1997).

On the other hand, Geary's cumulant estimators (and hence higher-order moments estimators) have received little attention until lately. This is partly because the early experiments indicated that the estimators based on these ideas have an unusually large variance (Madansky, 1959). Recently, Pal (1980), Pakes (1982), and van Montfort *et al.* (1987) advocated higher-order moment estimators. Van Montfort *et al.* (1987) obtained the optimal estimator among the third-order moment estimators (see Section 4.5) and their technique seems to be generalizable to fourth- or higher-order moments. In practice, Pakes (1982) pointed out that the data bases used in econometrics have become significantly larger in recent years. It is probably worthwhile to re-examine the precision of cumulant estimators. One thing that it is necessary to have in practice is the assurance that ξ is nonnormal and whether or not it is symmetrically distributed about 0. The former is crucial because one depends on nonnormality to make the model identifiable. The latter is to help decide whether to use the third- or fourth-order moment estimators. If the true ξ is symmetric, the third-order moment estimators are not consistent, and the fourth-order moment estimators should be used. On the other hand, if ξ is asymmetric, then using a fourth-order moment estimator is inferior to using a third-order moment estimator because of its higher variance.

In the 1950s, shortly after the basic identifiability results were proven by Reiersol (1950), there were a number of attempts to find consistent estimators of the slope under nonnormality. The most obvious approach is to use ML estimation. Unfortunately, this requires strong assumptions on the distribution, which one is generally unwilling to make once one leaves normality. Furthermore, maximum likelihood might well be mathematically intractable in the nonnormal case. Even further, in the functional model, the ML method suffers from the Neyman–Scott effect, as was seen in the normal case. It is also not clear how the presence of incidental parameters will affect ML estimation in the nonnormal functional model. Little has been done in ML estimation in either structural or functional models under nonnormality. Wolfowitz (1952; 1953; 1954; 1957) suggested the minimum distance method to estimate the intercept and slope. However, it is not clear how these estimators are to be computed. On the other hand, Neyman (1951) provided a consistent estimator for the nonnormal structural model for which explicit formulae are given, but they are complicated and lack obvious interpretation. Reiersol (1950) clearly characterizes the identification problem in the structural model, and this, together with the result of Wald (1949), solves the consistent estimation problem in the functional model as well (see Appendix A). These are nice results, but they do not tell us how to choose reasonable and practical estimates when the model is identifiable. It appears, judging from the existing results, that without additional knowledge, the estimation of β_1 in the ME model is not straightforward in most practical situations.

4.7 Exercises

4.1 *Show that the least-squares estimator is biased and inconsistent in model (4.1) when the independent variable, x, and the error, e, are correlated.*

4.2 *Show that the instrumental variable estimator (4.2) is consistent with asymptotic variance (4.3).*

4.3 *Suppose that in the structural ME model you have two instrumental variables, w and z, available with $\sigma_{xw} \geq \sigma_{xz}$. If σ_{wz} is known, is it possible to combine w and z into a new variable, v, which provides a better estimate of β_1 than w alone can provide? If so, how would you do so; and if not, why not?*

4.4 (*Wald, 1940*) *Show that the grouping estimator (4.15) is consistent if the grouping is independent of the error and (4.16) holds true.*

4.5 (*Theil, 1950*) *Let $\xi_1, \ldots, \xi_k$ be k random variables and $\eta = \beta_0 + \sum_{j=1}^{k} \beta_j \xi_j$. Observe $x_j = \xi_j + \delta_j$ and $y = \eta + \varepsilon$. It is assumed that the errors, δ_j, are independent and such that the order of the x_j is the same as that of the ξ_j ($x_{ji} < x_{li} \Leftrightarrow \xi_{ji} < \xi_{li}$, $i = 1, \ldots, n$). If the δ are such that for all $j \neq l$, $P\left(c_j < c_l\right) = 1/2$, where $c_i \equiv \varepsilon_i - \sum_{j=1}^{k} \beta_j \delta_{ji}$, then a confidence interval can be obtained in the manner of*

(4.32) for any of the β_j given the remainder of the β_j. Show also that a conservative confidence interval can be obtained for all the β_j by taking the union of individual intervals.

4.6 (Bartlett 1949) *You are given $n = 2l + 1$ observations $\binom{x_1}{y_1}, \ldots, \binom{x_n}{y_n}$ from the model*

$$x_i = i, \qquad y_i = \beta_0 + \beta_1 x_i + \varepsilon_i,$$
$$E\varepsilon_i = 0, \qquad \operatorname{var}(\varepsilon_i) = \sigma^2, \qquad 1 = 1, \ldots, n.$$

Thus, there is no error in x and one has a standard regression model. If β_1 is estimated by least squares, giving the minimum variance unbiased estimate, show that the variance of the estimate is $3\sigma^2/\{l(l+1)(2l+1)\}$. Show further that if the three-group estimate (4.25) is used then maximum efficiency is attained when $p_1 = p_3 = 1/3$ and this efficiency is at least $8/9$. Show that the efficiency for $p_1 = p_3 = 1/2$ is only $27/32$ of the maximum.

4.7 *Let*

$$\begin{pmatrix} x \\ y \end{pmatrix} \sim N\left(\begin{pmatrix} \mu_x \\ \mu_y \end{pmatrix}, \begin{bmatrix} \sigma_x^2 & \rho\sigma_x\sigma_y \\ \rho\sigma_x\sigma_y & \sigma_y^2 \end{bmatrix}\right)$$

and suppose that μ_x and μ_y are known. If

$$r^2 = \frac{\left(\sum(x_i - \mu_x)(y_i - \mu_y)\right)^2}{\sum(x_i - \mu_x)^2 \sum(y_i - \mu_y)^2},$$

what is the appropriate t statistic corresponding to (4.12) and how many degrees of freedom does it have?

4.8 *If the model has no intercept, that is, $\beta_0 = 0$, then show that $y - \beta_1 x$ is uncorrelated with the instrumental variable z and $y - \beta_1 x$ has mean zero. Give similar arguments to those used to obtain (4.14) to provide the following confidence interval for β_1:*

$$P\left(\frac{[\sum(z_i - \bar{z})y_i]^2 - 2\beta_1 \sum(z_i - \bar{z})y_i \sum(z_i - \bar{z})x_i + \beta_1^2[\sum(z_i - \bar{z})x_i]^2}{\sum(z_i - \bar{z})^2[\sum y_i^2 - 2\beta_1 \sum x_i y_i + \beta_1^2 \sum x_i^2]} \leq r_{1-\gamma}^2\right)$$
$$= 1 - \gamma,$$

where $r_{1-\gamma}^2$ is defined in (4.13). How many degrees of freedom does the t statistic have?

4.9 *Consider the estimator (4.15) on the data described in Example 4.4. Show that $E\hat{\beta}_{1G}$ does not exist. (Hint: use conditional arguments as in Exercise 2.3.)*

5

Linear Measurement Error Model with Vector Explanatory Variables

5.1 Introduction

This chapter discusses linear measurement error models with vector-valued explanatory variables, that is, with more than one x variable. This is an extension of the models discussed in the previous chapters. We focus on the point estimation of the parameters.

The models are again divided into two classes, namely, models with and without equation error. The no-equation-error model is

$$\eta_i = \beta_0 + \boldsymbol{\beta}'\boldsymbol{\xi}_i, \tag{5.1}$$

$$\mathbf{x}_i = \boldsymbol{\xi}_i + \boldsymbol{\delta}_i, \qquad y_i = \eta_i + \varepsilon_i, \qquad i = 1, \ldots, n, \tag{5.2}$$

where $\boldsymbol{\beta} = (\beta_1, \beta_2, \ldots, \beta_p)'$, $\boldsymbol{\xi}_i$, $\mathbf{x}_i$ and $\boldsymbol{\delta}_i$ are p-dimensional vectors, while β_0, η_i, y_i, and ε_i are scalars. The observed variables are $\mathbf{x}$ and y, the unobservable true variables are $\boldsymbol{\xi}$ and η, and the measurement errors are $\boldsymbol{\delta}$ and ε. The measurement errors $(\boldsymbol{\delta}_i', \varepsilon_i)$ are independent and identically distributed random vectors, which are independent of the true values $\boldsymbol{\xi}_i$. The equation error model is (5.2), together with

$$\eta_i = \beta_0 + \boldsymbol{\beta}'\boldsymbol{\xi}_i + q_i, \tag{5.3}$$

where q_i is the scalar-valued equation error, which is independent of both the measurement error and the true values, $\boldsymbol{\xi}_i$. This is frequently written as

$$\mathbf{x}_i = \boldsymbol{\xi}_i + \boldsymbol{\delta}_i, \qquad y_i = \beta_0 + \boldsymbol{\beta}'\boldsymbol{\xi}_i + e_i, \tag{5.4}$$

where $e_i = \varepsilon_i + q_i$.

As before, if the true value $\boldsymbol{\xi}_i$ are independent and identically distributed random vectors with mean $\boldsymbol{\mu}$ and covariance matrix $\boldsymbol{\Sigma}_{\xi\xi}$, the model is a structural model; while if the true values $\boldsymbol{\xi}_i$ are known constant vectors, the model is a functional model. The ultrastructural model assumes that the $\boldsymbol{\xi}_i$ are independent, each $\boldsymbol{\xi}_i$ with mean $\boldsymbol{\mu}_i$ and common covariance matrix $\boldsymbol{\Sigma}_{\xi\xi}$.

Some of the independent independent variables can be measured *without* error. In other words, some of the elements of the p-vector $\boldsymbol{\delta}_i$ can be identically zero, so that the corresponding explanatory variables are error-free. In this case, the error covariance matrix $\boldsymbol{\Sigma}_{\delta\delta}$ is singular. The row and column of $\boldsymbol{\Sigma}_{\delta\delta}$ corresponding to an error-free independent variable are zero vectors. Some authors include the intercept β_0 in $\boldsymbol{\beta}$, and set $x_{i0} = \xi_{i0} \equiv 1$ for all i, to account for the intercept, as is common practice in ordinary regression. In this formulation, x_0 is an error-free 'variable'.

5.2 Identifiability

The classical identifiability characterization of the simple structural model was extended to the multivariate structural model by Bekker (1986) as follows. Consider the model (5.4), and allow e_i to be either with or without equation error. Suppose that $(\boldsymbol{\delta}, e)$ is normally distributed with covariance $\boldsymbol{\Sigma}_{\delta e} = \mathbf{0}$; then the parameter vector, $\boldsymbol{\beta}$, is identified if and only if there does not exist a nonsingular matrix $\mathbf{A}^{(p\times p)} = (\mathbf{a}_1^{(p\times 1)}, \mathbf{A}_2^{(p\times(p-1))})$ such that $\boldsymbol{\xi}'\mathbf{a}_1$ is distributed normally and independently of $\boldsymbol{\xi}'\mathbf{A}_2$. The nonnormality of $\boldsymbol{\xi}$ ensures identification, and again, as in the simple ME model, consistent estimation of $\boldsymbol{\beta}$ is possible via higher-order moments; see, for example, Pal (1980). We will not discuss the nonnormal case here, but only point out that the identification characterization does serve as a warning that unless the 'distance to normality' of the distribution of $\boldsymbol{\xi}$ is large, any estimator will be unreliable unless some side condition, instrumental variable, or other additional information is available to support consistent estimation. If $\boldsymbol{\xi}$ is normal or near normal, then identifiability assumptions similar to those in the simple ME models can be used to ensure consistent estimates. Traditionally, for the equation error model, the typical identifiability assumption is that $\boldsymbol{\Sigma}_{\delta e}^{(p\times 1)} = \boldsymbol{\Sigma}_{\delta\varepsilon}^{(p\times 1)}$ $(= \text{cov}\,(\boldsymbol{\delta},\varepsilon))$ and $\boldsymbol{\Sigma}_{\delta\delta}$ are known. On the other hand, for the no-equation-error model, the typical identifiability assumption is that $\text{var}\left((\boldsymbol{\delta}', \varepsilon)\right) = \sigma^2\boldsymbol{\Omega}_0^{((p+1)\times(p+1))}$ with $\boldsymbol{\Omega}_0$ known. Gleser (1992) investigated the **reliability matrix** known case for the equation error model. This approach will be discussed in Section 5.3.2.

For the functional model, identifiability is again not the problem but the consistent estimation of $\boldsymbol{\beta}$ is; see Appendix A. Thus, the normal functional model still requires some form of additional knowledge in order to find consistent estimators of $\boldsymbol{\beta}$.

5.3 The Equation Error Model

In this section, we investigate the vector explanatory variable measurement error model with equation error, (5.4). We adopt the usual identifiability assumption, that is, that $\boldsymbol{\Sigma}_{\delta\delta}^{(p\times p)}$ and $\boldsymbol{\Sigma}_{\delta\varepsilon}^{(p\times 1)}$ are known but σ_e^2 is unknown. We also discuss the reliability matrix known case.

5.3.1 Maximum likelihood estimation

For the functional model, when $(\boldsymbol{\delta}_i', e_i)$ is normally distributed, ML estimation breaks down for the same reason as in the univariate explanatory variable model (see Section 1.4).

For the structural model, when $(\boldsymbol{\xi}_i', \boldsymbol{\delta}_i', e_i)$ is normally distributed, $(\mathbf{x}_i', y_i)$ is normal with mean $(\boldsymbol{\mu}_x', \mu_y)' = (\boldsymbol{\mu}', \beta_0 + \boldsymbol{\beta}'\boldsymbol{\mu})'$ and covariance matrix

$$\begin{bmatrix} \boldsymbol{\Sigma}_{\mathbf{xx}} & \boldsymbol{\Sigma}_{\mathbf{x}y} \\ \boldsymbol{\Sigma}_{y\mathbf{x}} & \sigma_y^2 \end{bmatrix} = \begin{bmatrix} \boldsymbol{\Sigma}_{\xi\xi} + \boldsymbol{\Sigma}_{\delta\delta} & \boldsymbol{\Sigma}_{\xi\xi}\boldsymbol{\beta} + \boldsymbol{\Sigma}_{\delta e} \\ \boldsymbol{\beta}'\boldsymbol{\Sigma}_{\xi\xi} + \boldsymbol{\Sigma}_{e\delta} & \boldsymbol{\beta}'\boldsymbol{\Sigma}_{\xi\xi}\boldsymbol{\beta} + \sigma_e^2 \end{bmatrix}.$$

It is well known that the ML estimators of $\boldsymbol{\mu}_{\mathbf{x}}$, μ_y, $\boldsymbol{\Sigma}_{\mathbf{xx}}$, $\boldsymbol{\Sigma}_{\mathbf{xy}}$, and σ_y^2 are the corresponding sample mean and elements of the sample covariance matrix. The functional invariance property of the ML estimation method gives

$$\hat{\boldsymbol{\mu}}_x = \bar{\mathbf{x}}, \qquad \hat{\mu}_y = \bar{y} = \hat{\beta}_0 + \hat{\boldsymbol{\beta}}'\bar{\mathbf{x}}, \qquad \hat{\boldsymbol{\Sigma}}_{\xi\xi} + \boldsymbol{\Sigma}_{\delta\delta} = \mathbf{S}_{\mathbf{xx}},$$
$$\hat{\boldsymbol{\Sigma}}_{\xi\xi}\hat{\boldsymbol{\beta}} + \boldsymbol{\Sigma}_{\delta e} = \mathbf{S}_{\mathbf{x}y}, \qquad \hat{\boldsymbol{\beta}}'\hat{\boldsymbol{\Sigma}}_{\xi\xi}\hat{\boldsymbol{\beta}} + \hat{\sigma}_e^2 = s_{yy};$$

and, therefore,

$$\begin{aligned} \hat{\boldsymbol{\mu}}_x &= \bar{\mathbf{x}}, \qquad \hat{\beta}_0 = \bar{y} - \hat{\boldsymbol{\beta}}'\bar{\mathbf{x}}, \\ \hat{\boldsymbol{\beta}} &= (\mathbf{S}_{\mathbf{xx}} - \boldsymbol{\Sigma}_{\delta\delta})^{-1}(\mathbf{S}_{\mathbf{x}y} - \boldsymbol{\Sigma}_{\delta e}), \qquad \hat{\boldsymbol{\Sigma}}_{\xi\xi} = \mathbf{S}_{\mathbf{xx}} - \boldsymbol{\Sigma}_{\delta\delta}, \qquad (5.5) \\ \hat{\sigma}_e^2 &= s_{yy} - \hat{\boldsymbol{\beta}}'\mathbf{S}_{\mathbf{xx}}\hat{\boldsymbol{\beta}} + \hat{\boldsymbol{\beta}}'\boldsymbol{\Sigma}_{\delta\delta}\hat{\boldsymbol{\beta}} = s_{yy} - 2\mathbf{S}_{y\mathbf{x}}\hat{\boldsymbol{\beta}} + \hat{\boldsymbol{\beta}}'\mathbf{S}_{\mathbf{xx}}\hat{\boldsymbol{\beta}} + 2\boldsymbol{\Sigma}_{e\delta}\hat{\boldsymbol{\beta}} - \hat{\boldsymbol{\beta}}'\boldsymbol{\Sigma}_{\delta\delta}\hat{\boldsymbol{\beta}}, \end{aligned}$$

provided $\hat{\sigma}_e^2 \geq \boldsymbol{\Sigma}_{e\delta}\boldsymbol{\Sigma}_{\delta\delta}^{+}\boldsymbol{\Sigma}_{\delta e}$, where $\boldsymbol{\Sigma}_{\delta\delta}^{+}$ denotes the Moore–Penrose generalized inverse of $\boldsymbol{\Sigma}_{\delta\delta}$ and $\hat{\boldsymbol{\Sigma}}_{\xi\xi} = \mathbf{S}_{\mathbf{xx}} - \boldsymbol{\Sigma}_{\delta\delta}$ is positive definite. If either of these provisos is violated the maximum occurs on the boundary of the parameter space (see Fuller, 1987, p. 105).

Because ε and q are independent in model (5.4), $\boldsymbol{\Sigma}_{\delta e} = \boldsymbol{\Sigma}_{\delta\varepsilon}$ and $\sigma_e^2 = \sigma_\varepsilon^2 + \sigma_q^2$. Knowledge of both $\boldsymbol{\Sigma}_{\delta\delta}$ and $\boldsymbol{\Sigma}_{\delta\varepsilon}$ is quite often obtained from replications. From these same replications one should also be able to obtain information about σ_ε^2. With the additional knowledge of σ_ε^2, one can also estimate the equation error variance σ_q^2 by $\hat{\sigma}_q = \hat{\sigma}_e^2 - \sigma_\varepsilon^2$.

If $\boldsymbol{\delta}$ and ε are independent and $\boldsymbol{\Sigma}_{\delta\delta}$ is known, the parameter estimates are, of course, (5.5) with $\boldsymbol{\Sigma}_{\delta e} = 0$. This situation is very common in practice, especially in the nonlinear multivariate model (Carroll *et al.*, 1995). Neither σ_ε^2 nor σ_q^2 is now identifiable.

The ML estimators of the parameters for the structural model above can be viewed as method of moments estimators in the corresponding functional and ultrastructural models. These estimates will be consistent under conditions corresponding to those for the structural model.

Example 5.1

We now generate an artificial set of data to demonstrate the use of the estimates (5.5) on a normal model with two explanatory variables and equation error. We performed the following experiment:

1. Generate 36 independent values, $\begin{pmatrix} \xi_{1,1} \\ \xi_{2,1} \end{pmatrix}, \ldots, \begin{pmatrix} \xi_{1,36} \\ \xi_{2,36} \end{pmatrix}$, of

$$\boldsymbol{\xi} \sim N\left(\mathbf{0}, \begin{bmatrix} 4 & 2 \\ 2 & 4 \end{bmatrix}\right).$$

2. Generate 36 independent values, $\begin{pmatrix} \delta_{1,1} \\ \delta_{2,1} \\ \varepsilon_1 \end{pmatrix}, \ldots, \begin{pmatrix} \delta_{1,36} \\ \delta_{2,36} \\ \varepsilon_{36} \end{pmatrix}$, of

$$\begin{pmatrix} \delta_1 \\ \delta_2 \\ \varepsilon \end{pmatrix} \sim N\left(\mathbf{0}, \begin{bmatrix} 1 & 0.5 & 0.5 \\ 0.5 & 1 & 0.5 \\ .5 & 0.5 & 1 \end{bmatrix}\right).$$

3. Generate 36 independent values, $q_1, \ldots, q_{36}$, of $q \sim N(0, 1)$.
4. Form the data

$$\begin{pmatrix} x_{1,i} \\ x_{2,i} \\ y_i \end{pmatrix} = \begin{pmatrix} \xi_{1,i} + \delta_{1,i} \\ \xi_{2,i} + \delta_{2,i} \\ 1 + 2\xi_{1,i} + 3\xi_{2,1} + q_i + \varepsilon_i \end{pmatrix}, \qquad i = 1, \ldots, 36.$$

Thus, $\beta_0 = 1$, $\beta_1 = 2$, $\beta_2 = 3$, ξ_1 and ξ_2 are correlated, and δ_1, δ_2, and ε are correlated. The data are given in Table 5.1.

Table 5.1 Thirty-six equation error model data with two explanatory variables: $\beta_0 = 1,\ \beta_1 = 2,\ \beta_2 = 3$

	x_1	x_2	y		x_1	x_2	y
1	0.21	−2.30	−3.51	19	2.27	0.60	7.15
2	−1.09	−1.89	−4.70	20	−6.53	−2.51	−16.83
3	−0.25	1.20	4.78	21	−0.21	1.98	2.26
4	−1.06	−2.46	−7.36	22	−2.21	−0.91	−2.56
5	1.24	−2.94	0.64	23	−2.15	−0.44	−3.45
6	−1.00	−2.76	−8.13	24	3.74	2.41	17.96
7	−4.37	−0.28	−9.34	25	−1.78	−0.54	0.27
8	−2.47	−2.98	−5.56	26	2.78	1.10	2.42
9	−0.51	0.50	7.95	27	−0.10	1.84	6.62
10	−0.64	0.77	−3.55	28	−1.74	−2.35	−11.76
11	1.55	−0.71	0.72	29	−1.34	−0.01	2.14
12	−0.57	1.38	3.24	30	−0.81	−1.86	−9.67
13	−1.19	−1.08	−2.81	31	1.21	1.58	7.39
14	4.13	0.05	12.79	32	2.59	3.68	15.88
15	2.46	4.00	17.67	33	0.18	0.51	2.97
16	2.87	1.19	5.78	34	1.08	3.64	12.32
17	2.04	0.85	0.63	35	−3.15	−2.94	−10.63
18	1.44	0.72	5.21	36	2.26	4.52	14.36

It is assumed that $\mathbf{\Sigma}_{\delta\delta} = \begin{bmatrix} 1 & 0.5 \\ 0.5 & 1 \end{bmatrix}$ and $\mathbf{\Sigma}_{\delta\varepsilon} = \begin{bmatrix} 0.5 \\ 0.5 \end{bmatrix}$ are known. From the data we obtain

$$\boldsymbol{\mu}_x = \begin{pmatrix} -0.03 \\ 0.10 \end{pmatrix}, \qquad \mu_y = 1.42,$$

$$\mathbf{S}_{xx} = \begin{bmatrix} 5.27 & 2.86 \\ 2.86 & 4.41 \end{bmatrix}, \qquad \mathbf{S}_{xy} = \begin{pmatrix} 16.15 \\ 15.35 \end{pmatrix},$$

and from (5.5)

$$\begin{aligned} \hat{\boldsymbol{\beta}} &= \left(\begin{bmatrix} 5.27 & 2.86 \\ 2.86 & 4.41 \end{bmatrix} - \begin{bmatrix} 1 & 0.5 \\ 0.5 & 1 \end{bmatrix} \right)^{-1} \left(\begin{pmatrix} 16.15 \\ 15.35 \end{pmatrix} - \begin{pmatrix} 0.5 \\ 0.5 \end{pmatrix} \right) \\ &= \begin{pmatrix} 2.04 \\ 2.94 \end{pmatrix} \end{aligned}$$

and $\hat{\beta}_0 = 1.20$, which are quite good. However, $s_{yy} = 74.50$ and

$$\begin{aligned} \hat{\sigma}_e^2 &= 74.50 - \begin{pmatrix} 2.04 \\ 2.94 \end{pmatrix}' \begin{pmatrix} 5.27 & 2.86 \\ 2.86 & 4.41 \end{pmatrix} \begin{pmatrix} 2.04 \\ 2.94 \end{pmatrix} \\ &\quad + \begin{pmatrix} 2.04 \\ 2.94 \end{pmatrix}' \begin{pmatrix} 1 & 0.5 \\ 0.5 & 1 \end{pmatrix} \begin{pmatrix} 2.04 \\ 2.94 \end{pmatrix} = -1.05, \end{aligned}$$

which is clearly a poor estimate of a variance. Because $\hat{\sigma}_e^2$ is negative, $\hat{\boldsymbol{\beta}}$ is not the ML estimate but only a method of moments estimate. With a sample size this small, the authors have found negative estimates for σ_ε^2 not uncommon. If the sample size is increased somewhat this problem tends to disappear.

Ordinary regression gives

$$\hat{\boldsymbol{\beta}}_R = \begin{pmatrix} 1.81 \\ 2.31 \end{pmatrix} \quad \text{and} \quad \hat{\beta}_{0R} = 1.25,$$

which are much less accurate than the ME model estimates.

5.3.2 The reliability matrix known

The extension of the reliability ratio known case to vector explanatory variable models was done by Gleser (1992). This extension is, in fact, for models with both **x** and **y** vector-valued. These results have the further advantage of establishing an additional useful connection between ordinary multivariate regression models and multivariate ME models. These connections are exploited, for example, in Cheng and Tsai (1996), where a score test for heteroscedasticity, a score test for autocorrelation in an autoregressive model of order 1, and the ML estimates and likelihood ratio tests for the Box–Cox transformation parameters are given for the structural model.

Assume that $(\boldsymbol{\xi}_i', \boldsymbol{\delta}_i', e_i)$ is jointly normally distributed and that $\boldsymbol{\Sigma}_{\delta e} = \mathbf{0}$. Let $\boldsymbol{\mu} = E\boldsymbol{\xi}_i = E\mathbf{x}_i$, $\mathbf{I}_p$ be the $p \times p$ identity matrix, and the reliability matrix $\mathbf{K}$ be defined as

$$\mathbf{K}^{(p\times p)} = \boldsymbol{\Sigma}_{\mathbf{xx}}^{-1}\boldsymbol{\Sigma}_{\xi\xi} = (\boldsymbol{\Sigma}_{\xi\xi} + \boldsymbol{\Sigma}_{\delta\delta})^{-1}\boldsymbol{\Sigma}_{\xi\xi}.$$

Assuming that $\mathbf{K}$ is known, then using the fact that

$$E(\boldsymbol{\xi}_i \mid \mathbf{x}_i) = (\mathbf{I}_p - \mathbf{K}')\boldsymbol{\mu} + \mathbf{K}'\mathbf{x}_i,$$

one can rewrite (5.4) as

$$\begin{aligned} y_i &= \beta_0 + \boldsymbol{\beta}'\boldsymbol{\xi}_i + e_i = \beta_0 + \boldsymbol{\beta}'E(\boldsymbol{\xi}_i \mid \mathbf{x}_i) + \boldsymbol{\beta}'\boldsymbol{\xi}_i - \boldsymbol{\beta}'E(\boldsymbol{\xi}_i \mid \mathbf{x}_i) + e_i \\ &= \beta_0^* + \boldsymbol{\beta}^{*\prime}\mathbf{x}_i + \varepsilon_i^*, \end{aligned} \tag{5.6}$$

where

$$\beta_0^* = \beta_0 + \boldsymbol{\beta}'(\mathbf{I}_p - \mathbf{K}')\boldsymbol{\mu}, \qquad \boldsymbol{\beta}^* = \mathbf{K}\boldsymbol{\beta}, \qquad \varepsilon_i^* = \boldsymbol{\beta}'\{\boldsymbol{\xi}_i - (\mathbf{I}_p - \mathbf{K}')\boldsymbol{\mu} - \mathbf{K}'\mathbf{x}_i\} + e_i.$$

Here, ε_i^* has mean zero and is uncorrelated with (and hence independent of) $\mathbf{x}_i$. The elements of the covariance matrix $\boldsymbol{\Omega}^*$ (with elements ω_{ij}^*) of ε_i^* are

$$\sigma_{\varepsilon^*}^2 = \omega_{ii}^* = \boldsymbol{\beta}'\boldsymbol{\Sigma}_{\delta\delta}\mathbf{K}\boldsymbol{\beta} + \sigma_e^2, \quad \omega_{ij}^* = 0, \qquad \text{for all } i \neq j. \tag{5.7}$$

Now one can separate the density function of the nuisance parameters in equation (5.6) from the joint density of $(\mathbf{x}_i', y_i)$. Specifically, one has $f(\mathbf{x}_i, y_i) = g(\mathbf{x}_i)h(\varepsilon_i^*)$, where f, g, and h are density functions of $(\mathbf{x}_i, y_i)$, $\mathbf{x}_i$, and ε_i^*, respectively. Hence, the log-likelihood function for the structural model is

$$\begin{aligned} L(y, \mathbf{X}; \beta_0, \boldsymbol{\beta}, \sigma_e^2) &= \sum \log g(\mathbf{x}_i) - \frac{n}{2}\log(2\pi) - \frac{1}{2}\log\left|\boldsymbol{\Omega}^*\right| \\ &\quad - \frac{1}{2}(\mathbf{y} - \beta_0^*\mathbf{1} - \mathbf{X}\boldsymbol{\beta}^*)'(\boldsymbol{\Omega}^*)^{-1}(\mathbf{y} - \beta_0^*\mathbf{1} - \mathbf{X}\boldsymbol{\beta}^*), \end{aligned}$$

where $\boldsymbol{\Omega}^*$ is defined in (5.7) and $\mathbf{y} = (y_1, \ldots, y_n)'$, $\mathbf{1} = (1, \ldots, 1)'$, $\mathbf{X} = (\mathbf{x}_1, \ldots, \mathbf{x}_n)'$. This explicit form of the log-likelihood function is quite useful (see, for example, Cheng and Tsai, 1996).

Another way to look at the argument above is to write

$$f(\mathbf{x}_i, y_i) = f(y_i \mid \mathbf{x}_i; \beta_0^*, \boldsymbol{\beta}^*, \sigma_\varepsilon^{*2})g(\mathbf{x}_i; \boldsymbol{\mu}, \boldsymbol{\Sigma}_{\xi\xi}, \boldsymbol{\Sigma}_{\delta\delta})$$

and treat the model in the following form:

$$y_i = E(y_i \mid \mathbf{x}_i) + \varepsilon_i^*.$$

It is easily seen that $E(y_i \mid \mathbf{x}_i) = \beta_0^* + \boldsymbol{\beta}^{*\prime}\mathbf{x}_i$.

Gleser (1992) showed that the ML estimators of β_0^*, $\boldsymbol{\beta}^*$, and σ_ε^{*2} are just the ordinary least-squares estimators, that is,

$$\hat{\beta}_0^* = \bar{y} - \hat{\boldsymbol{\beta}}^{*\prime}\bar{\mathbf{x}}, \qquad \hat{\boldsymbol{\beta}}^* = \mathbf{S}_{\mathbf{xx}}^{-1}\mathbf{S}_{\mathbf{x}y}, \tag{5.8}$$

provided

$$\hat{\sigma}_e^2 = \hat{\sigma}_\varepsilon^{*2} - \hat{\boldsymbol{\beta}}^{*\prime}\mathbf{K}^{-1}\boldsymbol{\Sigma}_{\delta\delta}\hat{\boldsymbol{\beta}}^* \geq 0,$$

where

$$\hat{\sigma}_\varepsilon^{*2} = n^{-1}\sum \hat{\varepsilon}^{*2} = n^{-1}\sum (y_i - \hat{\beta}_0^* - \hat{\boldsymbol{\beta}}^{*\prime}\mathbf{x}_i)^2$$

is the estimate of the error variance (without any adjustment for the degrees of freedom) in the regression model (5.6). It follows that the ML estimators of β_0, $\boldsymbol{\beta}$, and σ_e^2 are

$$\hat{\beta}_0 = \hat{\beta}_0^* - \hat{\boldsymbol{\beta}}'(\mathbf{I}_p - \mathbf{K}')\hat{\boldsymbol{\mu}}, \qquad \hat{\boldsymbol{\beta}} = \mathbf{K}^{-1}\hat{\boldsymbol{\beta}}^*, \quad \text{and} \quad \hat{\sigma}_e^2 = n^{-1}\sum \hat{\varepsilon}_i^{*2} - \hat{\boldsymbol{\beta}}'\boldsymbol{\Sigma}_{\delta\delta}\mathbf{K}\hat{\boldsymbol{\beta}}, \tag{5.9}$$

where

$$\hat{\boldsymbol{\mu}} = \bar{\mathbf{x}} = n^{-1}\sum \mathbf{x}_i, \quad \hat{\varepsilon}_i^* = y_i - \hat{\beta}_0^* - \hat{\boldsymbol{\beta}}^{*\prime}\mathbf{x}_i. \tag{5.10}$$

Gleser's results are also applicable to the case when $\boldsymbol{\Sigma}_{\delta\delta}$ is known as in Section 5.3.1 (recall that $\boldsymbol{\Sigma}_{\delta e} = \mathbf{0}$ is assumed). First observe that the ML estimators are (5.8) if $\mathbf{K}$ is known; however, if $\mathbf{K}$ is not known, then one can estimate the reliability matrix by

$$\hat{\mathbf{K}} = \mathbf{S}_{\mathbf{xx}}^{-1}\hat{\boldsymbol{\Sigma}}_{\xi\xi} = \mathbf{S}_{\mathbf{xx}}^{-1}(\mathbf{S}_{\mathbf{xx}} - \boldsymbol{\Sigma}_{\delta\delta}). \tag{5.11}$$

Then the ML estimators of β_0, $\boldsymbol{\beta}$ and σ_e^2 are

$$\hat{\beta}_0 = \hat{\beta}_0^* - \hat{\boldsymbol{\beta}}'(\mathbf{I}_p - \hat{\mathbf{K}}')\hat{\boldsymbol{\mu}}, \qquad \hat{\boldsymbol{\beta}} = \hat{\mathbf{K}}^{-1}\hat{\boldsymbol{\beta}}^* \quad \text{and} \quad \hat{\sigma}_e^2 = n^{-1}\sum \hat{\varepsilon}_i^{*2} - \hat{\boldsymbol{\beta}}'\boldsymbol{\Sigma}_{\delta\delta}\hat{\mathbf{K}}\hat{\boldsymbol{\beta}},$$

where $\hat{\boldsymbol{\mu}}$ and $\hat{\varepsilon}_i^*$ are defined by (5.10) and $\hat{\mathbf{K}}$ is defined by (5.11). Furthermore, the explicit forms of $\hat{\beta}_0$ and $\hat{\boldsymbol{\beta}}$ are

$$\hat{\beta}_0 = \bar{y} - \hat{\boldsymbol{\beta}}'\bar{\mathbf{x}}, \qquad \hat{\boldsymbol{\beta}} = (\mathbf{S}_{\mathbf{xx}} - \boldsymbol{\Sigma}_{\delta\delta})^{-1}\mathbf{S}_{xy},$$

provided that

$$\hat{\sigma}_e^2 = s_{yy} - \mathbf{S}_{y\mathbf{x}}(\mathbf{S}_{\mathbf{xx}} - \boldsymbol{\Sigma}_{\delta\delta})^{-1}\mathbf{S}_{\mathbf{x}y} \geq 0,$$

which are the same estimates as in Section 5.3.1. Note that the above restriction is the same as requiring $\mathbf{S}_{\mathbf{xx}} - \boldsymbol{\Sigma}_{\delta\delta}$ to be positive definite and $\hat{\sigma}_e^2$ in (5.5) to be nonnegative.

5.4 Maximum Likelihood for the No-Equation-Error Model

In this section, we consider the maximum likelihood estimation of the vector explanatory variable ME model with no equation error. The typical identifiability assumption is that the error covariance matrix is known up to a proportionality factor. If the measurement errors, $\delta_1, \ldots, \delta_p$ and ε, are independent, this assumption is the extension to the multivariate model of the ratio $\lambda = \sigma_\varepsilon^2/\sigma_\delta^2$ known case of the univariate model.

5.4.1 Functional model with uncorrelated errors

It is convenient to define $\zeta_0 \equiv 1$ and write the no equation error model, (5.1), in the symmetric form:

$$\sum_{j=0}^{k} \alpha_j \zeta_j = 0, \tag{5.12}$$

where $\boldsymbol{\zeta} = (\zeta_0, \ldots, \zeta_k)' = (1, \xi_1, \ldots, \xi_p, \eta)'$ and $(\beta_0, \boldsymbol{\beta}') = -(\alpha_0/\alpha_k, \ldots, \alpha_p/\alpha_k)$ (recall (1.16)). This model is not uniquely defined because one can multiply (5.12) by any nonzero constant. To make it unique, the condition $\sum_{j=1}^{k} \alpha_j^2 = 1$

will prove convenient. Note that α_0 does not appear in this condition (see, for example, Zamar, 1989). Equation (5.12) defines a hyperplane in the k-dimensional space of the data,

$$\begin{pmatrix} x_{1i} \\ x_{2i} \\ \vdots \\ x_{pi} \\ x_{ki} \end{pmatrix} = \begin{pmatrix} \zeta_{1i} + \delta_{1i} \\ \zeta_{2i} + \delta_{2i} \\ \vdots \\ \zeta_{pi} + \delta_{pi} \\ \zeta_{ki} + \delta_{ki} \end{pmatrix} = \begin{pmatrix} \xi_{1i} + \delta_{1i} \\ \xi_{2i} + \delta_{2i} \\ \vdots \\ \xi_{pi} + \delta_{pi} \\ \eta_i + \varepsilon_i \end{pmatrix}, \qquad i = 1, \ldots, n.$$

Except for ζ_0, the true variables are observed with error as described in (5.2). For maximum likelihood, the errors, $\delta_1, \ldots, \delta_k$, are assumed independent of each other and normal with zero means. The identifiability assumption used here is that $\lambda_1, \ldots, \lambda_k$ are known and that

$$\frac{\sigma_{\delta_1}^2}{\lambda_1} = \frac{\sigma_{\delta_2}^2}{\lambda_2} = \ldots = \frac{\sigma_{\delta_k}^2}{\lambda_k}.$$

Again the observables can be standardized by considering $x_1/\sqrt{\lambda_1}, \ldots, x_k/\sqrt{\lambda_k}$, and it can be assumed that all the λs equal one and all the errors have the same variance, say σ_δ^2.

The log-likelihood function is

$$\log L \propto -nk \log \sigma_\delta - \frac{1}{2\sigma_\delta^2} \sum_{j=1}^{k} \sum_{i=1}^{n} \left(x_{ji} - \zeta_{ji}\right)^2. \qquad (5.13)$$

Maximizing the log-likelihood is equivalent to minimizing the double sum in (5.13) and, as before, this is equivalent to minimizing the sum of the orthogonal distances squared from the data points to the hyperplane (5.12) (see Sections 1.2.3 and 3.4). Furthermore, as before, the estimated hyperplane, which minimizes this distance, must pass through the sample mean, that is,

$$\sum_{j=0}^{k} \hat{\alpha}_j \bar{x}_j = 0, \qquad (5.14)$$

where $\bar{x}_j = \sum_{i=1}^{n} x_{ji}/n$. Finding the minimizing hyperplane is a familiar mathematical problem. Using (5.14), the squared distance from a point, $\mathbf{x}_i$, to the estimated hyperplane is

$$\frac{\left(\sum_{j=0}^{k} \hat{\alpha}_j x_{ji}\right)^2}{\sum_{j=1}^{k} \hat{\alpha}_j^2} = \left(\sum_{j=0}^{k} \hat{\alpha}_j x_{ji}\right)^2 = \left(\sum_{j=1}^{k} \hat{\alpha}_j \dot{x}_{ji}\right)^2,$$

where the $\dot{x}_{ji} = x_{ji} - \bar{x}_j$, $j = 1, \ldots, k$, are the variables measured from their sample means. So the quantity to be minimized is

$$W = \sum_{i=1}^{n} \left(\sum_{j=1}^{k} \alpha_j \dot{x}_{ji}\right)^2 \qquad (5.15)$$

subject to the constraint $\sum_{j=1}^{k} \alpha_j^2 = 1$. This can be solved using Lagrange multipliers. One must minimize

$$\sum_{i=1}^{n} \left(\sum_{j=1}^{k} \alpha_j \dot{x}_{ji} \right)^2 + \tau \left(1 - \sum_{j=1}^{k} \alpha_j^2 \right),$$

where τ is the Lagrange multiplier. Differentiating with respect to the α_j gives

$$\sum_{i=1}^{n} \dot{x}_{li} \left(\sum_{j=0}^{k} \hat{\alpha}_j \dot{x}_{ji} \right) = \tau \hat{\alpha}_l, \qquad l = 1, \ldots, k. \tag{5.16}$$

Let $c_{lj} = \frac{1}{n} \sum_{i=1}^{n} \dot{x}_{li} \dot{x}_{ji}$ be the sample covariance between the lth and jth variables; then the equations become

$$\sum_{j=1}^{k} \hat{\alpha}_j c_{lj} - \frac{\tau}{n} \hat{\alpha}_l = 0, \qquad l = 1, \ldots, k. \tag{5.17}$$

Eliminating the α_j in these equations gives the following determinant equation for τ:

$$\begin{vmatrix} c_{11} - \frac{\tau}{n} & c_{12} & c_{13} & \cdots & c_{1k} \\ c_{12} & c_{22} - \frac{\tau}{n} & c_{23} & \cdots & c_{2k} \\ c_{13} & c_{23} & c_{33} - \frac{\tau}{n} & \cdots & c_{3k} \\ \vdots & \vdots & \vdots & \ddots & \vdots \\ c_{1k} & c_{2k} & c_{3k} & \cdots & c_{kk} - \frac{\tau}{n} \end{vmatrix} = 0. \tag{5.18}$$

This is a polynomial of degree k in τ. The k roots of this equation are all real because the matrix $\{c_{ij}\}$ is nonnegative definite. The desired root is the smallest one. To see this, multiply the lth equation in (5.16) by $\hat{\alpha}_l$ and add the last k equations; then the left-hand side of this sums to (5.15). Thus $W = \tau \sum_{j=1}^{k} \hat{\alpha}_j^2$, which was to be minimized.

Having τ, one can solve the system of linear equations, (5.17) for the $\hat{\alpha}_j$, and $\hat{\alpha}_0$ in the original formulation is found from (5.14).

This method gives estimates which can be obtained as special cases of the more general estimates, (5.19), given in the next subsection.

Example 5.2

We now generate an artificial set of data to demonstrate the use of the methods of this subsection on a normal functional model without equation error. We performed the following experiment.

1. Generate 36 fixed values, $\binom{\xi_{1,1}}{\xi_{2,1}}, \ldots, \binom{\xi_{1,36}}{\xi_{2,36}}$, where

$$\begin{pmatrix} \xi_{1,j} \\ \xi_{2,j} \end{pmatrix} = \begin{pmatrix} -5 + \frac{5}{3} (\lceil j/6 \rceil) \\ -5 + \frac{5}{3} (1 + (j-1) \pmod{6}) \end{pmatrix},$$

$\lceil b \rceil$ is the next greatest integer in b, and $(j-1) \pmod 6$ is $j-1$ modulo 6. This is just a 6×6 square lattice in steps of $5/3$ with, for example,

$$\begin{pmatrix} \xi_{1,1} \\ \xi_{2,1} \end{pmatrix} = \begin{pmatrix} -10/3 \\ -10/3 \end{pmatrix}, \quad \begin{pmatrix} \xi_{1,18} \\ \xi_{2,18} \end{pmatrix} = \begin{pmatrix} 0 \\ 5 \end{pmatrix}, \quad \text{and} \quad \begin{pmatrix} \xi_{1,36} \\ \xi_{2,36} \end{pmatrix} = \begin{pmatrix} 5 \\ 5 \end{pmatrix}.$$

2. Generate 36 independent values,

$$\begin{pmatrix} \delta_{1,1} \\ \delta_{2,1} \\ \varepsilon_1 \end{pmatrix}, \ldots, \begin{pmatrix} \delta_{1,36} \\ \delta_{2,36} \\ \varepsilon_{36} \end{pmatrix}, \quad \text{of} \quad \begin{pmatrix} \delta_1 \\ \delta_2 \\ \varepsilon \end{pmatrix} \sim N(\mathbf{0}, \mathbf{I}).$$

3. Form the data

$$\begin{pmatrix} x_{1,i} \\ x_{2,i} \\ y_i \end{pmatrix} = \begin{pmatrix} \xi_{1,i} + \delta_{1,i} \\ \xi_{2,i} + \delta_{2,i} \\ 1 + 2\xi_{1,i} + 3\xi_{2,i} + \varepsilon_i \end{pmatrix}, \qquad i = 1, \ldots, 36.$$

Thus, $\beta_0 = 1$, $\beta_1 = 2$, $\beta_2 = 3$; and δ_1, δ_2, and ε are independent. The data are given in Table 5.2.

Table 5.2 Thirty-six functional model data with two explanatory variables: $\beta_0 = 1$, $\beta_1 = 2$, $\beta_2 = 3$

	x_1	x_2	y
1	−2.60	−4.90	−16.56
2	−3.45	−2.36	−11.18
3	−3.72	0.81	−5.53
4	−4.03	0.87	0.60
5	−1.72	2.29	3.19
6	−3.55	3.75	9.53
7	−3.75	−3.17	−12.64
8	−1.04	−3.01	−10.32
9	−1.53	0.70	−3.54
10	−2.90	1.94	4.24
11	−0.65	2.52	7.48
12	2.33	5.92	13.07
13	−0.26	−3.28	−9.44
14	2.51	−2.05	−4.42
15	1.13	1.35	0.02
16	1.06	1.59	6.85
17	0.70	2.45	11.42
18	0.60	5.34	16.11

	x_1	x_2	y
19	2.99	−3.39	−5.91
20	−1.23	−1.92	−1.35
21	1.35	0.13	4.14
22	0.39	2.34	9.61
23	0.33	3.89	15.17
24	3.59	5.84	18.61
25	2.29	−2.94	−2.26
26	4.10	2.01	4.20
27	3.01	0.94	7.62
28	3.10	0.23	−12.28
29	3.00	4.40	17.18
30	3.33	3.65	22.46
31	5.42	−2.65	1.21
32	6.23	−0.27	5.26
33	5.01	−0.08	10.89
34	5.57	3.41	15.27
35	4.10	2.55	20.30
36	5.51	6.83	26.55

It is assumed that $\lambda_1 = \lambda_2 = \lambda_3 = 1$ are known. From the data we obtain

$$\bar{x}_1 = 0.90, \qquad \bar{x}_2 = 0.88, \qquad \bar{y} = \bar{x}_3 = 5.00.$$

Subtracting these means from each element in the appropriate columns of the data matrix yields the data matrix for $\dot{x}_1$, $\dot{x}_2$, and $\dot{x}_3$. From these centred data we calculate the matrix

$$\{c_{lj}\} = \begin{bmatrix} 9.283 & 1.261 & 16.756 \\ 1.261 & 9.141 & 27.356 \\ 16.756 & 27.356 & 109.751 \end{bmatrix}.$$

This matrix has eigenvalues 0.802, 8.172, and 119.202. Thus $\tau/n = 36 \times 0.802 = 28.872$. The equations (5.17) become

$$\sum_{j=1}^{k} \hat{\alpha}_j c_{lj} - \tau \hat{\alpha}_l = 0, \qquad l = 1, \ldots, k,$$

that is,

$$\begin{aligned} 9.283\hat{\alpha}_1 + 1.261\hat{\alpha}_2 + 16.756\hat{\alpha}_3 - 0.802\hat{\alpha}_1 &= 0, \\ 1.261\hat{\alpha}_1 + 9.141\hat{\alpha}_2 + 27.356\hat{\alpha}_3 - 0.802\hat{\alpha}_2 &= 0, \\ 16.756\hat{\alpha}_1 + 27.356\hat{\alpha}_2 + 109.751\hat{\alpha}_3 - 0.802\hat{\alpha}_3 &= 0. \end{aligned}$$

Solving the first two of these equations yields

$$\hat{\beta}_1 = -\frac{\hat{\alpha}_1}{\hat{\alpha}_3} = 1.52 \quad \text{and} \quad \hat{\beta}_2 = -\frac{\hat{\alpha}_2}{\hat{\alpha}_3} = 3.05.$$

Finally, the third equation gives $\hat{\beta}_0 = 0.94$. These are quite good estimates.

Ordinary regression gives

$$\hat{\boldsymbol{\beta}}_R = \begin{pmatrix} 1.43 \\ 2.80 \end{pmatrix} \quad \text{and} \quad \hat{\beta}_{0R} = 1.25,$$

which are much less accurate than the ME model estimates.

5.4.2 Functional and structural models with correlated errors

The above results can be generalized to a general error covariance structure; see, for example, Fuller (1987, p. 124). For linear ME models, it is frequently convenient to estimate the intercept separately. We change notation slightly and write the no-equation-error model in the symmetric form,

$$\sum_{j=0}^{k} \alpha_j \gamma_{ji} = 0, \qquad i = 1, \ldots, n,$$

where $k = p+1$, $\boldsymbol{\gamma}_i = (\gamma_{0i}, \ldots, \gamma_{ki})' = (\eta_i, 1, \xi_{1i}, \ldots, \xi_{pi})'$, $\boldsymbol{\alpha} = (1, -\beta_0, -\boldsymbol{\beta}')'$, and $\mathbf{z}_i = (y_i, 1, x_{1i}, \ldots, x_{pi})'$. It is assumed that the error covariance structure is known up to a scalar multiple. That is,

$$\begin{aligned}\boldsymbol{\Omega}_{error}^{((k+1)\times(k+1))} &= \operatorname{var}\left((\varepsilon, 1, \boldsymbol{\delta}')\right) = \begin{bmatrix} \sigma_\varepsilon^2 & 0 & \boldsymbol{\Sigma}_{\varepsilon\delta} \\ 0 & 0 & \mathbf{0}' \\ \boldsymbol{\Sigma}_{\delta\varepsilon} & \mathbf{0} & \boldsymbol{\Sigma}_{\delta\delta} \end{bmatrix} \\ &= \sigma_\delta^2 \boldsymbol{\Omega}_0 = \sigma_\delta^2 \begin{bmatrix} \lambda & 0 & \boldsymbol{\Sigma}_{\varepsilon\delta} \\ 0 & 0 & \mathbf{0}' \\ \boldsymbol{\Sigma}_{\delta\varepsilon} & \mathbf{0} & \boldsymbol{\Sigma}_{\delta\delta} \end{bmatrix}\end{aligned}$$

and $\boldsymbol{\Omega}_0$ is known. For maximum likelihood, the errors are again assumed normal with zero means. Then the ML estimators for the functional model are

$$\left(\hat{\beta}_0, \hat{\boldsymbol{\beta}}'\right)' = \left(\mathbf{M}_{\mathbf{xx}} - \hat{\tau} \begin{bmatrix} 0 & \mathbf{0}' \\ \mathbf{0} & \boldsymbol{\Sigma}_{\delta\delta} \end{bmatrix}\right)^{-1} \left(\mathbf{M}_{\mathbf{x}y} - \hat{\tau}\,(0, \boldsymbol{\Sigma}_{\varepsilon\delta})'\right) \tag{5.19}$$

and $\hat{\sigma}_\delta^2 = \hat{\tau}/(p+1)$, provided the matrix inverse exists, where

$$\mathbf{M}_{xx}^{((p+1)\times(p+1))} = \frac{1}{n}\sum \left(1, \mathbf{x}_i'\right)'\left(1, \mathbf{x}_i'\right), \qquad \mathbf{M}_{xy}^{((p+1)\times 1)} = \frac{1}{n}\sum \left(1, \mathbf{x}_i'\right)' y_i,$$

$\hat{\tau}$ is the smallest root of $|\mathbf{M}_{zz} - \tau\boldsymbol{\Omega}_0| = 0$, and $\mathbf{M}_{zz} = n^{-1}\sum \mathbf{z}_i'\mathbf{z}_i$. The ML estimator of $\boldsymbol{\gamma}_i$, $i = 1, \ldots, n$, is

$$\begin{aligned}\hat{\boldsymbol{\gamma}}_i &= \mathbf{z}_i - \left(\left(y_i - \hat{\beta}_0 - \hat{\boldsymbol{\beta}}'\mathbf{x}_i\right)\left[\left(1, -\hat{\beta}_0, -\hat{\boldsymbol{\beta}}'\right)\boldsymbol{\Omega}_0\left(1, -\hat{\beta}_0, -\hat{\boldsymbol{\beta}}'\right)'\right]\right. \\ &\quad \left.\times \left(1, -\hat{\beta}_0, -\hat{\boldsymbol{\beta}}'\right)\boldsymbol{\Omega}_0\right).\end{aligned}$$

The estimate of σ_δ^2 is not consistent: $\hat{\sigma}_\delta^2 \to \frac{1}{p+1}\sigma_\delta^2$ as $n \to \infty$. Some authors have suggested the estimate $\tilde{\sigma}_\delta^2 = n\hat{\tau}/(n-p-1)$ based on a degrees of freedom argument.

For the normal structural model, the ML estimates of β_0 and $\boldsymbol{\beta}$ are (5.19) and $\hat{\sigma}_\delta^2 = \hat{\tau}$. These estimates are consistent (see Chan and Mak, 1984; Amemiya and Fuller, 1984). The estimates (5.19) are also consistent for the ultrastructural model.

Note that, in the above formulation, we allowed some of the ξ_j to be measured without error, which implies that $\boldsymbol{\Sigma}_{\delta\delta}$ is singular. There appears to be no generally accepted definition of maximum likelihood in the situation. We interpret 'maximum likelihood' to mean maximizing the density on the appropriate subspace over which the density is defined. This can be done if this subspace does not depend on any unknown parameters, which is the case here.

5.5 Alternative Approaches to Estimating the Parameters

This section investigates some alternative approaches to estimating the parameters other than maximum likelihood and the method of moments, namely, instrumental variable methods, generalized least-squares methods and modified least-squares methods. Because the ideas are the same as those for the case of one explanatory variable models discussed in previous chapters, the discussion here will be brief. Recall that some of the explanatory variables can be measured without error. This can include variables besides the dummy variable, ξ_0.

5.5.1 Instrumental variables

Consider the **regression** model

$$\mathbf{y} = \mathbf{X}\boldsymbol{\theta} + \mathbf{e}, \tag{5.20}$$

where $\mathbf{y}$ is an $n \times 1$ vector, $\mathbf{X}$ is the $n \times (p+1)$ observations on independent variables $(1, x_1, \ldots, x_p)$ (including the dummy variable for the intercept), and $\mathbf{e}$ is the $n \times 1$ error vector ($n > k = p+1$). If the elements of $\mathbf{X}$ are not asymptotically uncorrelated with the errors, then the ordinary least-squares estimator of $\boldsymbol{\theta}$ is not consistent. If this happens, there might exist a vector instrumental variable $\mathbf{z} = (z_1, \ldots, z_k)'$, which provides the data matrix $\mathbf{Z}^{(n \times k)}$, and has the following properties:

(i) $\lim_{n \to \infty} n^{-1}\mathbf{Z}'\mathbf{e} = \mathbf{0}$ or $E\mathbf{Z}'\mathbf{e} = \mathbf{0}$, and
(ii) $\lim_{n \to \infty} n^{-1}\mathbf{Z}'\mathbf{X}$ or $E\mathbf{Z}'\mathbf{X}$ is a nonsingular matrix of constants.

Then the instrumental variable estimator of $\boldsymbol{\theta}$ defined by

$$\hat{\boldsymbol{\theta}} = (\mathbf{Z}'\mathbf{X})^{-1} Z'\mathbf{y}$$

is a consistent estimate of $\boldsymbol{\theta}$.

Now consider the ME model with possible equation error,

$$\mathbf{x}_i = \boldsymbol{\xi}_i + \boldsymbol{\delta}_i, \qquad y_i = \boldsymbol{\beta}'\boldsymbol{\xi}_i + e_i,$$

which can be written as a special case of (5.20) (see (5.21)). If there is an instrumental variable $\mathbf{z}$ such that

(i′) $E\mathbf{z}_i(\boldsymbol{\delta}_i', e_i) = \mathbf{0}$ and
(ii′) $\sum \mathbf{z}_i'\mathbf{x}_i$ has rank equal to the dimension of $\boldsymbol{\beta} = (\beta_0, \ldots, \beta_p)'$ (with probability one),

then the instrumental variable estimator

$$\hat{\boldsymbol{\beta}}_{\mathbf{IV}} = (\mathbf{Z}'\mathbf{X})^{-1}\mathbf{Z}'\mathbf{y}$$

will be a consistent estimator of $\boldsymbol{\beta}$.

It is possible for the dimension of the instrumental variable $\mathbf{z}$ to be higher than the dimension of $\mathbf{x}$. In this case, the instrumental variable estimator takes another form; see Fuller (1987, p. 148 ff.) and Bowden and Turkington (1984).

5.5.2 The generalized and modified least-squares estimators

The generalized least-squares and modified least-squares approaches described in Section 3.5 can be easily extended to vector explanatory variables models. Both Sprent (1966) and Gleser (1981) discussed generalized least-squares estimation for the multivariate functional model. The latter also investigated the large-sample properties of the generalized least-squares estimator. In this section, we treat the generalized least estimator in a slightly different manner and extend the modified least-squares method to the vector explanatory variable model.

Generalized least squares.

Consider model (5.1)–(5.2), and write

$$y_i = \beta_0 - \boldsymbol{\beta}'\mathbf{x}_i + (\varepsilon_i - \boldsymbol{\beta}'\boldsymbol{\delta}_i), \qquad i = 1, \ldots, n. \tag{5.21}$$

Let $v_i = \varepsilon_i - \boldsymbol{\beta}'\boldsymbol{\delta}_i$, $i = 1, \ldots, n$, be the **residuals**, which are independent and identically distributed with mean zero and variance $\sigma_v^2 = \sigma_\varepsilon^2 - 2\boldsymbol{\beta}'\boldsymbol{\Sigma}_{\delta\varepsilon} + \boldsymbol{\beta}'\boldsymbol{\Sigma}_{\delta\delta}\boldsymbol{\beta}$ regardless of whether the model is functional, structural or ultrastructural. The generalized least-squares estimator is defined to be the $(\hat{\beta}_0, \hat{\boldsymbol{\beta}}')$ that minimizes

$$Q(\beta_0, \boldsymbol{\beta}') = n^{-1} \sum \frac{(y_i - \beta_0 - \boldsymbol{\beta}'\mathbf{x}_i)^2}{\sigma_v^2}. \tag{5.22}$$

It is seen that one need only know the error covariance matrix up to a scalar multiple in order to solve this minimization problem. Gleser (1981) showed that the resulting estimator $(\hat{\beta}_0, \hat{\boldsymbol{\beta}}')$ is the same as the ML estimator (5.19) under the normality assumption (see Exercise 5.1).

Modified least squares.

Now consider the modified least-squares approach described in Section 3.5. Obviously, the modified least-squares estimator coincides with the generalized least squares under the assumption that one knows the error covariance matrix up to a proportionality factor in the no-equation-error model.

For the equation error model (5.4), which can possibly be degenerate ($q_i \equiv 0$), the modified least-squares estimator minimizes

$$Q(\beta_0, \boldsymbol{\beta}') = n^{-1} \sum \{(y_i - \beta_0 - \boldsymbol{\beta}'\mathbf{x}_i)^2 + 2\boldsymbol{\beta}'\boldsymbol{\Sigma}_{\delta\varepsilon} - \boldsymbol{\beta}'\boldsymbol{\Sigma}_{\delta\delta}\boldsymbol{\beta}\}. \tag{5.23}$$

Naturally, knowledge of $\boldsymbol{\Sigma}_{\delta\delta}$ and $\boldsymbol{\Sigma}_{\delta\varepsilon}$ is required. The estimator resulting from minimization of (5.23) is the same as the ML estimator (5.5) of the normal structural model.

The restriction, $Q(\hat{\beta}_0, \hat{\boldsymbol{\beta}}') \geq 0$, is required in both (5.22) and (5.23) because Q is estimating the error variance. It is seen that such a restriction is satisfied automatically in (5.22) but one needs $\mathbf{S}_{\mathbf{xx}} - \boldsymbol{\Sigma}_{\delta\delta}$ positive definite and $\hat{\sigma}_e^2 \geq \boldsymbol{\Sigma}_{e\delta}\boldsymbol{\Sigma}_{\delta\delta}^{+}\boldsymbol{\Sigma}_{\delta e}$ for (5.23) (see Section 5.3.1). The asymptotic properties are the same as those in Section 3.5 for the univariate model.

5.6 Asymptotic Properties of the Estimates

This section gives the asymptotic properties of the estimators for the vector explanatory variable ME model derived in the previous two sections. For simplicity, the model formulation is slightly different. Consider the vector explanatory variable, equation error, ultrastructural model,

$$\mathbf{x}_i = \boldsymbol{\xi}_i + \boldsymbol{\delta}_i, \qquad y_i = \boldsymbol{\theta}'\boldsymbol{\xi}_i + e_i, \tag{5.24}$$

with $e_i = \varepsilon_i + q_i$, and the no-equation-error ultrastructural model,

$$\mathbf{x}_i = \boldsymbol{\xi}_i + \boldsymbol{\delta}_i, \qquad y_i = \eta_i + \varepsilon_i, \qquad \eta_i = \boldsymbol{\theta}'\boldsymbol{\xi}_i, \tag{5.25}$$

where $\boldsymbol{\theta} = (\beta_0, \beta_1, \ldots, \beta_p)'$. Thus, if the model has an intercept, $\xi_{i0} \equiv 1$ and $\delta_{i0} \equiv 0$. The true variables $\boldsymbol{\xi}_i$ have means $\boldsymbol{\mu}_i$ and common covariance $\boldsymbol{\Sigma}_{\xi\xi}$. Assume that the limit $\mathbf{S}^*_{\mu\mu} = \lim_{n\to\infty} n^{-1} \sum \boldsymbol{\mu}_i \boldsymbol{\mu}_i'$ exists, and

$$\mathbf{S}_{\xi\xi} = \mathbf{S}^*_{\mu\mu} + \boldsymbol{\Sigma}_{\xi\xi}, \tag{5.26}$$

where $\mathbf{S}_{\xi\xi}$ is a positive definite matrix. Furthermore, let $v_i = e_i - \boldsymbol{\theta}'\boldsymbol{\delta}_i$ in the equation error model and $u_i = \varepsilon_i - \boldsymbol{\theta}'\boldsymbol{\delta}_i$ in the no-equation-error model; then

$$\sigma_v^2 = \sigma_q^2 + \sigma_u^2, \qquad \sigma_u^2 = \sigma_\varepsilon^2 - 2\boldsymbol{\Sigma}_{\varepsilon\delta}\boldsymbol{\theta} + \boldsymbol{\theta}'\boldsymbol{\Sigma}_{\delta\delta}\boldsymbol{\theta}, \qquad \boldsymbol{\Sigma}_{\delta v} = \boldsymbol{\Sigma}_{\delta u} = \boldsymbol{\Sigma}_{\delta\varepsilon} - \boldsymbol{\Sigma}_{\delta\delta}\boldsymbol{\theta}, \tag{5.27}$$

because q is independent of $\boldsymbol{\delta}$ and ε, which implies $\boldsymbol{\Sigma}_{\delta\varepsilon} = \boldsymbol{\Sigma}_{\delta e}$.

The estimator of $\boldsymbol{\theta}$ is

$$\hat{\boldsymbol{\theta}} = (\mathbf{M}_{\mathbf{xx}} - \boldsymbol{\Sigma}_{\delta\delta})^{-1}(\mathbf{M}_{\mathbf{x}y} - \boldsymbol{\Sigma}_{\delta e}), \tag{5.28}$$

where $\mathbf{M}_{\mathbf{xx}} = n^{-1} \sum \mathbf{x}_i \mathbf{x}_i'$ and $\mathbf{M}_{\mathbf{x}y} = n^{-1} \sum \mathbf{x}_i y_i$ in the equation error model and $\hat{\boldsymbol{\theta}}$ is defined by (5.19) in the no-equation-error model.

The following asymptotic results are due to Schneeweiss (1976) and Fuller (1980); see also Fuller (1987, Chapter 2).

Theorem 5.1 *Assume the equation error ultrastructural model (5.24) with normally distributed measurement errors* $(\boldsymbol{\delta}_i', \varepsilon_i)$ *and that the equation error* q_i *has finite fourth moments. Let* $\hat{\boldsymbol{\theta}}$ *be defined as in (5.28). Then* $n^{1/2}(\hat{\boldsymbol{\theta}} - \boldsymbol{\theta})$ *is asymptotically normal with mean vector zero and covariance matrix*

$$\boldsymbol{\Gamma} = \mathbf{S}_{\xi\xi}^{-1}\{(\mathbf{S}_{\xi\xi} + \boldsymbol{\Sigma}_{\delta\delta})\sigma_v^2 + \boldsymbol{\Sigma}_{\delta\mathbf{v}}\boldsymbol{\Sigma}_{\mathbf{v}\delta}\}\mathbf{S}_{\xi\xi}^{-1},$$

where $\mathbf{S}_{\xi\xi}$ *is defined in (5.26) and* $\boldsymbol{\Sigma}_{\delta\mathbf{v}}$ *and* σ_v^2 *are defined by (5.27).*

For the no-equation-error model, let $\hat{\boldsymbol{\theta}}$ be defined by (5.19); then the large sample results are provided by the following theorem.

Theorem 5.2 *Assume the no-equation-error ultrastructural model (5.25) with normally distributed measurement error* $(\boldsymbol{\delta}_i', \varepsilon_i)$. *Let* $\hat{\boldsymbol{\theta}}$ *be defined in (5.19). Then* $n^{1/2}(\hat{\boldsymbol{\theta}} - \boldsymbol{\theta})$ *is asymptotically normal with mean vector zero and covariance matrix*

$$\boldsymbol{\Gamma} = \mathbf{S}_{\xi\xi}^{-1}\{(\mathbf{S}_{\xi\xi} + \boldsymbol{\Sigma}_{\delta\delta})\sigma_u^2 - \boldsymbol{\Sigma}_{\delta\mathbf{u}}\boldsymbol{\Sigma}_{\mathbf{u}\delta}\}\mathbf{S}_{\xi\xi}^{-1},$$

where $\mathbf{S}_{\xi\xi}$ *is defined in (5.26) and* $\boldsymbol{\Sigma}_{\delta\mathbf{u}}$ *and* σ_u^2 *are defined by (5.27).*

Note that the normality of $\boldsymbol{\xi}$ is not required in the theorems above. Moreover, both theorems apply to both the structural and functional models. For the functional model, $\mathbf{S}_{\xi\xi} = \boldsymbol{\Sigma}_{\xi\xi}$ because $\mathbf{S}_{\mu\mu}^*$ equals zero. On the other hand, $\mathbf{S}_{\xi\xi} = \mathbf{S}_{\xi\xi}^* = \mathbf{S}_{\mu\mu}^*$ because $\boldsymbol{\Sigma}_{\xi\xi}$ equals zero and $\boldsymbol{\mu}_i = \boldsymbol{\xi}_i$.

Fuller (1980) and Amemiya and Fuller (1988) (see also Fuller 1987, Chapter 2) treated the case when the error covariance matrix is independently estimated instead of being given. The asymptotic distribution of $\hat{\boldsymbol{\theta}}$ is still normal, but with a more complicated asymptotic covariance matrix, under the condition that the estimated error covariance matrix is unbiased and follows a Wishart distribution with degrees of freedom nearly proportional to sample size n.

5.7 Bibliographic Notes and Discussion

The vector explanatory variable models are a direct extension of the univariate model. These models correspond to multiple regression models with measurement error in the independent variables. The ideas and methodologies are analogous to those in the univariate model so we have only stated the basic results. The models can be 'mixed' in the sense that some of the independent variables can be without measurement error.

The ML estimators and their large-sample properties for the multivariate normal no-equation-error functional and structural relationships were first obtained by Patefield (1981a) for the case in which the error covariance matrix is diagonal and known up to a proportionality factor. Gleser (1981) obtained the ML estimates and the associated asymptotic results for the full (both x and y) multivariate functional no-equation-error model when the error covariance is known up to a proportionality constant. Amemiya and Fuller (1984) discussed ML estimators for the full multivariate no-equation-error functional and structural normal models and the associated large-sample results with independently estimated error covariance matrix. Chan and Mak (1984) investigated the ML estimation of the multivariate structural model when the error covariance is known or known up to a proportionality constant. Vector explanatory variable models are a special case of the full multivariate model.

In the equation error normal functional relationship ML estimation breaks down, as happened with the univariate model in Section 1.3.2. Schneeweiss (1976) discussed the large-sample properties of the equation error model estimator (5.5). Fuller (1980) discussed the large-sample properties of the estimators (5.5) and (5.19) with or without equation error when the error covariance matrix is estimated.

For the multivariate normal ultrastructural model, it is believed that ML estimation parallels that for the functional model, just as in the univariate case (see Section 1.3.3).

Indeed, Tracy and Jinadasa (1987) showed that, in the no-equation-error multivariate normal ultrastructural model with the error covariance matrix known, the ML estimator of $\boldsymbol{\beta}$ occurs on the boundary of the parameter space where $\boldsymbol{\Sigma}_{\xi\xi} = \mathbf{0}$. Therefore, the ML estimate of $\boldsymbol{\beta}$ is the same as that of the corresponding functional model (see Section 5.4 and Exercise 5.7).

Gleser (1992) extended the reliability ratio known case to the multivariate structural model in which the reliability matrix is known. Because it is closely related to ordinary least squares (see Section 5.3), this approach allows for the use of least-squares methodology, and paves the way for adapting regression diagnostic methods to diagnostic methods for ME models (cf. Cheng and Tsai, 1996).

The generalized least-squares method was discussed in Sprent (1966) and more extensively for the multivariate model in Gleser (1981). The link between the generalized least-squares estimator and the ML estimator for the normal model was also established by Gleser (1981). The ML estimator of the transformed data discussed in Section 1.5 (see also Cox, 1993, p. 908) also makes this connection. There are other articles that investigate the properties of generalized least-squares estimators, for example, Höschel (1978a; 1978b), and Nussbaum (1976; 1977; 1979; 1984). Again, as in the univariate models, modified least-squares gives a unified approach for 'least-squares' methods in vector explanatory variable ME models. Note that modified and generalized least-squares methods provide simple unbiased estimating equations for the parameters. This is not only applied to nonnormal models (with suitable side assumptions), but also provides standard large-sample results. More significantly, such an approach allows one to investigate other statistical questions, for example, diagnostics for the ME model (Wellman and Gunst, 1991; Cheng and Tsai, 1993).

The instrumental variables estimator is widely discussed in the econometrics literature. Here we have given only a brief introduction to the subject. See Section 4.6 for a summary and Bowden and Turkington (1984) for a more complete treatment.

5.8 Exercises

5.1 *Show that the estimator* $(\hat{\beta}_0, \hat{\boldsymbol{\beta}})$ *from (5.22) is the same as the ML estimator (5.19) under the normality assumption.*

5.2 *Show that estimate of* $\boldsymbol{\beta}$ *in (5.5) does not have finite first moment for the normal model. What happens in the nonnormal model (cf. Exercise 2.1)?*

5.3 *Consider the normal multivariate structural model with* $\boldsymbol{\Sigma}_{\delta\varepsilon} = \mathbf{0}$ *and* $p = 2$. *Give an example with a nonsingular matrix* $\mathbf{A}^{(2\times2)}$, *as defined in Section 5.2, such that* $\boldsymbol{\xi}'\mathbf{a}_1^{(2\times1)}$ *is distributed independently of* $\boldsymbol{\xi}'\mathbf{A}_2^{(2\times1)}$ *thereby showing that the model is not identifiable.*

5.4 *Do a Monte Carlo study similar to Example 7.1 but replicated, say,* $m = 100$ *times. How many of the 100 replications give negative estimates for* σ_ε^2? *Try increasing the sample size from* $n = 36$ *to* $n = 100$ *and* $n = 200$. *Now how many times do negative estimates for* σ_ε^2 *occur?*

5.5 *Do a Monte Carlo example similar to Example 7.1 but under the conditions of Section 5.3.2. Compare on the same data your estimates with the ordinary least-squares estimates.*

5.6 *Evaluate (5.19) for the data of Example 1.11 and verify the estimates of* β_0 *and* β_1 *given there. That is, assume that* $\lambda = 1$ *and create the matrices given in Section 5.4.2 for the case* $p = 1$. *Your answers might differ slightly due to round-off errors.*

5.7 *Extend the results of Tracy and Jinadasa (1987) to ML estimators for the multivariate normal ultrastructural model with equation error assuming that both* $\mathbf{\Sigma}_{\delta\delta}$ *and* $\mathbf{\Sigma}_{\delta\varepsilon}$ *are known.*

5.9 Research Problem

5.8 *In the univariate no-equation-error normal model, the estimate (1.30) does not have finite first moment. But it is not known whether this property carries over to the vector explanatory variable no-equation-error model estimate (5.19). Prove or disprove that the estimate (5.19) possesses no finite first moment.*

6

Polynomial Measurement Error Models

The previous chapters have considered only linear relationships between the variables. Extensions to nonlinear relationships are not as straightforward for ME models as they are for ordinary regression. Indeed several fundamental questions concerning nonlinear ME models have successfully resisted attack hitherto.

We will only briefly mention the general nonlinear ME model. Our focus will be on a special case of nonlinear models, the polynomial ME model with univariate explanatory variable. Discussions will include polynomial ME models, with and without equation error, and the polynomial Berkson model.

6.1 Introduction

The general definition of a nonlinear ME model with no equation error and vector explanatory variable is

$$\mathbf{x}_i = \boldsymbol{\xi}_i + \boldsymbol{\delta}_i, \quad y_i = \eta_i + \varepsilon_i, \qquad i = 1, \ldots, n \tag{6.1}$$

and

$$\eta_i = g(\boldsymbol{\xi}_i; \boldsymbol{\beta}), \qquad i = 1, \ldots, n, \tag{6.2}$$

where g is a real-valued function that possesses continuous first and second derivatives with respect to the second argument, $\boldsymbol{\xi}_i$ is an $r \times 1$ vector, and $\boldsymbol{\beta} = (\beta_0, \beta_1, \ldots, \beta_p)'$ is the unknown parameter vector to be estimated. We adopt the usual definition of nonlinear ME models (cf. Fuller, 1987, p. 226), that is, the model is nonlinear if $g(\boldsymbol{\xi}; \boldsymbol{\beta})$ is nonlinear in either $\boldsymbol{\xi}$ or $\boldsymbol{\beta}$. Note that this definition of 'nonlinear' is different from that of ordinary regression, where a model which is nonlinear in ξ but linear in $\boldsymbol{\beta}$ is considered to be linear. For example, the polynomial regression model is considered to be linear in classical regression but it is a nonlinear ME model. This difference in the definitions of linear and nonlinear is a natural consequence of the fact that $\boldsymbol{\xi}$ is measured with error in the ME model.

In this chapter, we assume that the $(\boldsymbol{\delta}_i', \varepsilon_i)$ are independent and identically distributed vectors (ε_i and $\boldsymbol{\delta}_i$ can be correlated). In the structural model, the $\boldsymbol{\xi}_i$ are independent and identically distributed, with the $(\boldsymbol{\delta}_i', \varepsilon_i)$ independent of the $\boldsymbol{\xi}_i$.

The nonlinear ME model with equation error is (6.1) with

$$\eta_i = g(\boldsymbol{\xi}_i; \boldsymbol{\beta}) + q_i, \qquad i = 1, \dots, n, \tag{6.3}$$

so that

$$y_i = g(\boldsymbol{\xi}_i; \boldsymbol{\beta}) + e_i,$$

where ε is the measurement error, q is the equation error and $e = \varepsilon + q$ (see the linear equation error ME model in Section 1.4). The equation errors, q_i, are independent, identically distributed and independent of the $(\boldsymbol{\delta}_i', \varepsilon_i)$ and the $\boldsymbol{\xi}_i$.

Estimation of the unknown parameters in the general nonlinear ME model is much more complicated than that in the linear ME model. The primary interest is in estimating $\boldsymbol{\beta}$. Some of the difficulties encountered can be illustrated by considering the no-equation error quadratic functional relationship with univariate explanatory variable:

$$\eta_i = \beta_0 + \beta_1 \xi_i + \beta_2 \xi_i^2. \tag{6.4}$$

Assume that the errors, ε and δ, are uncorrelated and, furthermore, that $\lambda = \sigma_\varepsilon^2/\sigma_\delta^2$ is known and, without loss of generality, taken to be unity. Then the log-likelihood function of (6.4) with normal errors is

$$\begin{aligned} L(\beta_0, \beta_1, \beta_2, \sigma_\varepsilon^2, \xi_1, \dots, \xi_n) \quad \propto \quad & -n \log \sigma_\varepsilon^2 - \frac{1}{2\sigma_\varepsilon^2} \sum (x_i - \xi_i)^2 \\ & - \frac{1}{2\sigma_\varepsilon^2} \sum (y_i - \beta_0 - \beta_1 \xi_i - \beta_2 \xi_i^2)^2. \end{aligned} \tag{6.5}$$

Differentiating (6.5) gives

$$x_i - \xi_i + (y_i - \beta_0 - \beta_1 \xi_i - \beta_2 \xi_i^2)(\beta_1 + 2\beta_2 \xi_i) = 0, \qquad i = 1, \dots, n, \tag{6.6}$$

$$\sum (y_i - \beta_0 - \beta_1 \xi_i - \beta_2 \xi_i^2) = 0, \tag{6.7}$$

$$\sum (y_i - \beta_0 - \beta_1 \xi_i - \beta_2 \xi_i^2)\xi_i = 0, \tag{6.8}$$

$$\sum (y_i - \beta_0 - \beta_1 \xi_i - \beta_2 \xi_i^2)\xi_i^2 = 0, \tag{6.9}$$

$$-\frac{2n}{\sigma_\varepsilon} + \frac{1}{\sigma_\varepsilon^3}\left[\sum (x_i - \xi_i)^2 + \sum (y_i - \beta_0 - \beta_1 \xi_i - \beta_2 \xi_i^2)^2\right] = 0. \tag{6.10}$$

Summing (6.6) over i and using (6.7) and (6.8) gives

$$\sum \hat{\xi}_i = \sum x_i. \tag{6.11}$$

From (6.7)–(6.9) and (6.11),

$$\begin{aligned} n\beta_0 + \beta_1 \sum x_i + \beta_2 \sum \xi_i^2 &= \sum y_i, \\ \beta_0 \sum x_i + \beta_1 \sum \xi_i^2 + \beta_2 \sum \xi_i^3 &= \sum y_i \xi_i, \\ \beta_0 \sum \xi_i^2 + \beta_1 \sum \xi_i^3 + \beta_2 \sum \xi_i^4 &= \sum y_i \xi_i^2. \end{aligned} \tag{6.12}$$

Equations (6.12) have the same form as the corresponding equations in regression analysis, but here the ξ are not observable. In order to obtain the LE estimators one must solve the $n+3$ equations, (6.6) and (6.12), for the $n+3$ unknowns $\beta_0, \beta_1, \beta_2$, and $\xi_1, \ldots, \xi_n$. Note that (6.6) is a cubic in ξ_i:

$$x_i + (y_i - \beta_0)\beta_1 + [-1 + 2(y_i - \beta_0)\beta_2 - \beta_1^2]\xi_i - 3\beta_1\beta_2\xi_i^2 - 2\beta_2^2\xi_i^3 = 0, \\ i = 1, \ldots, n.$$

This can have either one real solution, two different real solutions, or three different real solutions. If there is more than one solution, one could choose that solution, ξ_i, closest to the corresponding x_i. The estimate of σ_ε^2 then follows from (6.10). One does not know that these LE estimators are ML estimators because it is not known that they actually maximize the likelihood function. It is not known whether these estimates are consistent even if they are ML estimators.

In practice, it appears that these equations could possibly be solved by iterative methods. The iterative procedure would proceed as follows. Choose some initial values, $\beta_{0,0}$, $\beta_{1,0}$, and $\beta_{2,0}$ for β_0, β_1, and β_2; Wolter and Fuller (1982b) suggest using the estimates obtained from ordinary regression as the initial values. For the first iteration, use (6.6) to obtain estimates, $\xi_{i,1}$, $i = 1, \ldots, n$, of the unobservable $\xi_1, \ldots, \xi_n$ and then use $\xi_{i,1}$, $i = 1, \ldots, n$, and (6.12) to obtain new estimates, $\beta_{0,1}$, $\beta_{1,1}$, and $\beta_{2,1}$. The second iteration repeats the procedure of the first iteration, but starting with $\beta_{0,1}$, $\beta_{1,1}$, and $\beta_{2,1}$. Iterations are continued until the estimates change by less than some prespecified tolerance.

Example 6.1

The authors experimented with the iterative scheme just described for solving (6.6)–(6.9) for the estimates of β_0, β_1, and β_2 in the normal quadratic functional model with λ known. In general, the results were not encouraging. Convergence was obtained only for very small σ_δ^2, and when convergence was obtained the estimates were not any better than the ordinary least-squares estimates.

Increasing the sample size did not help because with each new data point one gets another ξ_j parameter to estimate.

Replications immediately come to mind, but, if one has replications, these replications could be used via averaging to reduce σ_δ^2. That is, if there are m replications for each ξ_i, then just consider the data

$$\bar{x}_{i\cdot} = \frac{1}{m}\sum_{j=1}^{m} x_{ij}, \qquad \text{and} \qquad \bar{y}_{i\cdot} = \frac{1}{m}\sum_{j=1}^{m} y_{ij}, \qquad i = 1, \ldots, n,$$

which has errors whose variances are divided by m. The idea is to get to a small error variance for the explanatory variable, which is just leading back to least squares.

Other iterative schemes for the general nonlinear model have been given; see Wolter and Fuller (1982b), Amemiya and Fuller (1988), Moon and Gunst (1995), and Section 6.5.5. The Monte Carlo studies in these papers use replicated data with relatively small variances for the explanatory variable errors – a situation in which

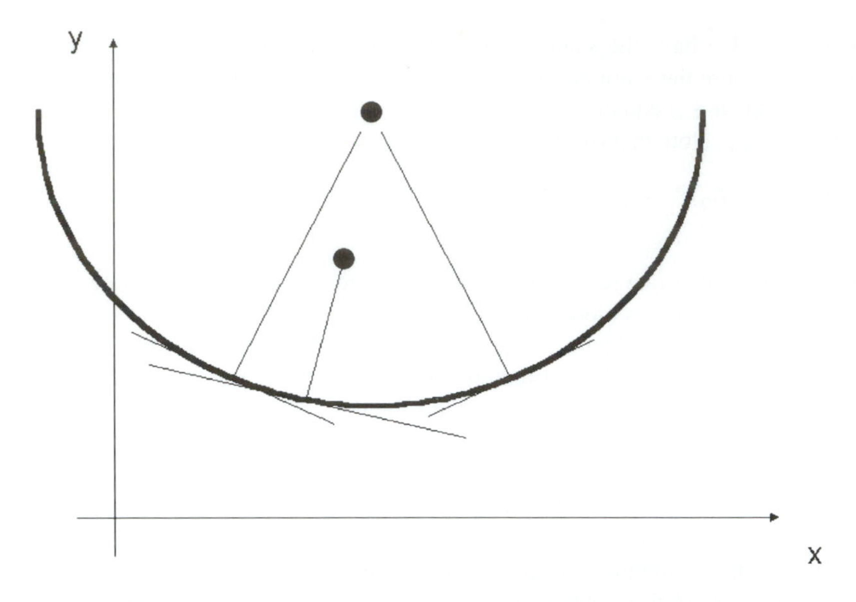

Figure 6.1: Non-unique and non-parallel distances to a curve.

one is close to the ordinary regression model. This chapter presents some alternative estimation procedures for the particular case of polynomial ME models.

The difficulties with estimating nonlinear ME models can also be seen from a geometrical point of view. From (6.5), one seeks to determine a quadratic curve such that the sum of squares of the distances of points to the curve is minimized. The joins of the data points to their corresponding closest points on the curve are not parallel and might not even be unique (see Figure 6.1). This means that it is going to be difficult to find analytically a curve which is 'closest' to the data. A solution, though possibly numerically attainable, is apparently not expressible in simple closed form.

Furthermore, if σ_δ^2 is assumed known or the equation error quadratic functional relationship is assumed, the problem is even more serious because the ML estimation approach breaks down just as it did in the linear case (see Section 1.3.2).

Some nonlinear relationships can be reduced to a linear relationship by a transformation. Consider, for example, a functional relationship of the type $\eta^\beta \xi^\gamma =$ constant. Here the obvious solution is to take logarithms. If the model can be transformed to linearity before the estimation begins, a great deal will be gained. A possible drawback is that, if errors in η and ξ are, say, normal and homoscedastic, those of the transforms log η and log ξ will not be. One should endeavour to obtain as much prior information as possible about the nature of the observational errors; and, when the errors are substantial and of unknown distribution, one should use methods of estimation that make as few assumptions as possible.

Nonlinear ME models are of both theoretical and practical interest and are currently areas of active research. Much of the work is summarized in Fuller (1987, Chapter 3) and Carroll *et al.* (1995).

In this chapter, we concentrate on polynomial ME models – perhaps the most natural extension of the linear ME model towards nonlinearity. Concise consistent estimators of the parameters of both the no-equation-error and equation error models will be given.

6.1.1 Calculating the fit of data to a nonlinear ME model

Suppose in an application one has data, $\binom{x_1}{y_2}, \ldots, \binom{x_n}{y_n}$, from an ME model and two prospective (linear or nonlinear) models, $\eta = g_1(\xi)$ and $\eta = g_2(\xi)$ for fitting the data. How can one decide which model fits better? First one needs to decide what one means by 'fits better'. The error structure has to enter into the problem because we only can measure x and y and not ξ and η. It is clear from Chapter 1 that one does not want to use least squares or any other measure, such as mean absolute error, that is based on the vertical distances from the data points to the line.

For polynomial models, it will be seen that, under certain conditions, one can find estimates of an unknown error variance and, in such cases, one can use these as a measure of the quality of fit. For example, in Section 6.4, it is shown that, for a polynomial of degree k, if $E\delta^r$, $r = 1, \ldots, 2k$, is known, then an estimate, (6.36), of error variance in y is available. If one is considering more than one model, then the one with smallest $\hat{\sigma}_e^2$ could be chosen.

If we assume that the error covariance matrix is known up to a scalar multiple,

$$\operatorname{cov}\left(\begin{pmatrix}\delta \\ \varepsilon\end{pmatrix}\right) = \sigma_0^2 \mathbf{\Omega}_0,$$

where $\mathbf{\Omega}_0$ is known, then a common way to measure fit is to rescale the data so that the error covariance is a scalar multiple of the identity matrix and measure the orthogonal distances from the data to the lines $y = g_1(x)$ and $y = g_2(x)$. This measure of fit is motivated by ML solutions to the nonlinear ME model with normal errors (see Section 6.5.4). If one knows $\mathbf{\Omega}_0$, then frequently one can estimate σ_0^2 given $g_1(x)$ and $g_2(x)$ and then choose the one that minimizes the estimate of σ_0^2.

Another method of finding the error variances is demonstrated here for the quadratic model. Suppose the proposed model values, β_0, β_1, and β_2, are given. Then one can use the estimates

$$\widehat{E\xi} = \hat{\mu} = \bar{x} \qquad \text{and} \qquad \widehat{Ey} = \bar{y}$$

and the fact that

$$Ey = \beta_0 + \beta_1 E\xi + \beta_2 E\xi^2$$

to obtain an estimate of $E\xi^2$,

$$\widehat{E\xi^2} = \frac{\bar{y} - \beta_0 - \beta_1 \bar{x}}{\beta_2}.$$

One now has an estimate of σ^2,

$$\hat{\sigma}^2 = \widehat{E\xi^2} - (\widehat{E\xi})^2,$$

and an estimate of σ_δ^2,

$$\hat{\sigma}_\delta^2 = s_{xx} - \hat{\sigma}^2.$$

If this is positive, it could be used as a measure of fit of the model, β_0, β_1, and β_2. If one prefers to use σ_ε^2 as a measure of fit, it can also be estimated by solving

$$Ey^2 = E(\beta_0 + \beta_1\xi + \beta_2\xi^2 + \varepsilon)^2$$

for $E\varepsilon^2$ and substituting estimates of the unknowns to get $\hat{\sigma}_\varepsilon^2$. This involves the moments $E\xi^3$ and $E\xi^4$. If ξ is assumed normal, one can obtain estimates of these moments from $\hat{\mu}$ and $\hat{\sigma}^2$ and the relationships between the higher and lower moments of a normal distribution. This method generalizes to higher-order polynomials for ξ normally distributed. It seems unreasonable to assume one knows the higher moments of ξ otherwise.

For the rest of this subsection, suppose that one has set the scale so that the error covariance matrix is a scalar multiple of the identity matrix. An example of the use of this approach is given in Example 6.3.

The distance from a point, $\binom{x_i}{y_i}$, to the curve, g, is (assume g is continuous)

$$d_i(g) = \min_x \sqrt{(y_i - g(x))^2 + (x_i - x)^2}. \tag{6.13}$$

If the curve is only to be considered only over the range of the x data, $[a, b]$, where $a = \min_i x_i$ and $b = \max_i x_i$, then (6.13) would become

$$d_i(g) = \min_{a \le x \le b} \sqrt{(y_i - g(x))^2 + (x_i - x)^2}. \tag{6.14}$$

Note that (6.13) is the radius of the smallest circle, centred at $\binom{x_i}{y_i}$, that touches the curve, g. The distance from the data to the curve would then be

$$d(g) = \sum_1^n d_i(g).$$

This distance can easily be calculated numerically. An algorithm for doing this using (6.14) would be as follows:

1. Form a lattice of m equally space points, $z_1, \ldots, z_m$, from a to b (m might be 1000 or 10 000).
2. For each data point, $\binom{x_i}{y_i}$, evaluate

$$d_{zi}(g) = \min_{j=1,\ldots,m} \sqrt{(y_i - g(z_j))^2 + (x_i - z_j)^2}.$$

3. Calculate

$$d(g) \simeq \sum_1^n d_{zi}(g).$$

Written in the statistics language S, this algorithm is very simple. For example, let $xdata$ be the vector of the n values of the x data, $ydata$ be the n values of the y data, $g(x) = \beta_0 + \beta_1 x + \beta_2 x^2$, and $m = 1000$; β_0, β_1, and β_2 are just numbers. Then the algorithm, **orthdist**, could be written as follows:

orthdist

```
z <- seq (min(xdata), max(xdata), length = 1000)
lat.y <- β0 + β1*z + β2*z ^2
sum <- 0
for (i in 1: length(xdata)) {
      dist <- sqrt (( lat.y - ydata[i] ) ^2 + ( z - xdata[i] ) ^2)
      sum <- sum + min(dist)
      }
```

(You should not use for loops in S, instead you should use arrays, but we write it this way for clarity.)

This algorithm is easily generalized to more that one x variable as long as one knows the covariance matrix of the errors up to an unknown multiplicative constant. In this case the distance (6.13) becomes the usual Euclidean distance and (6.14) is taken over the range of the data.

6.2 The Nonlinear Structural Model

For the nonlinear structural ME model, one could, in principle, use maximum likelihood to estimate the parameters if the joint distribution of $(\mathbf{x}', y)$ is known and tractable. So far this approach has not proved fruitful. For example, it is not clear how to do this even for the simple quadratic structural relation when $(\xi, \delta, \varepsilon)$ are normally distributed. There is very little known about ML estimators for the nonlinear structural model.

6.2.1 Gleser's conditional approach

Gleser (1990) proposed formulating the structural equation error model, (6.1) and (6.3) with $\boldsymbol{\delta}$ and ε independent, as

$$y_i = E(y_i \mid \mathbf{x}_i) + u_i, \tag{6.15}$$

where

$$u_i = g(\boldsymbol{\xi}_i; \boldsymbol{\beta}) - E(y_i \mid \mathbf{x}_i) + e_i = g(\boldsymbol{\xi}_i; \boldsymbol{\beta}) - E[g(\boldsymbol{\xi}_i; \boldsymbol{\beta}) \mid \mathbf{x}_i] + e_i.$$

It can be shown that $E(y \mid \mathbf{x})$ and u are uncorrelated and $Eu = 0$ (see Exercise 6.1). By this artifice, the nonlinear ME regression problem has been converted to a nonlinear ordinary regression problem with heteroscedastic error in general. That is, (6.15) gives a classical nonlinear regression model for the conditional distribution of y given $\mathbf{x}$. Note that the joint likelihood of $(\mathbf{x}', y)$ can be written as

$$f(\mathbf{x}, y) = f(y \mid \mathbf{x}) f(\mathbf{x}).$$

Thus, the information about $\boldsymbol{\beta}$ comes from the conditional distribution of y given $\mathbf{x}$. It should, however, be noted that the error u is a function of $\boldsymbol{\beta}$ and $\mathbf{x}_i$. There are many techniques for fitting such nonlinear ordinary regression models; see Carroll and Ruppert (1988) for a summary. Estimates of $\boldsymbol{\beta}$ in (6.15) will be consistent and efficient in many cases provided $\boldsymbol{\beta}$ is identifiable. The difficulty is that the conditional expectation $E(y \mid \mathbf{x})$ cannot be always be expressed in a convenient form for use in standard regression. Examples of this approach, involving exponential models, are given in Gleser (1990, p. 110).

6.2.2 The cumulant method

The cumulant method of Section 4.5.1 can be extended to the polynomial structural model. Consider the cubic model,

$$\eta_i = \beta_0 + \beta_1 \xi_i + \beta_2 \xi_i^2 + \beta_3 \xi_i^3, \tag{6.16}$$

where it is no longer necessary that the errors be normally distributed. The joint characteristic function of η and ξ is

$$\phi(t_1, t_2) = \int \mathrm{e}^{\theta_1 \xi + \theta_2 \eta} \, \mathrm{d}F(\xi, \eta),$$

where $\theta_j = \mathrm{i}t_j$. Then, by (6.16),

$$\frac{\partial \phi}{\partial \theta_2} - \beta_0 \phi - \beta_1 \frac{\partial \phi}{\partial \theta_1} - \beta_2 \frac{\partial^2 \phi}{\partial \theta_1^2} - \beta_3 \frac{\partial^3 \phi}{\partial \theta_1^3} \tag{6.17}$$

$$= \int (\eta - \beta_0 - \beta_1 \xi - \beta_2 \xi^2 - \beta_3 \xi^3) \mathrm{e}^{\theta_1 \xi + \theta_2 \eta} \, \mathrm{d}F(\xi, \eta) = 0. \tag{6.18}$$

Letting $\psi = \log\, \phi$ and using the relations

$$\frac{\partial \phi}{\partial \theta} = \phi \frac{\partial \psi}{\partial \theta}, \qquad \frac{\partial^2 \phi}{\partial \theta^2} = \phi \left\{ \frac{\partial^2 \psi}{\partial \theta^2} + \left(\frac{\partial \psi}{\partial \theta} \right)^2 \right\},$$

$$\frac{\partial^3 \phi}{\partial \theta^3} = \phi \left\{ \frac{\partial^3 \psi}{\partial \theta^3} + 3 \frac{\partial^2 \psi}{\partial \theta^2} \frac{\partial \psi}{\partial \theta} + \left(\frac{\partial \psi}{\partial \theta} \right)^3 \right\},$$

and (6.18) gives

$$\frac{\partial \psi}{\partial \theta_2} - \beta_0 - \beta_1 \frac{\partial \psi}{\partial \theta_1} - \beta_2 \left\{ \frac{\partial^2 \psi}{\partial \theta_1^2} + \left(\frac{\partial \psi}{\partial \theta_1} \right)^2 \right\}$$

$$-\beta_3 \left\{ \frac{\partial^3 \psi}{\partial \theta_1^3} + 3 \frac{\partial^2 \psi}{\partial \theta_1^2} \frac{\partial \psi}{\partial \theta_1} + \left(\frac{\partial \psi}{\partial \theta_1} \right)^3 \right\} = 0. \tag{6.19}$$

The next step is to substitute the power series in the cumulants (recall (4.38)) for ψ in (6.19) and equate the coefficients of the terms to zero. This yields a set of equations for $\boldsymbol{\beta} = (\beta_0, \beta_1, \beta_2, \beta_3)$, which is linear in the βs but not, in general, in the cumulants. Recall that the cumulants, $\kappa(r, s)$, of ξ and η are the same as those for x and y if r and s are both positive (see (4.40)).

The product cumulant method can be somewhat complicated and the nonlinearity in the cumulants can lead to difficulties. This can be demonstrated by again specializing to the quadratic model, that is, by setting $\beta_3 = 0$ in (6.16). The power series expansion is

$$\psi(t_1, t_2) = \sum_{r=0}^{\infty}\sum_{s=0}^{\infty} \kappa(r, s)\frac{\theta_1^r \theta_2^s}{r!s!}.$$

From (6.19)

$$\begin{aligned} 0 \; = \; & \sum_{r=1}^{\infty}\sum_{s=1}^{\infty} \kappa(r, s)\frac{\theta_1^{r-1}\theta_2^s}{(r-1)!s!} - \beta_0 - \beta_1 \sum_{r=0}^{\infty}\sum_{s=1}^{\infty} \kappa(r, s)\frac{\theta_1^r\theta_2^{s-1}}{r!(s-1)!} \\ & -\beta_2\left\{\sum_{r=0}^{\infty}\sum_{s=2}^{\infty} \kappa(r, s)\frac{\theta_1^r\theta_2^{s-2}}{r!(s-2)!} + \left(\sum_{r=0}^{\infty}\sum_{s=1}^{\infty} \kappa(r, s)\frac{\theta_1^r\theta_2^{s-1}}{r!(s-1)!}\right)^2\right\}. \end{aligned} \tag{6.20}$$

Equating coefficients to zero, in order, yields the following results:

(a) constant terms,
$\kappa(0, 1) - \beta_0 - \beta_1\kappa(1, 0) - \beta_2[\kappa^2(1, 0) + \kappa(2, 0)] = 0;$

(b) terms in θ_1,
$\kappa(1, 1) - \beta_1\kappa(2, 0) - \beta_2[\kappa(3, 0) + 2\kappa(1, 0)\kappa(2, 0)] = 0;$

(c) terms in θ_2,
$\kappa(0, 2) - \beta_1\kappa(1, 1) - \beta_2[\kappa(2, 1) + 2\kappa(1, 0)\kappa(1, 1)] = 0;$

(d) terms in θ_1^2,
$\kappa(2, 1) - \beta_1\kappa(3, 0) - \beta_2[\kappa(4, 0) + 2\kappa^2(2, 0) + 2\kappa(3, 0)\kappa(1, 0)] = 0;$

(e) terms in $\theta_1\theta_2$,
$\kappa(1, 2) - \beta_1\kappa(2, 1) - \beta_2[\kappa(3, 1) + 2\kappa(1, 1)\kappa(2, 0) + 2\kappa(1, 0)\kappa(2, 1)] = 0;$

(f) terms in θ_2^2,
$\kappa(0, 3) - \beta_1\kappa(1, 2) - \beta_2[\kappa(2, 2) + 2\kappa^2(1, 1) + 2\kappa(1, 0)\kappa(1, 2)] = 0.$

(6.21)

Observe that $\kappa(1, 0)$ can be estimated by $\bar{x}$ and $\kappa(0, 1)$ can be estimated by $\bar{y}$. Equation (6.21a) is the only equation involving β_0. In order to use it for estimating β_0 one must estimate not only β_2 but also $\kappa(2, 0)$, which depends on the variance of the error term and cannot be estimated from the observables. Equation (6.21c) involves $\kappa(0, 2)$, which does not occur again. Similarly, (6.21f) involves $\kappa(0, 3)$, and, in fact any equation from terms involving θ_1^r contains the term $\kappa(r, 0)$, which cannot be estimated from the observables unless one assumes that ε (not x and y) is normally distributed (with cumulants beyond the second equal to zero). If ε is not normally distributed, (6.21) cannot be used for estimating the βs. If ε and δ are normally distributed, on the other hand, then $\kappa(r, 0)$ and $\kappa(0, r)$ can be estimated from the data for $r > 2$ because the corresponding cumulants of the errors are zero. Then, using (6.21b) and (6.21e), the term $\kappa(2, 0)$ can be eliminated, and β_1 and β_2 can be estimated from (6.21b), (6.21e), and (6.21f) and the cumulant estimates, $K(0, 1)$, $K(1, 0)$, $K(1, 1)$, $K(1, 2)$, $K(2, 1)$, $K(3, 0)$, $K(0, 3)$, and $K(3, 1)$ (see Section 4.5 and Exercise 6.4).

If ε and δ are not normally distributed, then more equations are needed. (6.20) yields:

(a) terms in $\theta_1\theta_2^2$,
$$\kappa(1,3) - \beta_1\kappa(2,2) - \beta_2[\kappa(3,2) + 2\kappa(1,2)\kappa(2,0) + 4\kappa(1,1)\kappa(2,1) + 2\kappa(1,0)\kappa(2,2)] = 0;$$
(b) terms in $\theta_1^2\theta_2$,
$$\kappa(2,2) - \beta_1\kappa(3,1) - \beta_2[\kappa(4,1) + 2\kappa(1,1)\kappa(3,0) + 4\kappa(2,0)\kappa(2,1) + 2\kappa(1,0)\kappa(3,1)] = 0;$$
(c) terms in θ_1^3,
$$\kappa(3,1) - \beta_1\kappa(4,0) - \beta_2[\kappa(5,0) + 6\kappa(2,0)\kappa(3,0) + 2\kappa(1,0)\kappa(4,0)] = 0;$$
(d) terms in θ_2^3,
$$\kappa(0,4) - \beta_1\kappa(1,3) - \beta_2[\kappa(2,3) + 6\kappa(1,2)\kappa(1,1) + 2\kappa(1,0)\kappa(1,3)] = 0. \tag{6.22}$$

Equations (6.22a) and (6.22b) contain $\kappa(2, 0)$, $\kappa(3, 0)$, and $\kappa(2, 2)$, which one can eliminate using (6.21c) and (6.21e) and thereby estimate β_1 and β_2. Thus $\kappa(2, 0)$ can also be estimated and hence also, from (6.21a), the value of β_0. These equations are nonlinear and one might obtain more than one set of estimators. If there is more than one solution, one can, for example, calculate the orthogonal distance (see Section 6.1.1) of the data to each solution and choose that model with smallest orthogonal distance. If there is no estimate of λ available, then the mean absolute error would probably suffice to select the better solution.

Example 6.2

We performed several Monte Carlo experiments using the above cumulant method on the quadratic structural model. We assumed that ξ was distributed as χ_4^2 and that the errors were normally distributed with mean zero and variance one. This allowed us to use equations (6.21).

We found that a rather large sample size was required in order to get accurate estimates. This is not surprising because we have to estimate several fourth-order cumulants and a nonlinear system of equations. For smaller sample sizes, it was not uncommon for the least-squares estimates to be more accurate than the cumulant method estimates; but as the sample size increased, the cumulant method generally outperformed least squares. In fact, the cumulant method is a method of moments approach to estimation and has no general optimality properties. The only thing recommending the cumulant method is that it leads to consistent estimates. Remember that the cumulant method is not unique. We have used one of the obvious solutions, and different solutions might give different results.

The following is a typical Monte Carlo run. We took 350 observations from the quadratic model

$$\eta_i = \xi_i - \xi_i^2, \qquad i = 1, \ldots, 350.$$

Thus $\beta_0 = 0$, $\beta_1 = 1$, and $\beta_2 = -1$. We obtained the following cumulant estimates:

$$\begin{array}{lll} K(1,0) = 4.084, & K(2,1) = -537.624, & K(3,0) = 32.636, \\ K(0,1) = -20.516, & K(1,2) = 8807.754, & K(0,3) = -146\,388.17, \\ K(1,1) = -92.612, & K(2,2) = 63\,463.61, & K(3,1) = -3274.696. \end{array}$$

From (6.21), we obtain the following set of equations

$$\begin{array}{ll} \text{(a)} & -20.516 - \hat{\beta}_0 - 4.084\,\hat{\beta}_1 - [4.084^2 + K(2,0)]\,\hat{\beta}_2 = 0; \\ \text{(b)} & -92.612 - K(2,0)\,\hat{\beta}_1 - [32.636 + 2 \times 4.084\,K(2,0)]\,\hat{\beta}_2 = 0; \\ \text{(e)} & 8807.754 + 537.624\,\hat{\beta}_1 \\ & \quad -[-3274.696 - 2 \times 92.612\,K(2,0) - 4391.313]\,\hat{\beta}_2 = 0; \\ \text{(f)} & -146\,388.169 - 8807.754\,\hat{\beta}_1 \\ & \quad -[63\,463.606 + 2(-92.612)^2 + 71\,941.7]\,\hat{\beta}_2 = 0. \end{array}$$

Eliminating $K(2,0)$ from (b) and (e) gives

$$8807.75 + 537.624\,\hat{\beta}_1 - \left(\frac{17154. + 6045.0\hat{\beta}_2}{\hat{\beta}_1 + 8.168\,\hat{\beta}_2} - 7666.0 \right) \hat{\beta}_2 = 0. \qquad (6.23)$$

Solving (f) for $\hat{\beta}_1$, we obtain

$$\hat{\beta}_1 = -16.62 - 17.321\hat{\beta}_2, \qquad (6.24)$$

which, when substituted into (6.23), gives

$$2.65 + 14.216\hat{\beta}_2 + 11.278\hat{\beta}_2^2 = 0.$$

This has solutions $\hat{\beta}_{2,1} = -0.22745$ and $\hat{\beta}_{2,2} = -1.0331$. From (6.24), we obtain the corresponding estimates,

$$\hat{\beta}_{1,1} = -16.62 - 17.321(-0.22745) = -12.68,$$
$$\hat{\beta}_{1,2} = -16.62 - 17.321\,(-1.0331) = 1.2743.$$

Finally, from (b) we obtain

$$K(2,0) = \frac{-92.612 - 32.636\,\hat{\beta}_2}{\hat{\beta}_1 + 8.168\,\hat{\beta}_2},$$

and substituting this in (a) and solving gives $\hat{\beta}_{0,1} = 36.38$ and $\hat{\beta}_{0,2} = 0.01$. There are two solutions so we calculate the sample mean absolute errors

$$\frac{1}{350} \sum_{i=1}^{350} |y_i - 36.38 + 12.68x_i + 0.22745x_i^2| = 17.16$$

and

$$\frac{1}{350}\sum_{i=1}^{350}|y_i - 0.01 - 1.2743\,x_i + 1.0331x_i^2| = 5.94$$

and choose the model

$$\eta_i = 0.01 + 1.2743\,\xi_i - 1.0331\xi_i^2.$$

Least squares gives the estimates, $\hat{\beta}_{0R} = -0.100$, $\hat{\beta}_{1R} = 1.465$, and $\hat{\beta}_{2R} = -0.987$, which are not as accurate as the cumulant method estimates.

Note that in the above example no side conditions were required. Recall that the linear structural model was identifiable if ξ was nonnormal (see Section 1.2.1). This subsection indicates that this result continues to hold for the quadratic model. In fact, all that is required for this method is that (x, y) not be normally distributed. If ξ is normally distributed, then the βs are identifiable because then y cannot be normal because it is the sum of two independent variables, η and ε, and if y were normal so then must be η; but η cannot be normal if ξ is. This argument applies to any polynomial structural model with degree $k > 1$. This result contrasts with the linear structural model where β_1 is identifiable if ξ is not normally distributed. The difference is that the linear model preserves normality while the polynomial model destroys it.

Van Montfort (1988, Chapter 2) investigated the estimation of the polynomial structural model via higher-order moments. In particular, he derived a consistent estimate of the βs in the quadratic model when $(\xi, \delta, \varepsilon)$ are jointly normal. The resulting estimator crucially depends on the 'nonlinearity' coefficient, β_2. For example, under the assumption that $E\xi = 0$, the estimate of β_1 is $\hat{c}_{12}/2\hat{c}_{21}$, where $\hat{c}_{ij}$ is defined by (4.36). It is easy to see that $\hat{c}_{12}$ converges to $4\beta_1\beta_2\sigma^4$ and $\hat{c}_{21}$ converges to $2\beta_2\sigma^4$, where σ^2 is the variance of ξ. Consequently this estimate is very unstable when β_2 is close to zero. This is not surprising since the model is not identifiable if $\beta_2 = 0$.

Of course there are many estimates possible for the polynomial model using higher-order moments. They all bear similar properties to those discussed earlier (see Section 4.5).

As mentioned earlier, the primary merit of the cumulant method is the consistency of the estimates. The above example suggests that this consistency requires a high price – large-sample sizes.

Example 6.3

We performed the method of this section on the corn yield versus rainfall data available in the statistics package, S (Becker *et al.*, 1988; see also Ezekiel and Fox, 1959). Table 6.1 gives 38 observations of corn yield in bushels per acre and average seasonal rainfall in inches for six Corn Belt states. Because neither rainfall nor corn yield can be measured precisely and both can be considered random, the structural model is appropriate.

Table 6.1 Corn yield (y) versus rainfall (x) in six Corn Belt states

year	x	y	year	x	y	year	x	y
1890	9.6	24.5	1903	14.1	30.2	1916	9.3	29.7
1891	12.9	33.7	1904	10.6	32.4	1917	9.4	35.0
1892	9.9	27.9	1905	10.0	36.4	1918	8.7	29.9
1893	8.7	27.2	1906	11.5	36.9	1919	9.5	35.2
1894	6.8	21.7	1907	13.6	31.5	1920	11.6	38.3
1895	12.5	31.9	1908	12.1	30.5	1921	12.1	35.2
1896	13.0	36.8	1909	12.0	32.3	1922	8.0	35.5
1897	10.1	29.9	1910	9.3	34.9	1923	10.7	36.7
1898	10.1	30.2	1911	7.7	30.1	1924	13.9	26.8
1899	10.1	32.0	1912	11.0	36.9	1925	11.3	38.0
1900	10.8	34.0	1913	6.9	26.8	1927	11.6	31.7
1901	7.8	19.4	1914	9.5	30.5	1928	10.4	32.6
1902	16.2	36.0	1915	16.5	33.3			

The sample cumulants are given by

$$\begin{array}{lll} K(1,0) = 10.784, & K(2,1) = -6.117, & K(3,0) = 6.138, \\ K(0,1) = 31.916, & K(1,2) = -23.339, & K(0,3) = -68.268, \\ K(1,1) = 3.876, & K(2,2) = 6.244, & K(3,1) = -3.528. \end{array}$$

Using (4.47), the linear cumulant method gives

$$\hat{\beta}_1 = \frac{K(1,2)}{K(2,1)} = \frac{-23.339}{-6.117} = 3.815, \qquad \hat{\beta}_0 = -9.228.$$

This is obviously a very poor fit of the data (see Figure 6.2), which is to be expected since these data do not fit a line and since the sample size is too small to give good estimates of the cumulants. Turning to the quadratic model, we solve the system of equations corresponding to those in Example 6.2 and obtain

$$\hat{\beta}_{0q} = -46.075, \qquad \hat{\beta}_{1q} = 12.568 \qquad \text{and} \qquad \hat{\beta}_{2q} = -0.482,$$

which is a much better fit (see Figure 6.2). The data are plotted in Figure 6.2 along with the linear and quadratic least-squares regressions and the linear and quadratic cumulant method estimates. The quadratic cumulant method gives a curve which looks quite good even compared to the quadratic least-squares curve. Note that one cannot measure the relative merit of a curve, $\hat{y}$, by calculating say the sum of the absolute differences of the vertical distances of the data points to the curve, $MAE = \sum |y_i - \hat{y}(x_i)|$, because this is an ME model and the vertical distances are not being minimized. The quadratic least-squares line uses vertical distances and has smaller MAE than the quadratic ME model line. However, if one uses the orthogonal distance measure of fit in Section 6.1.1 (the **orthdist** algorithm), then the quadratic

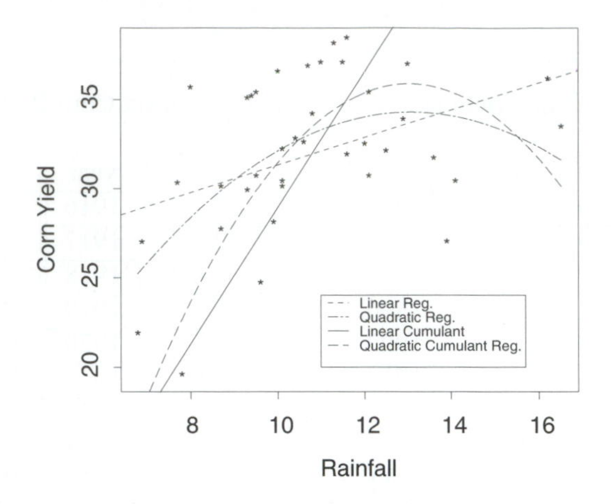

Figure 6.2: Corn yield versus rainfall. Linear and quadratic regression lines obtained via the least-squares and cumulant methods.

least-squares curve has an orthogonal distance of 95.3 and the quadratic ME model curve has an orthogonal distance of 82.1 and the latter curve fits better. Unfortunately, in order to use **orthdist,** we had to pick a value for the relative scale of the errors. We assumed that the variations of the errors were proportional to the scale of the variables and used the very crude estimate, $\sqrt{s_{yy}/s_{xx}} = \sqrt{3.7} = 1.9$.

6.3 Identifiability in Nonlinear ME Models

There are no broad results available on identifiability for general nonlinear structural models. As seen in Section 6.2.2, the βs in the polynomial structural model will be identifiable for polynomials of degree 2 or higher no matter what the distribution of ξ and the errors. Model (6.15) sheds some light on the identifiability of $\boldsymbol{\beta}$ when $\mathbf{x}$ is normally distributed and the error covariance matrix $\boldsymbol{\Sigma}_{\delta\delta}$ is known. If $\mathbf{x}$ is normal, then this implies that both $\boldsymbol{\xi}$ and $\boldsymbol{\delta}$ are also normal and so is the distribution of $\boldsymbol{\xi}$ given $\mathbf{x}$. Under these assumptions, the conditional distribution, $F(\boldsymbol{\xi} \mid \mathbf{x})$, can be estimated from the data. Then

$$E(y \mid \mathbf{x}) = E(g(\boldsymbol{\xi}, \boldsymbol{\beta}) \mid \mathbf{x}) = \int g(\boldsymbol{\xi}, \boldsymbol{\beta})\, \mathrm{d}F(\boldsymbol{\xi} \mid \mathbf{x}) = h(\mathbf{x}, \boldsymbol{\beta}), \qquad (6.25)$$

which has as its only unknown the parameter vector, $\boldsymbol{\beta}$. So now one can fit

$$y_i = h(\mathbf{x}_i, \boldsymbol{\beta}) + u_i \qquad (6.26)$$

using least squares. If h in (6.26) is of a form that permits a consistent estimate of $\boldsymbol{\beta}$, then clearly $\boldsymbol{\beta}$ is identifiable. Thus, the question of identifiability for the nonlinear normal structural model with $\boldsymbol{\Sigma}_{\delta\delta}$ known has been reduced to a question of the identifiability of a corresponding nonlinear ordinary regression model. Of course, as was pointed out earlier, the conditional expectation of y given x might not have an explicit or computable form even when x is normal. The previous argument does not, therefore, completely resolve the issue of the consistent estimation of $\boldsymbol{\beta}$ for this model.

As for the nonlinear functional model, the situation is complicated by the presence of incidental parameters whose number increases with sample size. We have seen in the normal linear functional model that identifiability does not ensure the existence of consistent estimates of the parameters. Hence, attention should focus on consistent estimation rather than identifiability. Because the true values, ξ, are nonstochastic, the distribution of $(\mathbf{x}', y)$ is the same as that of $(\boldsymbol{\delta}', \varepsilon)$ but with a different mean. Again, in principle, one could use ML estimation if one assumes knowledge of the distribution of the errors. The quadratic normal functional relation discussed earlier exemplifies this situation. Because of the presence of incidental parameters, however, there is no guarantee that the solution of the ML equations is the ML estimate. Even if it is, one is still not sure the estimate is consistent. In the quadratic functional model discussed earlier, the likelihood equations were derived and an iterative method for their solution was suggested. The consistency and efficiency of the resulting estimates are still under question, however.

Remember that identifiability is similar to an existence result. That is, just knowing a model is identifiable does not tell one how to estimate its parameters. In fact, if a model is 'close to' a nonidentifiable model, then parameter estimates might be very unstable. On the other hand, if a parameter is not identifiable, a consistent estimate of that parameter cannot exist no matter what procedure is used. The following comment is made in Carroll *et al.* (1995, p. 143): 'Many problems of practical importance actually are identifiable, but only barely so, and estimation without additional data is not practical.'

6.4 Polynomial Model with Equation Error

This section investigates the general polynomial model with equation error; that is, model (6.1) and (6.3) with

$$\eta_i = \beta_0 + \beta_1\xi_i + \beta_2\xi^2 + \ldots + \beta_k\xi_i^k + q_i, \qquad i = 1, \ldots, n. \tag{6.27}$$

We focus our discussion on the functional model; however, the resulting estimators are also consistent estimators for the corresponding structural and ultrastructural models. Thus, the ξs are assumed to be fixed unknown constants and the $(\delta_i, \varepsilon_i)$ are assumed independent of the q_j for all i and j. It is also assumed that $E\delta^r$ is known for $r = 1, \ldots, 2k$.

We follow the treatment of Cheng and Schneeweiss (1996; 1998a). The goal is to find unbiased estimating equations for the parameter vector, $\boldsymbol{\beta} = (\beta_0, \beta_1, \ldots, \beta_k)'$,

$$\sum_{i=1}^{n} \Psi_i(x_i, y_i, \hat{\boldsymbol{\beta}}) = 0 \tag{6.28}$$

(recall Section 3.5).

The model can be written as

$$x_i = \xi_i + \delta_i, \qquad y_i = \beta_0 + \beta_1 \xi_i + \beta_2 \xi_i^2 + \ldots + \beta_k \xi_i^k + e_i,$$

where $e_i = \varepsilon_i + q_i$. If the ξ_i, $i = 1, \ldots, n$, were observable, one could simply employ ordinary least squares and minimize

$$\sum (y_i - \beta_0 - \beta_1 \xi_i - \beta_2 \xi_i^2 - \ldots - \beta_k \xi_i^k)^2.$$

Taking derivatives with respect to the βs leads to the equations

$$\sum_{i=1}^{n} (y_i \xi_i^j - \beta_0 \xi_i^j - \beta_1 \xi_i^{j+1} - \beta_2 \xi_i^{j+2} - \ldots - \beta_k \xi_i^{j+k}) = 0, \qquad j = 0, 1, \ldots, k. \tag{6.29}$$

These will yield unbiased estimating equations if one can find unbiased estimates t_{ir}, say, for each ξ_i^r, $i = 1, \ldots, n$ and $r = 0, 1, \ldots, 2k$. One is, therefore, seeking a variable t_{ir} to be computed from the observable x_i such that $Et_{ir} = \xi_i^r$. For simplicity, we omit the index i when there is no possible confusion. The obvious choice for t_0 is 1 and for t_1 is just x, but the choices for the higher-order t_r are less obvious. The relation $x = \xi + \delta$ suggests a way to look for suitable t_r. Indeed, because $x^r = (\xi + \delta)^r$,

$$Ex^r = \sum_{j=0}^{r} \binom{r}{j} \xi^j \, E\delta^{r-j} = \sum_{j=0}^{r} c_{rj} \xi^j,$$

which is a polynomial in ξ with coefficients $c_{rj} = \binom{r}{j} E\delta^{r-j}$. If Ex^r is replaced by x^r, and ξ^j by a new variable t_j, then the resulting recursive system of equations

$$x^r = \sum_{j=0}^{r} c_{rj} t_j, \qquad r = 0, \ldots, 2k, \tag{6.30}$$

can be easily solved, one at a time, for the t_j, yielding

$$t_r = \sum_{j=0}^{r} a_{rj} x^j, \tag{6.31}$$

say, where the a_{rj} are functions of $E\delta^l$, $l = 0, \ldots, r$, which were assumed known. Obviously $Et_r = \xi^r$ by construction. Using this definition of the t_r, the first five t_r are $t_0 = 1$, $t_1 = x$, $t_2 = x^2 - \sigma_\delta^2$, $t_3 = x^3 - 3x\sigma_\delta^2 - E\delta^3$, and $t_4 = x^4 - 6x^2\sigma_\delta^2 - 4xE\delta^3 - E\delta^4 + 6\sigma_\delta^4$. It can be shown that the polynomial, t_r, in x of degree r, $r = 0, \ldots, 2k$, such that $Et_r = \xi^r$, is uniquely defined (see Exercise 6.2).

6.4.1 The case when δ and ε are independent

If δ and ε are independent, then what is called an **adjusted least-squares estimator** of $\boldsymbol{\beta}$ can be obtained via (6.29) by replacing ξ_i^r by t_{ir}. One can write down a more familiar matrix form as follows. Let $\boldsymbol{\zeta}_i' = (\xi_i^0, \xi_i^1, \ldots, \xi_i^k)$. If the ξ_i are known, then the normal equation of the least-squares estimator is

$$\left(\sum_{i=1}^n \zeta_i \zeta_i'\right) \hat{\beta}_{OLS} = \sum_{i=1}^n \zeta_i y_i.$$

Now let $\mathbf{H}_i = \mathbf{H}_i(x)$ be a $(k+1) \times (k+1)$ matrix, the (p, q)th element of which is $t_{i,p+q}$, $p + q = 0, \ldots, 2k$, that is,

$$\mathbf{H}_i = \begin{pmatrix} t_{i0} & t_{i1} & \cdots & t_{i,k} \\ t_{i1} & t_{i2} & \cdots & t_{i,k+1} \\ \vdots & \vdots & & \vdots \\ t_{i,k} & t_{i,k+1} & \cdots & t_{i,2k} \end{pmatrix}.$$

Obviously $E\mathbf{H}_i = \boldsymbol{\zeta}_i \boldsymbol{\zeta}_i'$. Then the normal equation for the adjusted least-squares estimator $\hat{\boldsymbol{\beta}} = \hat{\boldsymbol{\beta}}_{ALS}$ is

$$\left(\sum_{i=1}^n \mathbf{H}_i\right) \hat{\boldsymbol{\beta}} = \sum_{i=1}^n \mathbf{h}_i, \tag{6.32}$$

where $\mathbf{h}_i = (h_{i0}, h_{i1}, \ldots, h_{ik})'$ with $h_{ir} = t_{ir} y_i$. Thus, if $\sum_{i=1}^n \mathbf{H}_i$ is of full rank (which it usually is in applications), then (6.32) provides an explicit noniterative solution for $\boldsymbol{\beta}$ in a polynomial ME model. No assumption of normality or any other distribution was required. Furthermore, many of the known techniques for solving for ordinary least-squares estimates can be employed here.

In the next subsection, we will discuss the asymptotic behaviour of the adjusted least-squares estimator.

6.4.2 The case when δ and ε are dependent

The general case when δ and ε are dependent needs more care. Observe that

$$Et_{ir} y_i = Et_{ir}\eta_i + Et_{ir}\varepsilon_i \qquad \text{and} \qquad Et_{ir}\eta_i = \xi_i^r(\beta_0 + \ldots + \beta_k \xi_i^k)$$

because q is independent of δ. Note that the term $Et_{ir}\varepsilon_i$ in these equations does not vanish because t_{ir} is function of x (and hence of δ), and one must find its unbiased estimate. Again, we omit the index i in the following discussion. By (6.31),

$$E(t_r \varepsilon) = E\left[\sum_{j=0}^r a_{rj}(\xi + \delta)^j \varepsilon\right] = \sum_{j=0}^r b_{rj} \xi^j$$

with coefficients

$$b_{rj} = \sum_{s=j}^r a_{rs} \binom{s}{j} E\left(\delta^{s-j} \varepsilon\right), \tag{6.33}$$

which depend on the $E\delta^l$ and $E(\delta^l \varepsilon)$, $l = 0, \ldots, r$, whose values are assumed known a priori. The coefficients, a_{rs}, are determined via (6.33). A natural estimate of $E(t_r\varepsilon)$ is, therefore, $\hat{E}(t_r\varepsilon) = \sum_{j=0}^{r} b_{rj}t_j$. Hence, if one defines

$$h_r^* = t_r y - \hat{E}(t_r\varepsilon) = t_r y - \sum_{j=0}^{r} b_{rj}t_j, \tag{6.34}$$

then $Eh_r^* = \xi^r(\beta_0 + \beta_1\xi + \ldots + \beta_k\xi^k)$.

The first three elements of $\mathbf{h}^*$ are $h_0^* = y$, $h_1^* = xy - \sigma_{\delta\varepsilon}$, and $h_2^* = (x^2 - \sigma_\delta^2)y - E(\delta^2\varepsilon) - 2\sigma_{\delta\varepsilon}x$.

The normal equation for the adjusted least-squares estimator when δ and ε are dependent becomes

$$\left(\sum_{i=1}^{n} \mathbf{H}_i\right)\hat{\boldsymbol{\beta}} = \sum_{i=1}^{n} \mathbf{h}_i^*, \tag{6.35}$$

where $\mathbf{h}_i^* = (h_{i0}^*, h_{i1}^*, \ldots, h_{ik}^*)'$. Obviously, if δ and ε are independent, then (6.35) reduces to (6.32).

The unknown equation error variance, σ_q^2, is not identifiable unless σ_ε^2 is known, but $\sigma_e^2 = \sigma_\varepsilon^2 + \sigma_q^2$ can be estimated by the least-squares estimate,

$$\hat{\sigma}_{e,OLS}^2 = \frac{1}{n}\left(\sum y_i^2 - \sum y_i \boldsymbol{\zeta}_i' \hat{\boldsymbol{\beta}}_{OLS}\right).$$

Replacing $\sum y_i \boldsymbol{\zeta}_i'$ and $\hat{\boldsymbol{\beta}}_{OLS}$ by $\mathbf{h}_i^{*\prime}$ and $\hat{\boldsymbol{\beta}}$ respectively, one obtains the adjusted least-squares estimate,

$$\hat{\sigma}_e^2 = \frac{1}{n}\sum (y_i^2 - \mathbf{h}_i^{*\prime}\hat{\boldsymbol{\beta}}). \tag{6.36}$$

If σ_ε^2 is known, then obviously $\hat{\sigma}_q^2 = \hat{\sigma}_e^2 - \sigma_\varepsilon^2$.

In order to compute the estimators $\hat{\boldsymbol{\beta}}$ and $\hat{\sigma}_e^2$ one needs $E\delta^r$ for $r = 1, \ldots, 2k$ and $E(\delta^r\varepsilon)$ for $r = 1, \ldots, k$. The adjusted least-squares estimators $\hat{\beta}$ and $\hat{\sigma}_e^2$ are consistent provided that $\lim_{n\to\infty} n^{-1}\sum \xi_i^r$ exists for $r = 0, \ldots, 2k$ and that $\lim_{n\to\infty} n^{-1}\sum \mathbf{H}_i$ is nonsingular. Moreover, under similar conditions (see Huber, 1967; or Carroll *et al.*, 1995, p. 262), $\sqrt{n}(\hat{\boldsymbol{\beta}} - \boldsymbol{\beta})$ is asymptotically normal with mean zero and covariance matrix $\boldsymbol{\Sigma}_\beta$, which can be estimated by

$$\hat{\boldsymbol{\Sigma}}_\beta = \frac{1}{n}\bar{\mathbf{H}}^{-1}\hat{\mathbf{V}}\bar{\mathbf{H}}^{-1},$$

with

$$\hat{\mathbf{V}} = \frac{1}{n}\sum_{i=1}^{n}(\mathbf{H}_i\hat{\boldsymbol{\beta}} - \mathbf{h}_i^*)(\mathbf{H}_i\hat{\boldsymbol{\beta}} - \mathbf{h}_i^*)',$$

where $\bar{\mathbf{H}} = n^{-1}\sum \mathbf{H}_i$.

6.4.3 The adjusted least-squares estimator under normality

If (δ, ε) are jointly normally distributed, the adjusted least-squares estimator can be simplified considerably and only $\sigma_{\delta\varepsilon}$ and σ_δ^2 need be known. In this case, simple formulae for the computation of t_r and $\hat{E}(t_r\varepsilon)$ and hence of $\mathbf{H}$ and $\mathbf{h}^*$ can be derived. In particular, the t_r, $r = 0, 1, 2, \ldots, 2k$, can be computed by a simple recursive formula.

Let $H_r(z)$ be the rth Hermite polynomial (Stuart and Ord, 1984, pp. 226 ff.):

$$H_0(z) = 1, \qquad H_1(z) = z, \qquad H_r(z) = zH_{r-1}(z) - (r-1)H_{r-2}(z), \quad r \geq 2.$$

Stulajter (1978) proved that, if z is normally distributed with mean μ and variance σ^2, then $E\{P_r(z)\} = \mu^r$, where $P_r(z) = \sigma^r H_r(z/\sigma)$. This fact can be used to obtain a recursive formula for t_r when x is normally distributed with mean ξ and variance σ_δ^2. A simpler proof of Stulajter's result is in Cheng and Schneeweiss (1996), which is given in the following lemma.

Lemma 6.1 *If δ is normally distributed with mean zero and variance σ_δ^2, then*

$$t_{j+1} = xt_j - \sigma_\delta^2 j t_{j-1} \tag{6.37}$$

for $j = 0, 1, 2, \ldots, 2k-1$, with $t_0 = t_{-1} = 1$.

Proof. By (6.30) and the definition of the coefficients c_{rj},

$$x^r = \sum_{j=0}^{r} \binom{r}{j} t_j E\delta^{r-j}.$$

Computing x^{r+1} using this expression, it follows that

$$\sum_{j=0}^{r+1} \binom{r+1}{j} t_j E\delta^{r+1-j} = \sum_{j=0}^{r} \binom{r}{j} x t_j E\delta^{r-j}. \tag{6.38}$$

On the other hand, under normality $\sigma_\delta^2 E\delta^{r-j-1} = E\delta^{r-j+1}/(r-j)$, so that

$$\begin{aligned}
&\sum_{j=0}^{r} \binom{r}{j} (t_{j+1} + \sigma_\delta^2 j t_{j-1}) E\delta^{r-j} \\
&= \sum_{j=1}^{r+1} \binom{r}{j-1} t_j E\delta^{r-j+1} + \sigma_\delta^2 \sum_{j=0}^{r-1} (j+1) \binom{r}{j+1} t_j E\delta^{r-j-1} \\
&= \sum_{j=1}^{r+1} \binom{r}{j-1} t_j E\delta^{r-j+1} + \sum_{j=0}^{r-1} \binom{r}{j} t_j E\delta^{r-j+1} = \sum_{j=0}^{r+1} \binom{r+1}{j} t_j E\delta^{r+1-j}
\end{aligned}$$

because $\binom{r}{j-1} + \binom{r}{j} = \binom{r+1}{j}$.

Together with (6.38), this implies

$$\sum_{j=0}^{r} \binom{r}{j} (t_{j+1} + \sigma_\delta^2 j t_{j-1}) E\delta^{r-j} = \sum_{j=0}^{r} \binom{r}{j} x t_j E\delta^{r-j}. \tag{6.39}$$

Now suppose (6.37) is true for $j < r$ (it is trivially true for $j = 0$); then substituting (6.37) into the left-hand side of (6.39) for $j < r$ and subtracting equal terms on both sides yields $t_{r+1} + \sigma_\delta^2 r t_{r-1} = x t_r$, which is (6.37) for $j = r$. By induction, this proves the assertion. □

For example, using Lemma 6.1, the first six values for t_{ir} are

$$t_{i1} = x_i, \qquad t_{i2} = x_i^2 - \sigma_\delta^2, \qquad t_{i3} = x_i^3 - 3x_i\sigma_\delta^2, \qquad t_{i4} = x_i^4 - 6x_i^2\sigma_\delta^2 + 3\sigma_\delta^4,$$

$$t_{i5} = x_i^5 - 10x_i^3\sigma_\delta^2 + 15x_i\sigma_\delta^4, \qquad t_{i6} = x_i^6 - 15x_i^4\sigma_\delta^2 + 45x_i^2\sigma_\delta^4 - 15\sigma_\delta^6.$$

As for the estimate of $E(t_r\varepsilon)$, one can prove the following.

Lemma 6.2 *If* (δ, ε) *are jointly normal, then* $\hat{E}(t_r\varepsilon) = \sigma_{\delta\varepsilon} r t_{r-1}$.

Proof. First note that, using (6.31), $E(t_r\varepsilon) = \sum_{j=0}^{r} a_{rj} E[(\xi + \delta)^j \varepsilon]$ can be written as $E(t_r\varepsilon) = \sum_{l=0}^{r} d_{rl}(\xi) E(\delta^l \varepsilon)$, where the $d_{rl}(\xi)$ are polynomials in ξ. But also

$$E(t_r\delta) = \Sigma a_{rj} E(\xi + \delta)^j \delta = \Sigma d_{rl}(\xi) E\delta^{l+1}.$$

Under the normality assumption $E(\delta^l\varepsilon) = \sigma_{\delta\varepsilon} E\delta^{l+1}/\sigma_\delta^2$ and hence $E(t_r\varepsilon) = \sigma_{\delta\varepsilon} E(t_r\delta)/\sigma_\delta^2$. Now, by (6.37),

$$\begin{aligned} E(t_r\delta) &= E(t_r x - t_r\xi) = E(t_{r+1} + \sigma_\delta^2 r t_{r-1} - t_r\xi) \\ &= \xi^{r+1} + \sigma_\delta^2 r \xi^{r-1} - \xi^{r+1} = \sigma_\delta^2 r \xi^{r-1}. \end{aligned}$$

It follows that $E(t_r\varepsilon) = \sigma_{\delta\varepsilon} r \xi^{r-1}$. □

Alternatively, Chan and Mak (1985) (see also Exercise 6.3) used series expansions and term-by-term differentiations to obtain a general formula for t_r, namely

$$t_r = r! \sum\sum (-\tfrac{1}{2}\sigma_\delta^2)^l \frac{x^m}{l!m!} \tag{6.40}$$

where the summation is taken over all nonnegative integers l, m with $2l + m = r$.

Example 6.4
Consider the quadratic functional relationship. No distributional assumptions are made on the errors. The unbiased estimating equations have the form

$$\begin{gathered} \beta_0 + \beta_1\bar{x} + \beta_2(m_{20} - \sigma_\delta^2) = \bar{y}, \\ \beta_0\bar{x} + \beta_1(m_{20} - \sigma_\delta^2) + \beta_2(m_{30} - 3\bar{x}\sigma_\delta^2 - E\delta^3) = m_{11} - \sigma_{\delta\varepsilon}, \\ \beta_0(m_{20} - \sigma_\delta^2) + \beta_1(m_{30} - 3\bar{x}\sigma_\delta^2 - E\delta^3) \\ +\beta_2(m_{40} - 6m_{20}\sigma_\delta^2 - 4\bar{x}E\delta^3 - E\delta^4 + 6\sigma_\delta^4) \\ = m_{21} - \bar{y}\sigma_\delta^2 - E(\delta^2\varepsilon) - 2\sigma_{\delta\varepsilon}\bar{x}, \end{gathered}$$

where $m_{kl} = n^{-1} \sum x_i^k y_i^l$ for nonnegative integers k and l and we use the traditional notation $\bar{x}$ and $\bar{y}$ instead of m_{10} and m_{01}, respectively.

When $\beta_2 = 0$, that is, in the linear case, the adjusted least-squares estimator reduces to the familiar estimators

$$\hat{\beta}_0 = \bar{y} - \hat{\beta}_1 \bar{x}, \qquad \hat{\beta}_1 = \frac{s_{xy} - \sigma_{\delta\varepsilon}}{s_{xx} - \sigma_\delta^2}.$$

This is derived as a modified least squares estimator in Section 3.5 and is also the ML estimator for the normal structural model provided the constraints following (1.64) are satisfied.

Example 6.5

Consider the cubic functional relationship. Assuming that δ is normal and independent of ε, the unbiased estimating equations become

$$\begin{aligned}
\beta_0 + \beta_1\bar{x} + \beta_2(m_{20} - \sigma_\delta^2) + \beta_3(m_{30} - 3\bar{x}\sigma_\delta^2)\} &= \bar{y}, \\
\beta_0\bar{x} + \beta_1(m_{20} - \sigma_\delta^2) + \beta_2(m_{30} - 3\bar{x}\sigma_\delta^2) + \beta_3(m_{40} - 6m_{20}\sigma_\delta^2 + 3\sigma_\delta^4)\} &= m_{11}, \\
\beta_0(m_{20} - \sigma_\delta^2) + \beta_1(m_{30} - 3\bar{x}\sigma_\delta^2) + \beta_2(m_{40} - 6m_{20}\sigma_\delta^2 + 3\sigma_\delta^4) & \\
+\beta_3(m_{50} - 10m_{30}\sigma_\delta^2 + 15\bar{x}\sigma_\delta^4) &= m_{21} - \bar{y}\sigma_\delta^2, \\
\beta_0(m_{30} - 3\bar{x}\sigma_\delta^2) + \beta_1(m_{40} - 6m_{20}\sigma_\delta^2 + 3\sigma_\delta^4) + \beta_2(m_{50} - 10m_{30}\sigma_\delta^2 + 15\bar{x}\sigma_\delta^4) & \\
+\beta_3(m_{60} - 15m_{40}\sigma_\delta^2 + 45m_{20}\sigma_\delta^4 - 15\sigma_\delta^6) &= m_{31} - 3m_{11}\sigma_\delta^2.
\end{aligned}$$

Note that, there is no need to impose any distributional assumption on ε and q except for the existence of σ_ε^2 and σ_q^2. A similar property is true for the classical regression model. Moreover, the adjusted least-squares estimator reduces to the ordinary least-squares estimator when $\sigma_\delta^2 = 0$.

6.5 The Polynomial Functional Relationship without Equation Error

This section investigates the estimation of the parameters of the general polynomial functional relation without equation error. The model under investigation is (6.1)–(6.2) with

$$\eta_i = \beta_0 + \beta_1\xi_i + \beta_2\xi^2 + \ldots + \beta_k\xi_i^k, \qquad i = 1, \ldots, n,$$

which is a generalization of (6.4) and (6.16).

If the $E\delta^r$, $r = 1, \ldots, 2k$ are known, then the methods of Section 6.4.2 can be used here. We will now introduce another estimator, which requires knowledge of $E\delta^r$, $r = 1, \ldots, 2k - 1$, and σ_ε^2. This new approach will lead to a method for the case in which the error covariance matrix known up to a scalar multiple in the polynomial normal ME model with no equation error.

6.5.1 A method of moments estimator

First consider the unbiased estimating equation approach similar to that used in Section 6.4. The goal is to find a system of $k + 1$ unbiased estimating equations for

$\boldsymbol{\beta} = (\beta_0, \beta_1, \ldots, \beta_k)'$. Assume that $(\delta_i, \varepsilon_i)$ are independent identically distributed with zero mean and that all the involved moments of the errors are known and finite. We follow the treatment of Cheng and Schneeweiss (1996; 1998a).

Equation (6.29) suggests a way to find such a system. In fact, keeping the first k equations in (6.29) and replacing the last equation by a suitable alternative provides the unbiased estimating equations. More precisely, consider the following set of equations:

$$\sum_{i=1}^{n}(y_i\xi_i^j - \beta_0\xi_i^j - \beta_1\xi_i^{j+1} - \beta_2\xi_i^{j+2} - \ldots - \beta_k\xi_i^{j+k}) = 0, \qquad j = 0, 1, \ldots, k-1, \tag{6.41}$$

together with

$$\sum(y_i - \beta_0 - \beta_1\xi_i - \ldots - \beta_k\xi_i^k)\eta_i = 0. \tag{6.42}$$

For the time being, assume that δ and ε are independent. Obviously, (6.41)–(6.42) are not solvable because ξ_i^r and η_i are all unknown. But, one can replace ξ_i^r by t_{ir}, the unbiased estimate of ξ_i^r defined by (6.31), and η_i by y_i, respectively. This is only a first step because $Ey_i^2 = \eta_i^2 + \sigma_\varepsilon^2$ and one, therefore, needs to subtract σ_ε^2 in order to make the last equation unbiased. Hence, the system of $k+1$ unbiased estimating equations becomes

$$\sum(y_i t_{ij} - \beta_0 t_{ij} - \beta_1 t_{i,j+1} - \ldots - \beta_k t_{i,j+k}) = 0, \qquad j = 0, 1, \ldots, k-1,$$

and

$$\sum(y_i^2 - \beta_0 y_i - \beta_1 t_{i1} y_i - \ldots - \beta_k t_{ik} y_i) = n\sigma_\varepsilon^2.$$

When δ and ε are dependent, the situation can be treated in a manner similar to that of Section 6.4. More specifically, it is necessary to subtract the bias $\hat{E}t_{ij}\varepsilon_i$ for $j = 0, \ldots, k-1$ from each of the first k equations and $\hat{E}t_{ir}\varepsilon_i$ from each term, t_{ir}, for $r = 1, \ldots, k$ in the $(k+1)$th equation. The resulting unbiased estimating equations become

$$\sum\{(y_i t_{ij} - \hat{E}(t_{ij}\varepsilon_i)) - \beta_0 t_{ij} - \beta_1 t_{i,j+1} - \ldots - \beta_k t_{i,j+k}\} = 0, \qquad j = 0, 1, \ldots, k-1, \tag{6.43}$$

and

$$\sum\{y_i^2 - \beta_0 y_i - \beta_1(t_{i1}y_i - \hat{E}(t_{i1}\varepsilon_i)) - \ldots - \beta_k(t_{ik}y_i - \hat{E}(t_{ik}\varepsilon_i))\} = n\sigma_\varepsilon^2, \tag{6.44}$$

where $\hat{E}(t_{ir}\varepsilon_i)$ is defined in Section 6.4.

The proposed method saves having to know $E\delta^{2k}$, but compensates by having to know σ_ε^2. This is the main reason why this method only applies to the no-equation-error model in which knowledge of $\sigma_e^2 = \sigma_\varepsilon^2 + \sigma_q^2$ is needed. It does not seem to be realistic in practice to assume the equation error variance σ_q^2 is known.

Example 6.6 (*Example 6.4 continued*)
Again consider the quadratic functional relation with no equation error. We do not make any distributional assumption on the errors. The unbiased estimating equations

of the proposed method of moments estimator have the following form:

$$\beta_0 + \beta_1\bar{x} + \beta_2(m_{20} - \sigma_\delta^2) = \bar{y},$$
$$\beta_0\bar{x} + \beta_1(m_{20} - \sigma_\delta^2) + \beta_2(m_{30} - 3\bar{x}\sigma_\delta^2 - E\delta^3) = m_{11} - \sigma_{\delta\varepsilon},$$
$$\beta_0\bar{y} + \beta_1(m_{11} - \sigma_{\delta\varepsilon}) + \beta_2(m_{21} - \sigma_\delta^2\bar{y} - E(\delta^2\varepsilon) - 2\bar{x}\sigma_{\delta\varepsilon}) = m_{02} - \sigma_\varepsilon^2.$$

Comparing this to Example 6.4 in the previous section, one finds that knowledge of $E\delta^4$ is not needed but information about σ_ε^2 is.

If $\beta_2 = 0$, then

$$\hat{\beta}_0 = \bar{y} - \hat{\beta}_1\bar{x} \qquad \text{and} \qquad \hat{\beta}_1 = (s_{yy} - \sigma_\varepsilon^2)/(s_{xy} - \sigma_{\delta\varepsilon}),$$

which, if $\sigma_{\delta\varepsilon} = 0$, is just (1.38).

Chan and Mak (1985, p. 512) obtained an estimator for $\boldsymbol{\beta}$ for the quadratic functional relation by a quite different approach and claimed that the resulting estimator is consistent without assuming any distributional assumption except that the error covariance matrix is known. In fact, the consistency holds only if both $E\delta^2\varepsilon$ and $E\delta^3$ are zero. Furthermore, it can be shown that this estimator is equivalent to the estimator derived in the previous example when both $E\delta^2\varepsilon$ and $E\delta^3$ are zero.

Example 6.7 (*Example 6.5 continued*)
Consider the cubic functional relation with no equation error. Assuming that δ is normal and independent of ε, then the unbiased estimating equations for the cubic functional relation become

$$\beta_0 + \beta_1\bar{x} + \beta_2(m_{20} - \sigma_\delta^2) + \beta_3(m_{30} - 3\bar{x}\sigma_\delta^2)\} = \bar{y},$$
$$\beta_0\bar{x} + \beta_1(m_{20} - \sigma_\delta^2) + \beta_2(m_{30} - 3\bar{x}\sigma_\delta^2) + \beta_3(m_{40} - 6m_{20}\sigma_\delta^2 + 3\sigma_\delta^4)\} = m_{11},$$
$$\beta_0(m_{20} - \sigma_\delta^2) + \beta_1(m_{30} - 3\bar{x}\sigma_\delta^2) + \beta_2(m_{40} - 6m_{20}\sigma_\delta^2 + 3\sigma_\delta^4)$$
$$+\beta_3(m_{50} - 10m_{30}\sigma_\delta^2 + 15\bar{x}\sigma_\delta^4) = m_{21} - \bar{y}\sigma_\delta^2,$$
$$\beta_0\bar{y} + \beta_1 m_{11} + \beta_2(m_{21} - \sigma_\delta^2\bar{y}) + \beta_3(m_{31} - 3m_{11}\sigma_\delta^2) = m_{02} - \sigma_\varepsilon^2.$$

6.5.2 Extended generalized least squares

Wolter and Fuller (1982a) proposed an estimator for β in the quadratic functional relation when (δ, ε) are normally distributed with known covariance matrix. Because this estimator reduces to the generalized least-squares estimator proposed by Sprent (1966) when $\beta_2 = 0$ and because it is essentially a weighted least-squares method (for a polynomial model of any degree k), we shall call it the extended generalized least squares or simply the generalized least-squares estimator.

Moon and Gunst (1995) extended the Wolter–Fuller's generalized least-squares estimator to the polynomial functional relation of any degree but they still retain the crucial assumption of normality. Cheng and Schneeweiss (1998b) extended this generalized least-squares estimator to a general distributional assumption.

6.5.3 The error covariance matrix known up to a proportionality factor

Recall that in the linear no equation error ME model, the typical identifiability assumption is that the error covariance matrix is known up to a proportionality factor instead of known completely. Up to now, in Section 6.5, it was assumed that the error covariance matrix was known completely (plus certain higher-order moments of the errors). In the polynomial model, it is also possible to relax the assumption to knowing the error covariance matrix up to a proportionality factor, but only if normality is assumed. The basic idea is as follows. In Section 6.5.1 a method of moments estimator was proposed, which requires the solution of $k+1$ linear equations in the βs, (6.43) and (6.44). Now suppose that one knows the error covariance matrix only up to a scalar multiple. There is no loss in generality in assuming that the unknown; quantity is either σ_δ^2 or σ_ε^2. Suppose it is assumed that σ_ε^2 is unknown; then one retains the $k+1$ equations, (6.43) and (6.44), and adds another equation to estimate σ_ε^2. Because

$$\sigma_\varepsilon^2 = n^{-1} \sum (y_i - \beta_0 - \beta_1 \xi_i - \ldots - \beta_k \xi_i^k)^2, \tag{6.45}$$

one can use this to estimate σ_ε^2. There are a few points that need to be mentioned. First one needs to expand (6.45) and then replace the ξ_i^r by their corresponding unbiased estimates, t_{ir}, defined by (6.31), and the $y_i \xi_i^r$ by their unbiased estimators, $y_i t_{ir} - \hat{E}(t_{ir}\varepsilon_i)$, defined in Section 6.4.2. Second, because σ_ε^2 is unknown in (6.43) and (6.44), one needs to solve (6.43) and (6.44) and for the estimate of σ_ε^2. Finally, an explicit solution is not available so one needs to use an iterative method to find the estimators of the βs and σ_ε^2. Note that a similar device works for the Wolter–Fuller extended generalized least-squares estimator, as reported in Cheng and Schneeweiss (1998b).

6.5.4 Approximation to the maximum likelihood estimator

In the introduction to this chapter, we discussed the ML estimation of the normal quadratic functional relationship. Now we briefly investigate the ML estimator of the general normal functional model. The following discussions are based on Wolter and Fuller (1982b) and Amemiya and Fuller (1988). We adopt their notation, which is slightly different from ours.

Wolter and Fuller (1982b) derived a one-step iterative approximation to the ML estimator of the general nonlinear functional ME model under the normality assumption. Suppose that the no-equation-error functional model,

$$\begin{aligned} &\text{(a)} \quad \eta_t = g(\boldsymbol{\xi}_t; \boldsymbol{\beta}), \\ &\text{(b)} \quad y_t = \eta_t + \varepsilon_t, \qquad \mathbf{x}_t = \xi_t + \delta_t, \end{aligned} \tag{6.46}$$

holds, where $\boldsymbol{\xi}_t$, $\mathbf{x}_t$ and $\boldsymbol{\delta}_t$ are $p \times 1$ vectors and $\boldsymbol{\beta} = (\beta_0, \beta_1, \ldots, \beta_k)'$. Let $\{a_n; n = 1, \ldots, \infty\}$ and $\{b_n; n = 1, \ldots, \infty\}$ be sequences of positive real numbers such that $n = a_n b_n$ for all n and assume that a sequence of experiments indexed by n exists. The observations $(y_{nt}, \mathbf{x}_{nt})$ are such that

$$y_{nt} = \eta_t + \varepsilon_{nt}, \qquad \mathbf{x}_{nt} = \xi_t + \delta_{nt}, \qquad t = 1, 2, \ldots, b_n.$$

Assume that the $(\varepsilon_{nt}, \boldsymbol{\delta}_{nt})$ are independent, normal random variables with mean zero and known covariance matrix $\boldsymbol{\Omega}_n = a_n^{-1}\boldsymbol{\Phi}$, where $\boldsymbol{\Phi}$ is a fixed positive definite matrix. The ML estimators of $\boldsymbol{\xi}_t$ and $\boldsymbol{\beta}$ minimize

$$\begin{aligned}\sum_{t=1}^{b_n} Q(\boldsymbol{\beta}, \boldsymbol{\xi}_t; y_{nt}, \mathbf{x}_{nt}) &= \sum_{t=1}^{b_n}\{(\mathbf{x}_{nt} - \boldsymbol{\xi}_t)', y_{nt} - g(\boldsymbol{\xi}_t; \boldsymbol{\beta})\}\boldsymbol{\Omega}_n^{-1} \\ &\quad \times\{(\mathbf{x}_{nt} - \boldsymbol{\xi}_t)', y_{nt} - g(\boldsymbol{\xi}_t; \boldsymbol{\beta})\}'. \end{aligned} \tag{6.47}$$

Wolter and Fuller then propose an iterative procedure for minimizing (6.47), which begins with an initial value for $\boldsymbol{\beta}$ equal to the ordinary least-squares solution, $\boldsymbol{\beta}_R$, of the nonlinear model.

The geometrical interpretation of (6.47) is as follows. Suppose that $\boldsymbol{\beta}$ is *fixed* at some value $\bar{\boldsymbol{\beta}}$, such as a preliminary estimate of $\boldsymbol{\beta}$. Let $\boldsymbol{\Phi}$ be the identity matrix, and let $\hat{\boldsymbol{\xi}}_t$ minimize $Q(\bar{\boldsymbol{\beta}}, \boldsymbol{\xi}_t; \mathbf{y}_{nt}, \mathbf{x}_{nt})$. Then $\hat{\boldsymbol{\xi}}_t$ is the point on $g(\boldsymbol{\xi}; \bar{\boldsymbol{\beta}})$ that is the minimum distance from $(y_t, \mathbf{x}_t)$. In other words, it is sort of an orthogonal regression estimate for the nonlinear model.

Amemiya and Fuller (1988) considered the ML solution of the normal implicit functional relationship.

$$f(\eta_t, \boldsymbol{\xi}_t; \boldsymbol{\beta}) = 0, \tag{6.48}$$

for $t = 1, 2, \ldots, b_n$. Note that the explicit model is a special case of the implicit model. Again both a_n and b_n tend to infinity and $n = a_n b_n$. For example, if a_n replicate observations are made at each b_n points, then the total number of observations is n. Large-sample results are proved under the assumption that a_n tends to infinity faster than b_n. Because $\boldsymbol{\Omega}_n = a_n^{-1}\boldsymbol{\Phi}$, the error covariance matrix tends to zero faster than the number of the incidental parameters tends to infinity, in order for the asymptotic properties to hold. Such large-sample, small error variance consistency is different from the usual definition of consistency. Amemiya and Fuller (1988) also showed that the assumption of $\boldsymbol{\Omega}_n$ is known can be relaxed to knowing $\boldsymbol{\Omega}_n$ up to a scalar multiple.

Hsiao (1989, p. 160) comments that the method given in this subsection 'yields useful approximations to the properties of estimators when the error variances are small and when the sample size is large', but they are 'not always applicable to data sets commonly encountered by economists'. He states that while replications do occur in economics, they do not usually increase in number with sample size. Note, further, that $\boldsymbol{\Omega}_n = a_n^{-1}\boldsymbol{\Phi}$ implies that the measurement error variance is going to zero, which means that the model is getting close to an ordinary regression model.

6.5.5 An iterative algorithm

The one-step approximation to the ML estimator, described in Wolter and Fuller (1982b), was applied to the polynomial normal no-equation-error model in Monte Carlo simulations by Moon and Gunst (1995). The implementation of the one-step approximation to the ML estimator is equivalent to the algorithm described in Britt and

Luecke (1973) for a normal polynomial functional model. We now briefly describe this algorithm.

Assume the model (6.48) and (6.46b) $(\boldsymbol{\delta}', \varepsilon)' \sim N(\mathbf{0}, \boldsymbol{\Omega})$, $\boldsymbol{\Omega}^{((p+1)\times(p+1))}$ known, and that $f(\eta_t, \boldsymbol{\xi}_t; \boldsymbol{\beta})$ has continuous first and second derivatives. The estimators are obtained by minimizing

$$Q = \sum_{i=1}^{n} \left(\begin{pmatrix} \mathbf{x}_i \\ y_i \end{pmatrix} - \begin{pmatrix} \boldsymbol{\xi}_i \\ \eta_i \end{pmatrix} \right)' \boldsymbol{\Omega}^{-1} \left(\begin{pmatrix} \mathbf{x}_i \\ y_i \end{pmatrix} - \begin{pmatrix} \boldsymbol{\xi}_i \\ \eta_i \end{pmatrix} \right) + \sum_{i=1}^{n} \tau_i f(\eta_i, \boldsymbol{\xi}_i; \boldsymbol{\beta}) \quad (6.49)$$

with respect to $\boldsymbol{\beta} = (\beta_0, \beta_1, \ldots, \beta_k)'$ and $\boldsymbol{\xi}_1, \boldsymbol{\xi}_2, \ldots, \boldsymbol{\xi}_n, \eta_1, \ldots, \eta_n$, where the τ_i are Lagrangian multipliers. Let $\boldsymbol{\beta}_0^{((k+1)\times 1)}, \boldsymbol{\xi}_{10}^{(p\times 1)}, \boldsymbol{\xi}_{20}^{(p\times 1)}, \ldots, \boldsymbol{\xi}_{n0}^{(p\times 1)}, \eta_{10}, \ldots, \eta_{n0}$ be initial values for the iterative scheme. Expanding f as a Taylor series about these initial values gives

$$\begin{aligned} f(\eta_t, \boldsymbol{\xi}_t; \boldsymbol{\beta}) \;\cong\; & f(\eta_{t0}, \boldsymbol{\xi}_{t0}; \boldsymbol{\beta}_0) + f_{\beta}(\eta_{t0}, \boldsymbol{\xi}_{t0}; \boldsymbol{\beta}_0)'(\boldsymbol{\beta} - \boldsymbol{\beta}_0) \\ & + f_{\xi}(\eta_{t0}, \boldsymbol{\xi}_{t0}; \boldsymbol{\beta}_0)'(\boldsymbol{\xi}_i - \boldsymbol{\xi}_{i0}) + f_{\eta}(\eta_{t0}, \boldsymbol{\xi}_{t0}; \boldsymbol{\beta}_0)(\eta_i - \eta_{i0}), \end{aligned} \quad (6.50)$$

where $f_{\beta}(\eta_{t0}, \boldsymbol{\xi}_{t0}; \boldsymbol{\beta}_0)$ is the vector of the partial derivatives of $f(\eta_t, \boldsymbol{\xi}_t; \boldsymbol{\beta})$ with respect to $\boldsymbol{\beta}$ evaluated at $\boldsymbol{\beta}_0, \boldsymbol{\xi}_{10}, \boldsymbol{\xi}_{20}, \ldots, \boldsymbol{\xi}_{n0}, \eta_0, \ldots, \eta_{n0}$, $f_{\xi}(\eta_{t0}, \boldsymbol{\xi}_{t0}; \boldsymbol{\beta}_0)$ is the vector of partials with respect to $\boldsymbol{\xi}$ evaluated at the initial values, and $f_{\eta}(\eta_{t0}, \boldsymbol{\xi}_{t0}; \boldsymbol{\beta}_0)$ is defined analogously. Replacing $f(\eta_t, \boldsymbol{\xi}_t; \boldsymbol{\beta})$ by the right-hand side of (6.50) in (6.49) and then minimizing Q by setting its derivatives equal to zero gives the updated values of the estimates:

$$\begin{aligned} \hat{\boldsymbol{\beta}} \;=\; & \boldsymbol{\beta}_0 - \left\{ \sum_{i=1}^{n} f_{\beta}(\eta_{i0}, \boldsymbol{\xi}_{i0}; \boldsymbol{\beta}_0) v_{i0}^{-1} f_{\beta}(\eta_{i0}, \boldsymbol{\xi}_{i0}; \boldsymbol{\beta}_0)' \right\}^{-1} \\ & \times \left\{ \sum_{i=1}^{n} f_{\beta}(\eta_{i0}, \boldsymbol{\xi}_{i0}; \boldsymbol{\beta}_0) v_{i0}^{-1} [f_{\beta}(\eta_{i0}, \boldsymbol{\xi}_{i0}; \boldsymbol{\beta}_0) + b_{i0}] \right\}, \end{aligned}$$

where

$$b_{i0} = [f_{\xi}(\eta_{i0}, \boldsymbol{\xi}_{i0}; \boldsymbol{\beta}_0)', \; f_{\eta}(\eta_{t0}, \boldsymbol{\xi}_{t0}; \boldsymbol{\beta}_0)] \left[\begin{pmatrix} \mathbf{x}_i \\ y_i \end{pmatrix} - \begin{pmatrix} \boldsymbol{\xi}_0 \\ \eta_0 \end{pmatrix} \right]$$

and

$$v_{i0} = (f_{\xi}(\eta_{i0}, \boldsymbol{\xi}_{i0}; \boldsymbol{\beta}_0)', f_{\eta}(\eta_{t0}, \boldsymbol{\xi}_{t0}; \boldsymbol{\beta}_0)) \boldsymbol{\Omega} (f_{\xi}(\eta_{i0}, \boldsymbol{\xi}_{i0}; \boldsymbol{\beta}_0)', f_{\eta}(\eta_{t0}, \boldsymbol{\xi}_{t0}; \boldsymbol{\beta}_0))'.$$

The new estimates of the true variables are

$$\begin{aligned} \begin{pmatrix} \hat{\boldsymbol{\xi}}_i \\ \hat{\eta}_i \end{pmatrix} \;=\; & \begin{pmatrix} \mathbf{x}_i \\ y_i \end{pmatrix} - v_{i0}^{-1} \left\{ [f_{\xi}(\eta_{i0}, \boldsymbol{\xi}_{i0}; \boldsymbol{\beta}_0)', \; f_{\eta}(\eta_{t0}, \boldsymbol{\xi}_{t0}; \boldsymbol{\beta}_0)] \left[\begin{pmatrix} \mathbf{x}_i \\ y_i \end{pmatrix} - \begin{pmatrix} \boldsymbol{\xi}_0 \\ \eta_0 \end{pmatrix} \right] \right. \\ & \left. + f(\eta_{t0}, \boldsymbol{\xi}_{t0}; \boldsymbol{\beta}_0) + f_{\beta}(\eta_{i0}, \boldsymbol{\xi}_{i0}; \boldsymbol{\beta}_0)'(\hat{\boldsymbol{\beta}} - \boldsymbol{\beta}_0) \right\} \boldsymbol{\Omega} \, f_{\beta}\left(\eta_{i0}, \boldsymbol{\xi}_{i0}; \boldsymbol{\beta}_0\right). \end{aligned}$$

These equations are iterated using the latest values for $\hat{\boldsymbol{\beta}}$, and $\hat{\boldsymbol{\xi}}_1, \hat{\boldsymbol{\xi}}_2, \ldots, \hat{\boldsymbol{\xi}}_n, \hat{\eta}_1, \ldots, \hat{\eta}_n$ as the initial values for the next iteration.

6.6 Polynomial Berkson Model

The linear Berkson model was introduced in Section 1.6. The analysis of the nonlinear Berkson model is not as straightforward as is in the linear case. The linear case can be reduced to regression analysis, whereas the nonlinear case cannot. Standard least-squares estimators for the nonlinear Berkson model can be inconsistent. The following results are from Cheng (1997).

Consider the simplest nonlinear Berkson model, the quadratic model:

$$x_i = \xi_i + \delta_i, \quad y_i = \eta_i + \varepsilon_i = \beta_0 + \beta_1 \xi_i + \beta_2 \xi_i^2 + \varepsilon_i, \qquad i = 1, \ldots, n. \tag{6.51}$$

The observed values are (x_i, y_i), the ξ_i are the unobserved true values, and the x_i are fixed and called controlled variables. In other words, the experimenter fixes the values, x_i, and then $\xi_i = x_i - \delta_i$, where the random errors $(\delta_i, \varepsilon_i)$ are independent and identically distributed random variables with mean zero. The quantity η_i depends only on the unobserved true values, ξ_i. Recall that in the usual ME model, one assumes that ξ_i is independent of δ_i, but in the Berkson model, ξ_i is a linear function of δ_i because x_i is fixed. The quadratic Berkson model (6.51) can be rewritten as

$$\begin{aligned} y_i &= \beta_0 + \beta_1 x_i + \beta_2 x_i^2 + u_i, \\ u_i &= \varepsilon_i - \beta_1 \delta_i - 2\beta_2 (x_i - \delta_i)\delta_i - \beta_2 \delta_i^2. \end{aligned} \tag{6.52}$$

Note that $Eu_i = \beta_2 \sigma_\delta^2 \neq 0$. This means that least-squares estimation cannot be indiscriminately applied to the quadratic Berkson model. It turns out, as seen below, that the measurement error, δ, causes the least-squares estimator for β_0 to be biased and inconsistent. The least-squares estimates for β_1 and β_2 are consistent, however. In fact, without additional knowledge, the quadratic Berkson model is not identified so the estimates cannot all be consistent.

In order to apply the least-squares approach to (6.52), one needs to remove the bias $\beta_2 \sigma_\delta^2$ from u_i; that is, to minimize

$$Q(\beta_0, \beta_1, \beta_2) = \sum (y_i - \beta_0 - \beta_1 x_i - \beta_2 x_i^2 - \beta_2 \sigma_\delta^2)^2, \tag{6.53}$$

where σ_δ^2 is assumed known a priori. Taking partial derivatives of (6.53) with respect to β_0, β_1, and β_2 yields

$$\sum (y_i - \beta_0 - \beta_1 x_i - \beta_2 x_i^2 - \beta_2 \sigma_\delta^2) = 0, \tag{6.54}$$

$$\sum (y_i - \beta_0 - \beta_1 x_i - \beta_2 x_i^2 - \beta_2 \sigma_\delta^2) x_i = 0, \tag{6.55}$$

and

$$\sum (y_i - \beta_0 - \beta_1 x_i - \beta_2 x_i^2 - \beta_2 \sigma_\delta^2) \left(x_i^2 + \sigma_\delta^2\right) = 0. \tag{6.56}$$

Using (6.55), (6.56) can be reduced to

$$\sum (y_i - \beta_0 - \beta_1 x_i - \beta_2 x_i^2 - \beta_2 \sigma_\delta^2) x_i^2 = 0. \tag{6.57}$$

Interestingly, if one were blindly to apply standard least squares and minimize

$$Q_L(\beta_0, \beta_1, \beta_2) = \sum (y_i - \alpha_0 - \beta_1 x_i - \beta_2 x_i^2)^2$$

with respect to α_0, β_1, and β_2, one would also obtain (6.55), (6.55), and (6.57) with β_0 replaced by $\alpha_0 = \beta_0 + \beta_2 \sigma_\delta^2$. This means that the estimates for β_1 and β_2, obtained by minizing (6.53), are just the ordinary least-squares estimators and β_1 and β_2 are identifiable even without knowledge of σ_δ^2. One cannot get at a consistent estimate of β_0 without some additional assumption on the model.

The resulting estimators from (6.53) are consistent but not efficient because the variances of the 'errors', u_i, are heteroscedastic. Assuming that δ is normally distributed and independent of ε, it can be shown that

$$\sigma_{u_i}^2 = \sigma_\varepsilon^2 + \beta_1^2 \sigma_\delta^2 + 2\beta_2^2 (2x_i^2 \sigma_\delta^2 + \sigma_\delta^4) + 4\beta_1 \beta_2 x_i \sigma_\delta^2.$$

Hence, if σ_ε^2 is also known, one might consider generalized (iterated weighted) least squares as a possible estimation procedure. Basically, one uses the least-squares estimator $(\hat{\beta}_0, \hat{\beta}_1, \hat{\beta}_2)$ derived by (6.53) as an initial estimator, then minimizes the weighted least squares

$$\hat{Q}(\beta_0, \beta_1, \beta_2) = \sum w_i(x_i, \hat{\beta}_0, \hat{\beta}_1, \hat{\beta}_2)(y_i - \beta_0 - \beta_1 x_i - \beta_2 x_i^2 - \beta_2 \sigma_\delta^2)^2,$$

where w_i is inversely proportional to σ_{u_i}. This cycle can be repeated using updated weights; see Carroll and Ruppert (1988, Chapter 2) for details.

Now we investigate the general polynomial Berkson model, with or without equation error, which is

$$\begin{aligned} x_i &= \xi_i + \delta_i, \qquad y_i = p(\xi_i, \boldsymbol{\beta}) + \varepsilon_i, \\ p(t_i, \boldsymbol{\beta}) &= \beta_0 + \beta_1 t_i + \ldots + \beta_k t_i^k + q_i, \qquad i = 1, \ldots, n, \end{aligned}$$

where $\boldsymbol{\beta} = (\beta_0, \beta_1, \ldots, \beta_k)$. The equation error, q_i, is assumed to have mean zero and variance, σ_q^2. If $\sigma_q^2 = 0$, the model is the no-equation-error model. As before, $(\delta_i, \varepsilon_i)$ and q_j are independent for all i and j; however, δ_i and ε_i are not necessarily independent. Note that

$$y_i = p(x_i, \boldsymbol{\beta}) + p(\xi_i, \boldsymbol{\beta}) - p(x_i, \boldsymbol{\beta}) + \varepsilon_i + q_i \equiv p(x_i, \boldsymbol{\beta}) + u_i, \tag{6.58}$$

where $u_i = p(\xi_i, \boldsymbol{\beta}) - p(x_i, \boldsymbol{\beta}) + \varepsilon_i + q_i$. Equation (6.58) looks like a classical polynomial regression model except that, as before, the expectation of u_i is not zero. In order to compute the bias of u_i, recall that x_i is fixed whereas ξ_i is not; therefore,

$$\begin{aligned} Eu_i &= E\{p(\xi_i, \boldsymbol{\beta}) - p(x_i, \boldsymbol{\beta}) + \varepsilon_i + q_i\} = E\{p(x_i - \delta_i, \boldsymbol{\beta})\} - p(x_i, \boldsymbol{\beta}) \\ &= E\{\beta_0 + \beta_1(x_i - \delta_i) + \beta_2(x_i - \delta_i)^2 + \ldots + \beta_k(x_i - \delta_i)^k\} - p(x_i, \boldsymbol{\beta}) \end{aligned}$$

and

$$E(x_i - \delta_i)^r = \sum_{j=0}^{r} \binom{r}{j} x_i^j (-1)^{r-j} E\delta_i^{r-j}, \qquad r = 1, \ldots, k.$$

The least-squares estimator $\hat{\boldsymbol{\beta}}$ for (6.58) results from minimizing

$$\begin{aligned} Q(\beta) &= \sum (y_i - p(x_i, \boldsymbol{\beta}) - Eu_i)^2 = \sum [y_i - E\{p(x_i - \delta_i)\}]^2 \\ &= \sum \{y_i - \beta_0 - \beta_1 E(x_i - \delta_i) - \beta_2 E(x_i - \delta_i)^2 - \\ &\quad \ldots - \beta_k E(x_i - \delta_i)^k\}^2. \end{aligned}$$

In order to find $\hat{\boldsymbol{\beta}}$, one needs to assume knowledge of $E\delta^r$ for $r = 1, \ldots, k$. Note that neither the dependency between measurement errors δ and ε nor the presence of equation error affects these estimators. This is analogous to the least-squares approach for the linear Berkson model (cf. Section 1.6). As in the quadratic model, it can be shown that the resulting estimators of β_{k-1} and β_k are ordinary least-squares estimators and are unbiased and consistent (see Exercise 6.13); but, without knowledge of $E\delta^r$ for $r = 1, \ldots, k$, $\beta_0, \ldots, \beta_{k-2}$ are not even identifiable and the least-squares estimates cannot be consistent because they do not use knowledge of $E\delta^r$ for $r = 1, \ldots, k$.

Note again that a major difference between the normal polynomial Berkson model and the normal polynomial ME model is that the estimates of the βs in the Berkson model apparently have finite variances (recall Section 1.6).

Again, if one wishes to use weighted least squares, one needs to compute the variances, $\sigma_{u_i}^2$. Note that $\sigma_{u_i}^2$ is a function of x_i, $\boldsymbol{\beta}$, and certain moments between the errors, in particular, σ_ε^2 and σ_q^2. Because the equation error, σ_q^2, is usually impossible to obtain a priori, one must rule out the equation error model if one wishes to use iterated weighted least squares. For the no-equation-error model, one needs to assume knowledge of σ_ε^2, $E\delta^r$ for $r = 1, \ldots, 2k$, and $E\delta^r\varepsilon$ for $r = 1, \ldots, k$. When (δ, ε) are normally distributed, the formulae for these moments can be simplified considerably.

Example 6.8

Consider the cubic Berkson model,

$$x_i = \xi_i + \delta_i, \qquad y_i = \eta_i + \varepsilon_i, \qquad \eta_i = \beta_0 + \beta_1\xi_i + \beta_2\xi_i^2 + \beta_3\xi_i^3 + q_i. \quad (6.59)$$

Rewrite (6.59) as

$$y_i = \beta_0 + \beta_1 x_i + \beta_2 x_i^2 + \beta_3 x_i^3 + u_i,$$

where

$$\begin{aligned} u_i &= q_i + \varepsilon_i - \beta_1\delta_i - 2\beta_2(x_i - \delta_i)\delta_i - \beta_2\delta_i^2 \\ &\quad -3\beta_3(x_i - \delta_i)^2\delta_i - 3\beta_3(x_i - \delta_i)\delta_i^2 - \beta_3\delta_i^3. \end{aligned}$$

Assuming $E\delta^3 = 0$, it is easy to see that $Eu_i = \beta_2\sigma_\delta^2 + 3\beta_3 x_i\sigma_\delta^2$. The least-squares estimator $(\hat{\beta}_0, \hat{\beta}_1, \hat{\beta}_2, \hat{\beta}_3)$ is obtained by minimizing

$$Q(\boldsymbol{\beta}) = \sum (y_i - \beta_0 - \beta_1 x_i - \beta_2 x_i^2 - \beta_3 x_i^3 - \beta_2\sigma_\delta^2 - 3\beta_3 x_i\sigma_\delta^2)^2.$$

Taking partial derivatives with respect to β_0, β_1, β_2, and β_3 gives

$$\begin{aligned} \hat{\beta}_0 + \hat{\beta}_1\bar{x} + \hat{\beta}_2(m_{20} + \sigma_\delta^2) + \hat{\beta}_3(m_{30} + 3\bar{x}\sigma_\delta^2) &= \bar{y}, \\ \hat{\beta}_0\bar{x} + \hat{\beta}_1 m_{20} + \hat{\beta}_2(m_{30} + \ barx\sigma_\delta^2) + \hat{\beta}_3(m_{40} + 3m_{20}\sigma_\delta^2) &= m_{11}. \\ \hat{\beta}_0 m_{20} + \hat{\beta}_1 m_{30} + \hat{\beta}_2(m_{40} + m_{20}\sigma_\delta^2) + \hat{\beta}_3(m_{50} + 3m_{30}\sigma_\delta^2) &= m_{21}, \\ \hat{\beta}_0 m_{30} + \hat{\beta}_1 m_{40} + \hat{\beta}_2(m_{50} + m_{30}\sigma_\delta^2) + \hat{\beta}_3(m_{60} + 3m_{40}\sigma_\delta^2) &= m_{31}. \end{aligned} \tag{6.60}$$

If the equation error vanishes and it is assumed that (δ, ε) are jointly normal with $\sigma_{\delta\varepsilon} = 0$, then lengthy calculation yields

$$\begin{aligned} \sigma_{u_i}^2 &= \sigma_\varepsilon^2 + \beta_1^2\sigma_\delta^2 + 2\beta_2^2(2x_i^2\sigma_\delta^2 + \sigma_\delta^4) + 3\beta_3^2(3x_i^4\sigma_\delta^2 + 12x_i^2\sigma_\delta^4 - 40\sigma_\delta^6) \\ &\quad + 4\beta_1\beta_2 x_i\sigma_\delta^2 + 6\beta_1\beta_3(x_i^2\sigma_\delta^2 + \sigma_\delta^4) + 12\beta_2\beta_3(x_i^3\sigma_\delta^2 + 2x_i\sigma_\delta^4). \end{aligned}$$

Thus the variance of u_i depends on x_i as well as the unknown parameters.

The solutions of the equations (6.60), $\hat{\boldsymbol{\beta}}$, can be used as the starting values in the following weighted least squares:

$$\begin{aligned} \hat{Q}(\boldsymbol{\beta}) &= \sum w_i(x_i, \hat{\beta}_0, \hat{\beta}_1, \hat{\beta}_2, \hat{\beta}_3) \\ &\quad \times \left(y_i - \beta_0 - \beta_1 x_i - \beta_2 x_i^2 - \beta_3 x_i^3 - \beta_2\sigma_\delta^2 - 3\beta_3 x_i\sigma_\delta^2\right)^2 \end{aligned}$$

with weights inversely proportional to σ_{u_i}.

Geary (1953) considered the cubic Berkson model of Example 6.8 in which replications of the observations over the same set of x_i are made under the assumption that

$$\sum x_i^{2k-1} = 0, \tag{6.61}$$

where k is a positive integer. He used a different technique to obtain four equations for the estimator of $\boldsymbol{\beta}$. These equations are the same as (6.60) except for the terms involving y_i where one needs to take averages because of the replications.

As noted earlier, even without (6.61), one can solve for $\hat{\beta}_2$ and $\hat{\beta}_3$, but not $\hat{\beta}_0$ and $\hat{\beta}_1$, without knowing σ_δ^2 (see Exercise 6.13).

In summary, it has been shown that classical least-squares estimation is no longer applicable to the nonlinear Berkson model without modification. Moreover, the polynomial Berkson model can be written as a polynomial regression model with heteroscedastic errors. Thus, the modified least-squares estimator has large variance and the iterated weighted (generalized) least-squares method should be considered as an alternative. One needs more information on the measurement error in order to apply least squares and/or iterated weighted least squares. A similar analysis can also be made for the general nonlinear Berkson model, but the algebra can be more cumbersome and there are no explicit formulae for Eu_i and $\sigma_{u_i}^2$.

6.7 Bibliographic Notes and Discussion

Nonlinear measurement error modelling has been an area of active research only in recent years, although it received attention long before this, dating back to the early 1930s (Deming, 1931). The major problem is that nonlinear ME models are very difficult to work with as compared to nonlinear classical regression models. The rapid development of ME models in the past several years has dramatically improved the situation, however. Early work in nonlinear ME models is summarized in Fuller (1987). Carroll *et al.* (1995) take great care to summarize the very diversified approaches developed in the past decade. Our aim in this chapter is quite simple – to provide a unified approach to *polynomial* ME models. Explicit formulae are given for consistent estimators in all types of polynomial models: structural, functional, and ultrastructural; with or without equation error.

Polynomial ME models are only a subclass of general nonlinear ME models. Thus, most of the techniques obtained for the general nonlinear ME model can be applied to the polynomial ME model. Practically these techniques might not yield results specific enough for direct use on polynomial ME models because the general nonlinear model is too general. It is important to note that the resulting estimators from most general nonlinear ME model methods are not necessarily consistent. Wolter and Fuller (1982b) and a later sequel, by Amemiya and Fuller (1988), focused on the large-sample, small (error) variance consistency (see Section 6.5.4). A majority of the nonlinear ME model literature deals with approximately consistent estimators such as regression calibration and SIMEX (simulation-extrapolation); see Carroll *et al.* (1995) for a summary.

Stefanski (1989) and Nakamura (1990) independently developed an approach employing an unbiased estimating function called the corrected score function. Consider a general linear or nonlinear ME model. Suppose that there exists an unbiased (multivariate) estimating equation

$$\boldsymbol{\psi}(\boldsymbol{\xi}_i, y_i, \boldsymbol{\beta}) = \mathbf{0}. \tag{6.62}$$

Further, suppose one can find a $\tilde{\boldsymbol{\psi}}$ such that

$$E\{\tilde{\boldsymbol{\psi}}(\mathbf{x}_i, y_i, \boldsymbol{\beta}) \mid y_i, \boldsymbol{\xi}_i)\} = \boldsymbol{\psi}(\boldsymbol{\xi}_i, y_i, \boldsymbol{\beta})$$

for all $(y_i, \boldsymbol{\xi}_i)$. Because $\mathbf{x}_i = \boldsymbol{\xi}_i + \boldsymbol{\delta}_i$ and taking expectation with respect to $\boldsymbol{\xi}_i$ and y_i again, one has

$$E\{\tilde{\boldsymbol{\psi}}(\mathbf{x}_i, y_i, \boldsymbol{\beta})\} = \mathbf{0}, \tag{6.63}$$

provided $\boldsymbol{\delta}_i$ is independent of ε_i. Then $\boldsymbol{\beta}$ is defined as the solution to $\sum \tilde{\boldsymbol{\psi}}(\mathbf{x}_i, y_i, \hat{\boldsymbol{\beta}}) = \mathbf{0}$. M-estimation theory implies that the estimator obtained by solving (6.63) is consistent and asymptotically normal. Obtaining (6.62) is quite easy because one can use ordinary regression pretending that $\boldsymbol{\xi}$ is observable. The general difficulty is in finding $\tilde{\boldsymbol{\psi}}$, if it exists. The corrected score function approach yields useful estimators in the generalized linear model with measurement error. For the polynomial model, Stefanski (1989) showed that the resulting estimator is the same as the adjusted least-squares estimator (when δ is normally distributed) obtained in Section 6.4. The solution to the general case when $\boldsymbol{\delta}$ and ε are dependent and/or the model is nonnormal is not known. On the other hand, the Stefanski–Nakamura approach allows the presence of an equation error.

The paper by O'Neill *et al.* (1969) seems to be one of the first to focus on the polynomial ME model. Their results concern the computational investigation of the no-equation-error normal functional polynomial relation with known error covariance matrix. As was pointed out in the introduction, large-sample results do not follow from the usual ML estimation theory because of the incidental parameters, ξ_i. Neither is it clear that the resulting estimator is the desired ML estimator. Wolter and Fuller (1982a) proposed an estimator for the quadratic no-equation-error functional model under the normality assumption. Moon and Gunst (1995) extended this to the polynomial functional relation of any degree. Cheng and Schneeweiss (1998b) extended the Wolter–Fuller results to the general nonnormal polynomial functional model. Furthermore, under the normality assumption, this method can be extended to the case in which the error covariance matrix is known up to a multiplicative scalar, as discussed in Section 6.5.3. Chan and Mak (1985) derived an estimator for the no-equation-error polynomial normal functional relation, which coincides with that in Section 6.4.3.

Moon and Gunst (1995) performed simulations on the normal no-equation-error polynomial functional model estimators, (6.32), and the extension of the Wolter–Fuller estimator, described in this chapter. Their overall recommendation is the one-step ML estimator (see Section 6.5.4) because it had the smallest bias and mean squared error in their simulations. One should note, however, that the one-step ML estimator is not consistent in the usual sense and the Britt–Luecke algorithm has serious convergence difficulties even for the quadratic model, contrary to what is indicated in Moon and Gunst (1995, p. 18). Our simulations indicate that these convergence problems do not get better as the sample size is increased. On the other hand, the estimator (6.32) is explicit (not iterative) and consistent. Thus, it should perform well for large-sample sizes. The estimators presented in this chapter are not likely to have finite moments, and therefore finite-sample modifications of these estimators are called for (see Cheng *et al.*, 1998).

The nonlinear Berkson model has received less attention than the nonlinear ME model. An early article by Geary (1953) considered the replicated cubic functional Berkson model mentioned in Section 6.6. Fedorov (1974) studied the nonlinear Berkson models using iterated weighted least squares. Burr (1988) investigated the Berkson model with binary variables. Rudemo *et al.* (1989) considered the nonlinear Berkson model using transformation and weighting techniques. Finally Carroll and Stefanski (1990) developed the full general theory for several (linear and nonlinear) measurement error models, including the Berkson model. Section 6.6 concentrated on the polynomial Berkson model and showed that it can be treated as a polynomial regression model with heteroscedastic errors whose error variances depend on both the observed valued x_i and the unknown parameters. Fitting techniques of heteroscedastic polynomial regression models can, therefore, be applied (Carroll and Ruppert, 1988).

The estimators for polynomial models derived in this chapter are also suitable for structural and ultrastructural models, under mild conditions on the moments of the true values, except for the one-step approximation to the ML estimator described in Section 6.5.5. The stochastic properties of the estimators, namely consistency and asymptotic normality, will not change.

6.8 Exercises

6.1 *For the nonlinear structural ME model (6.15), of Section 6.2 show that* $E(y \mid x)$ *and* u *are uncorrelated and* $Eu = 0$.

6.2 (*Cheng and Schneeweiss, 1996*) *In Section 6.4, show that the polynomials* t_r *in* x *of degree* r, $r = 0, \ldots, k$, *such that* $Et_r = \xi^r$, *are uniquely defined.*

6.3 (*Chan and Mak, 1985*) *Derive (6.40). (Hint: use the fact that*

$$\mathrm{e}^{\theta\xi_i} = \mathrm{e}^{-\theta^2\sigma_\delta^2/2}E\mathrm{e}^{\theta x_i}, \qquad \text{for all } \theta \in \Re,$$

then use series expansions and term-by-term differentiation.)

6.4 *For the nonnormal structural model with normal errors, solve (6.21) for* $\boldsymbol{\beta}$ *in terms of the estimated cumulants.*

6.5 (*Gleser, 1990*) *Let* $g(\xi; \beta_0, \beta_1) = \beta_0 \mathrm{e}^{\beta_1 \xi}$ *in the structural model, (6.1) and (6.3), with* x *normal and* δ *and* ε *independent. Suppose that reliability ratio,* κ_ξ, *is known. Find the least-squares estimates for* β_0 *and* β_1 *using* $y_i = g(\hat{x}_i; \beta_0, \beta_1) + \text{error}$, *where* $\hat{x}_i = E(\xi_i | x_i)$. *Are these estimators consistent?*

6.6 *In Example 6.3, find the unbiased estimating equations without assuming normality or assuming that the errors,* δ *and* ε *are independent.*

6.7 *Perform the Monte Carlo experiment of Exercise 3.4 with*

$$y_i = \beta_1\xi_i + \beta_2\xi_i^2 + \varepsilon_i$$

and

$$\xi_i = \frac{10i}{n} - 5.$$

Assume σ_δ^2 *is known. Try the different values for the sample size and* β_1 *given in Example 3.3. Try* $\beta_2 = 1,\ 2,$ *and 5. Add the parameter value* $\sigma_\delta^2 = 0.01$ *to those used in Exercise 3.4.*

6.8 *Perform the Monte Carlo experiment of Exercise 3.3 with*

$$y_i = \beta_1\xi_i + \beta_2\xi_i^2 + \varepsilon_i.$$

Assume σ_δ^2 *is known. Try the different values for the sample size and* β_1 *given in Example 3.3. Try* $\beta_2 = 1,\ 2,$ *and 5. Add the parameter value* $\sigma_\delta^2 = 0.01$ *to those used in Exercise 3.3.*

6.9 *Perform the Monte Carlo experiment of Exercise 3.7 with*

$$y_i = \beta_1\xi_i + \beta_2\xi_i^2 + \varepsilon_i.$$

Assume σ_δ^2 *is known. Try the different values for the sample size and* β_1 *given in Example 3.3. Try* $\beta_2 = 1,\ 2,$ *and 5. Add the parameter value* $\sigma_\delta^2 = 0.01$ *to those used in Exercise 3.7.*

6.10 *Perform the Monte Carlo experiment of Exercise 3.4 with*

$$y_i = \beta_1 \xi_i + \beta_2 \xi_i^2 + \beta_3 \xi_i^3 + \varepsilon_i$$

and

$$\xi_i = \frac{10\,i}{n} - 5.$$

Assume σ_δ^2 *is known. Try the different values for the sample size and* β_1 *given in Example 3.3. Try* $\beta_2 = 1,\ 2,$ *and* 5, *and* $\beta_3 = 1,\ 3$. *Add the parameter value* $\sigma_\delta^2 = 0.01$ *to those used in Exercise 3.4.*

6.11 *Perform the Monte Carlo experiment of Exercise 3.3 with*

$$y_i = \beta_1 \xi_i + \beta_2 \xi_i^2 + \beta_3 \xi_i^3 + \varepsilon_i.$$

Assume σ_δ^2 *is known. Try the different values for the sample size and* β_1 *given in Example 3.3. Try* $\beta_2 = 1,\ 2,$ *and* 5, *and* $\beta_3 = 1,\ 3$. *Add the parameter value* $\sigma_\delta^2 = 0.01$ *to those used in Exercise 3.3.*

6.12 *Perform the Monte Carlo experiment of Exercise 3.7 with*

$$y_i = \beta_1 \xi_i + \beta_2 \xi_i^2 + \beta_3 \xi_i^3 + \varepsilon_i.$$

Assume σ_δ^2 *is known. Try the different values for the sample size and* β_1 *given in Example 3.3. Try* $\beta_2 = 1,\ 2,$ *and* 5, *and* $\beta_3 = 1,\ 3$. *Add the parameter value* $\sigma_\delta^2 = 0.01$ *to those used in Exercise 3.7.*

6.13 *For the polynomial Berkson model of degree k, with normal errors,* δ, *show that the least-squares estimators of* β_{k-1} *and* β_k *are unbiased and consistent while the least squares estimators of* $\beta_0, \ldots, \beta_{k-2}$ *are biased and inconsistent. Moreover, with the additional knowledge of* $E\delta^r$ *for* $r = 1, \ldots, k$, $\beta_0, \ldots, \beta_{k-2}$ *are identifiable. What happens if the normality assumption on* δ *is dropped?*

6.14 *The following example is from Bal and Ojha (1975). A grain called paddy is grown in India. It is harvested after it 'flowers'. An experiment is conducted to study how long one should wait after flowering before harvesting the grain. Sixteen plots of paddy are harvested at two-day intervals beginning 16 days after flowering. The days after harvesting, x, and the yield in kilograms per hectare, y, are given in Table 6.2.*

Table 6.2 *Harvest day and yield of paddy*

	x	*y*		*x*	*y*		*x*	*y*
1	16	2508	7	28	3500	12	38	3333
2	18	2518	8	30	3883	13	40	3517
3	20	3304	9	32	3823	14	42	3241
4	22	3423	10	34	3646	15	44	3103
5	24	3057	11	36	3708	16	46	2776
6	26	3190						

(a) Give arguments why a Berkson model might fit these data.

(b) Assuming that the actual flowering could occur any time during the two-day interval with equal probability, use a uniform distribution on this interval to calculate the variance of ε.

(c) Assume that from previous experience it is known that $\sigma_\delta = 100$ *kg/ha. Estimate the quadratic Berkson model for these data using iterated weighted least squares.*

(d) Estimate the least-squares quadratic model for these data. Do your results agree with the theory on the Berkson model given in this chapter?

7

Robust Estimation in Measurement Error Models

In this chapter we consider robust estimation in linear ME models. The usual ME model estimates presented heretofore are known to be highly nonrobust (Carroll and Gallo, 1982; Brown, 1982; Ammann and Van Ness, 1988; 1989; Cheng and Van Ness, 1990; 1992). This means that if there is contamination in the data ('bad' data points) or if the assumptions used to derive the estimation procedures are not actually satisfied by the data (the model generating the data is not the model used to derive the estimators), then the estimates can behave very poorly. Thus, robust ME model estimation procedures are important for anyone concerned about contaminated data or violations of the model assumptions. Robust statistical procedures sacrifice some of the efficiency of the estimators at the ideal model in order to reduce the sensitivity of the resulting robust estimators to contamination or violation of the assumptions.

The parameter estimates for the ME model are even less robust than the least-squares estimates are for the ordinary regression model, and the latter are known to be quite nonrobust (see, for example, Ammann and Van Ness, 1988; 1989). One of the topics discussed later (see Section 7.4) is an algorithm for converting any robust ordinary regression procedure into a corresponding robust orthogonal regression procedure. In many ways, the methods of obtaining robust estimates in ME modelling are analogous to corresponding methods developed for ordinary regression.

7.1 Introduction

We first give a brief introduction to robust estimation as commonly applied in ME models. Our treatment of robustness follows the ideas developed in Huber (1981) and Hampel *et al.* (1986). Huber (1981, p. 1) states that the word 'robust' has many different meanings. In his book (and in Hampel *et al.*), robustness signifies insensitivity to small deviations from the assumptions. Hampel *et al.* (1986, p. 21) list four basic types of deviations from models:

1. occasional gross errors in the data;
2. round-off errors and grouping;

3. an approximation such as the central limit theorem was used in selecting the model;
4. an assumption of independence or some other specific correlation structure is only approximate.

The latter two types can be considered special cases of *distributional robustness*. Type (2) commonly occurs because the data are rounded to a limited accuracy and are, in fact, basically discrete. Sometimes observations are grouped or classified even more coarsely. Type (3) is also quite common because one is not certain of the underlying distribution of the data set under investigation. The derivation of the estimation procedure, however, might assume a specific distribution such as the normal distribution. A robust statistical procedure, such as an estimator, should work reasonably well even if the true distribution deviates moderately from the distribution assumed in deriving the procedure. Type (1) is also referred to as *contamination robustness*; however, it can also be viewed as a special case of distributional robustness. Contamination of the data can take several forms. One common form is the occurrence of 'outliers'. Outliers, occasional bad data points that are far from the main bulk of the data, are of major concern in regression procedures. They are caused by misrecorded data, measuring instrument failures, and the like. Figure 7.1 shows how drastic the difference can be between standard ME regression and robust ME regression in the presence of outliers (see Example 7.10 for a description of the data). One can view contaminated data as data coming from a mixture of the assumed distribution, F, and some other distribution, H, possibly quite different from the assumed distribution. In terms of distribution functions this can be written

$$F_{\text{data}} = \alpha F_{\text{assumed}} + (1-\alpha) H_{\text{contamination}}, \qquad 0 \le \alpha \ll 1. \tag{7.1}$$

Another important property of robust procedures is that they be reasonably *efficient*, that is, they should perform reasonably well on uncontaminated data. This means that the performance of the robust procedure is reasonably close to that of the best known nonrobust procedure when they are both applied to uncontaminated data.

7.1.1 The influence function

The measure of the robustness of an estimator that is most used in this chapter is the **influence function** introduced by Hampel (1968; 1974). The **breakdown point** is the other measure of robustness that will be mentioned in this chapter. We first define the influence function for the general estimation problem.

Let $\mathcal{F} = \{F_\theta; \boldsymbol{\theta} \in \boldsymbol{\Theta}\}$ be some parameterized family of distributions, where $\boldsymbol{\Theta}$ is a subset of $\Re^p$, and let F_n be the empirical distribution function of an independent and identically distributed sample, $\mathbf{x}_1, \ldots, \mathbf{x}_n$, distributed according to some unknown member of $\mathcal{F}$:

$$F_n(\mathbf{z}) \equiv \frac{1}{n} \sum_{i=1}^{n} I_{(-\infty,z_1]\otimes(-\infty,z_2]\otimes\ldots\otimes(-\infty,z_p]}(\mathbf{x}_i),$$

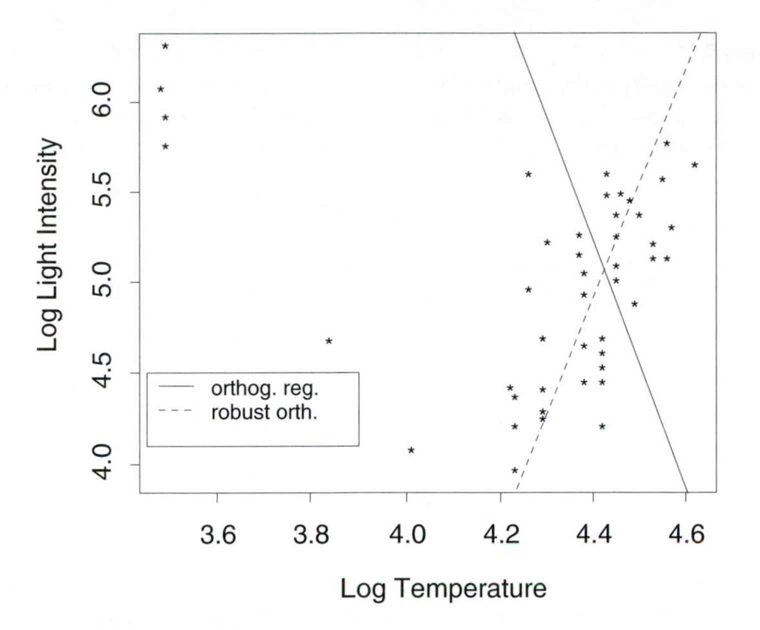

Figure 7.1: Orthogonal regression on the Hertzsprung–Russell data.

where I_A is the indicator function of the set A, p is the dimension of $\mathbf{x}$ and $\mathbf{z}$, and $\otimes$ indicates the direct product of sets. Thus $F_n(\mathbf{z})$ is $1/n$ times the number of data points with the property that all of their coordinates are less than or equal to the corresponding coordinates of $\mathbf{z}$. $F_n(\mathbf{z})$ carries all the relevant information about $\boldsymbol{\theta}$ that is in the independent and identically distributed data, that is, it is a sufficient statistic. The Glivenko–Cantelli theorem states that $F_n(\mathbf{z})$ converges in the strongest sense one could wish for; it converges uniformly with probability one:

$$\sup_{\mathbf{z}} |F_n(\mathbf{z}) - F_{\boldsymbol{\theta}}(\mathbf{z})| \to 0, \qquad \text{with probability 1 as } n \to \infty.$$

One can then consider an estimator to be a functional, $\mathbf{T} : \mathcal{D} \to \Re^p$, where the domain, $\mathcal{D}$, of $\mathbf{T}$ is the set of all distribution functions for which the functional $\mathbf{T}$ is defined. The estimator of the unknown $\boldsymbol{\theta}$ is then written $\hat{\boldsymbol{\theta}} = \hat{\boldsymbol{\theta}}_n = \mathbf{T}(F_n)$. All estimators considered in this chapter are assumed to be **Fisher consistent**, that is $\mathbf{T}(F_{\boldsymbol{\theta}}) = \boldsymbol{\theta}$ for all $\boldsymbol{\theta}$, which means that at the model the estimator measures the right quantity. Furthermore, when $p = 1$, if $\mathbf{T}$ is both Fisher consistent and weakly continuous at $F_{\boldsymbol{\theta}}$, then $\hat{\boldsymbol{\theta}}_n = \mathbf{T}(F_n)$ is a consistent estimator of $\boldsymbol{\theta}$ by the Glivenko–Cantelli theorem (Huber, 1981, p. 41).

In the one-dimensional case, that is, when x_i is scalar,

$$F_n(z) \equiv \frac{1}{n} \sum_{i=1}^{n} I_{(-\infty,z]}(x_i),$$

is the number of data points less than or equal to z divided by n.

Example 7.1

Suppose the family is the family of univariate normal distributions with mean θ and variance 1. Then $\mathcal{F} = \{F_\theta; \theta \in \Re\}$, where F_θ is the distribution function of the $N(\theta, 1)$ distribution and $\Re$ is the real line. Suppose the functional T is defined by

$$T(F) = \int_{-\infty}^{\infty} x \mathrm{d}F(x); \tag{7.2}$$

then $T(F_n)$ is the sample mean,

$$T(F_n) = \int_{-\infty}^{\infty} x \, \mathrm{d}F_n(x) = \frac{1}{n}\sum_{i=1}^{n} x_i.$$

The latter equality is true because F_n jumps by $1/n$ at each data point. This estimator is Fisher consistent because

$$T(F_\theta) = \int_{-\infty}^{\infty} x \, \mathrm{d}F_\theta(x) = \theta.$$

Next suppose that

$$T(F) = \tfrac{1}{2}\left[\sup\left(x : F(x) \le \tfrac{1}{2}\right) + \inf\left(x : F(x) \ge \tfrac{1}{2}\right)\right]; \tag{7.3}$$

then $T(F_n)$ is the sample median. This estimator is also Fisher consistent because, for the normal distribution, the median is equal to θ.

A useful concept in defining and working with influence functions is the Gâteaux derivative (see Hampel *et al.*, 1986, p. 83). A functional, $\mathbf{T}$, is Gâteaux differentiable at the distribution, F, in the domain of $\mathbf{T}$, if there exists a function, $\mathbf{k}$, such that for all G in the domain of $\mathbf{T}$,

$$\lim_{t\to 0} \frac{\mathbf{T}((1-t)F + tG) - \mathbf{T}(F)}{t} = \int \mathbf{k}(\mathbf{x}) \, \mathrm{d}G(\mathbf{x}).$$

Note that the Gâteaux derivative can be written as

$$\frac{\partial}{\partial t}[\mathbf{T}((1-t)F + tG)]_{t=0} = \int \mathbf{k}(\mathbf{x}) \, \mathrm{d}G(\mathbf{x}), \tag{7.4}$$

which is just the ordinary partial derivative of $\mathbf{T}((1-t)F + tG)$ with respect to the parameter t. Putting $F = G$ in (7.4) gives

$$\int \mathbf{k}(\mathbf{x}) \, \mathrm{d}F(\mathbf{x}) = 0. \tag{7.5}$$

Example 7.2

Consider the functional, (7.2), of the previous example. Then

$$\lim_{t\to 0} \frac{T((1-t)F + tG) - T(F)}{t} = \lim_{t\to 0} \frac{1}{t}\left[\int x \, \mathrm{d}\left((1-t)F - tG\right) - \int x \, \mathrm{d}F\right]$$

$$= \lim_{t\to 0} \frac{1}{t}\left[(1-t)\int x \, \mathrm{d}F - t\int x \, \mathrm{d}G - \int x \, \mathrm{d}F\right] = \int (-T(F) + x) \, \mathrm{d}G$$

because $T(F)$ is just a constant here and $\int \mathrm{d}G = 1$. So the Gâteaux derivative of T at F is

$$-T(F) + x. \tag{7.6}$$

Let $F_{t,\mathbf{z}}(\mathbf{x}) = (1-t)F(\mathbf{x}) + t\Delta_{\mathbf{z}}(\mathbf{x})$ where $0 \leq t \leq 1$ and $\Delta_{\mathbf{z}}$ is the distribution function of a unit mass at $\mathbf{z}$; then the influence function for $\mathbf{T}$ at $\mathbf{x}$ is defined to be

$$\mathbf{IF}(\mathbf{z}; \mathbf{T}, F) \equiv \mathbf{IF}(\mathbf{z}, \mathbf{T}(F)) = \lim_{t \to 0^+} \frac{\mathbf{T}(F_{t,\mathbf{z}}) - \mathbf{T}(F)}{t},$$

for those $\mathbf{z}$ where this limit exists. In other words, let $G = \Delta_{\mathbf{z}}$ in (7.4). Then it is clear that the influence function is just $\mathbf{k}(\mathbf{z})$ because

$$\int \mathbf{k}(\mathbf{x})\,\mathrm{d}\Delta_{\mathbf{z}}(\mathbf{x}) = \mathbf{k}(\mathbf{z}).$$

Also, by (7.5),

$$\int \mathbf{IF}(\mathbf{x}; \mathbf{T}, F)\,\mathrm{d}F\,(\mathbf{x}) = 0. \tag{7.7}$$

A heuristic interpretation of the influence function is as follows. The influence function, $\mathbf{IF}(\mathbf{x}; \mathbf{T}, F)$, measures the effect of an infinitesimal contamination at the point $\mathbf{x}$ on the estimate, standardized by the mass of the contamination. The influence function has two important uses. First, it allows one to assess the relative influence of an individual observation on the value of the estimate. If this influence is high, then this is bad because an outlier can have a large effect on the estimator and the estimator is, therefore, not robust. One type of robustness requirement that is sometimes required, in order to call an estimate 'robust', is that its influence function be bounded over all $\mathbf{x}$. This leads to the following definition. The **gross error sensitivity** of $\mathbf{T}$ at F is defined by

$$\gamma^*(\mathbf{T}, F) = \sup_{\mathbf{x}} \|\mathbf{IF}(\mathbf{x}; \mathbf{T}, F)\|, \tag{7.8}$$

which is the supremum (in the Euclidean norm) taken over all $\mathbf{x}$ where the influence function exists. The gross error sensitivity measures the approximate worst influence that a small amount of contamination can have on the value of the estimate no matter where the contaminated data lie. It can be regarded as an upper bound on the standardized asymptotic bias of the estimate: 'standardized' because one divides by the contamination mass, t; 'asymptotic' because γ^* is usually evaluated at the assumed model, $F_{\boldsymbol{\theta}}$, and not at the finite-sample estimate, F_n. Second, the influence function can be used to calculate the asymptotic covariance matrix of the estimate, $\mathbf{T}(F_n)$. Suppose that $\mathbf{T}$ is sufficiently regular and that G is 'near' F; then the first-order von Mises expansion (which is derived from a Taylor series) is

$$\mathbf{T}(G) = \mathbf{T}(F) + \int \mathbf{IF}(\mathbf{x}; \mathbf{T}, F)\,\mathrm{d}(G - F)(\mathbf{x}) + \mathbf{R}, \tag{7.9}$$

where $\mathbf{R}$ is the remainder term of the Taylor expansion. Substituting the empirical distribution F_n for G in (7.9), and using (7.7), yields

$$\begin{aligned}\sqrt{n}(\mathbf{T}(F_n) - \mathbf{T}(F)) &= \sqrt{n}\int \mathbf{IF}(\mathbf{x}; \mathbf{T}, F)\, \mathrm{d}F_n(\mathbf{x}) + \sqrt{n}\mathbf{R} \\ &= \frac{1}{\sqrt{n}}\sum \mathbf{IF}(\mathbf{x}_i; \mathbf{T}, F) + \sqrt{n}\mathbf{R}. \qquad (7.10)\end{aligned}$$

The first term of (7.10) is asymptotically normal with mean zero by the central limit theorem, and, if the remainder term is negligible, then $\sqrt{n}(\mathbf{T}(F_n) - \mathbf{T}(F))$ is asymptotically normal with mean zero and variance

$$\mathbf{V}(\mathbf{T}, F) = \int \mathbf{IF}(\mathbf{x}; \mathbf{T}, F)\, \mathbf{IF}(\mathbf{x}; \mathbf{T}, F)'\, \mathrm{d}F(\mathbf{x}). \qquad (7.11)$$

Of course, the argument above is not rigorous. More regularity conditions are needed for a mathematical proof; see, for example, Boos and Serfling (1980) and Fernholz (1983). In almost all practical situations (7.11) gives the correct asymptotic variance of the estimator. In fact, we used (7.11) in Sections 3.5 and 6.4 because the unbiased estimating equation is closely related to influence function (see the discussion below).

Example 7.3 (*Example 7.1 continued.*)
Again consider the sample mean, (7.2):

$$T(F_t) = \int_{-\infty}^{\infty} x\, \mathrm{d}F_t(x) = (1-t)\int_{-\infty}^{\infty} x\, \mathrm{d}F(x) + tx = (1-t)T(F) + tx$$

and

$$\begin{aligned}IF(x; T, F) &= \lim_{t\to 0^+} \frac{T(F_t(x)) - T(F)}{t} \\ &= \lim_{t\to 0^+} \frac{(1-t)T(F) + tx - T(F)}{t} \\ &= \lim_{t\to 0^+} \frac{-tT(F) + tx}{t} = -T(F) + x, \qquad (7.12)\end{aligned}$$

which agrees with (7.6). Obviously γ^* is ∞ and the sample mean is not a robust estimator of the mean. Note further that (7.11) gives the correct asymptotic variance of $\sqrt{n}(\bar{X} - \theta)$:

$$\int IF(x; T, F_\theta)^2\, \mathrm{d}F_\theta(x) = \int (x-\theta)^2\, \mathrm{d}F_\theta(x) = n\, \mathrm{var}(\bar{X}).$$

One the other hand, for the sample median, (7.3), suppose that F is strictly monotone increasing and has a derivative everywhere. Then $\frac{\mathrm{d}}{\mathrm{d}y}F^{-1}(y)|_{y=1/2} = \frac{1}{F'(\text{median})} = a$, say, and

$$IF(x; T, F) = \lim_{t\to 0^+} \frac{T(F_t(x)) - T(F)}{t}$$

$$= \lim_{t\to 0^+} \frac{\text{sign}(x)\left(F^{-1}\left(\frac{1}{2(1-t)}\right) - F^{-1}\left(\frac{1}{2}\right)\right)}{t}$$

$$= \text{sign}(x)\left[\lim_{t\to 0^+} \frac{a\left[\frac{1}{2(1-t)} - \frac{1}{2}\right]}{t}\right] = \text{sign}(x)\,\frac{a}{2}, \quad (7.13)$$

and $\gamma^* = a/2$. Thus the sample median is robust in the sense that its gross error sensitivity is bounded.

7.1.2 Breakdown point

The influence function describes the infinitesimal stability of the asymptotic value of an estimator. It is a local concept because it is evaluated at the model, F_θ. An important global robustness concept is the **breakdown point**, which is also an important criterion for defining a 'robust' estimator. Roughly speaking, the breakdown point is the smallest amount of contamination that can cause the estimator to become essentially useless. There are several definitions of breakdown point. The oldest definition was given by Hodges (1967). It was restricted to the one-dimensional location estimation problem. Hampel (1968; 1971) generalized this idea and gave a general definition of asymptotic breakdown point (see Hampel *et al.*, 1986, p. 97). This, rather complicated, definition is as follows.

Definition 7.1 *The asymptotic breakdown point, ε^*, of the sequence of estimators $\{\mathbf{T}_n; n \geq 1\}$ at F is defined by*

$$\begin{aligned}\varepsilon^* \;\equiv\; & \sup\{\varepsilon \leq 1;\ \textit{such that there is a compact set } K_\varepsilon \subsetneq \mathbf{\Theta} \textit{ such that} \\ & d_p(F, G) < \varepsilon \textit{ implies } G(\{T_n \in K_\varepsilon\}) \to 1 \textit{ as } n \to \infty\}, \end{aligned} \quad (7.14)$$

where d_p is the Prohorov distance (Prohorov, 1956) defined by

$$d_p(F, G) \equiv \inf\{\varepsilon;\ F(A) \leq G(A^\varepsilon + \varepsilon) \textit{ for all events } A\},$$

and A^ε is the set of all points whose distance from A is less than ε.

The idea is that as long as G does not wander too far from F (in the sense that the Prohorov distance is less than ε^*), then $T_n(G)$ cannot wander off without bound (get outside of any compact set). The larger ε^* is, the more robust the sequence $\{T_n\}$ is . Note that the breakdown point depends on the estimator $\mathbf{T}$ and the distribution F, and the formal notation should be $\varepsilon^*(\mathbf{T}, F)$; however, it usually does not depend on F.

Another form of breakdown point that is much easier to use is the **finite-sample breakdown point** introduced by Donoho and Huber (1983). There are three versions of finite-sample breakdown point. We will discuss the **ε-replacement breakdown point** (cf. Rousseeuw and Leroy, 1987, p. 9). Let $\mathbf{Z}$ represent the sample $(\mathbf{z}_1, \ldots, \mathbf{z}_n)$. Consider all possible corrupted samples $\mathbf{Z}'$ that are obtained by substituting any subset of size m of the original data points by arbitrary values. The maximum bias that can result from such contamination is denoted by

$$b(m, \mathbf{T}, \mathbf{Z}) \equiv \sup_{\mathbf{Z}'} \|\mathbf{T}(\mathbf{Z}) - \mathbf{T}(\mathbf{Z}')\|,$$

where the supremum is taken over all possible $\mathbf{Z}'$. If the maximum bias is infinite, this means that m outliers can have an arbitrarily large effect on the estimate, $\mathbf{T}$. In this case, one can define the estimate as having 'broken down'. This leads to the following definition of the finite-sample breakdown point of the estimate of $\mathbf{T}$ at the sample $\mathbf{Z}$.

Definition 7.2 *The finite sample breakdown point is*

$$\varepsilon_n^* = \min\left\{\frac{m}{n}; b(m, \mathbf{T}, \mathbf{Z}) = \infty\right\}.$$

Clearly, the finite-sample breakdown point does not depend on the underlying distribution. One might conjecture that $\lim_{n\to\infty} \varepsilon_n^* = \varepsilon^*$. In many cases this is true, as we shall see below; but for more complicated situations, it is not clear whether it is true or not. In fact, neither the asymptotic nor the finite-sample breakdown point is easy to calculate in complicated situations such as the GM-estimators for robust regression (Maronna and Yohai, 1981). Thus, the ε-replacement breakdown point gives the fraction of bad data an estimator can cope with before it becomes completely useless. Again, one wants ε_n^* to be large for robustness.

In the following two examples, assume one is given a sample, $\{x_1, \ldots, x_n\}$, of one-dimensional observations.

Example 7.4
Consider the sample mean, (7.2): $T_n = T(F_n) = \bar{x}_n = \sum x_i/n$. The asymptotic breakdown point is zero (Hampel *et al.*, 1986, p. 99). For contradiction, suppose $\varepsilon^* > 0$ and then there exists an $\varepsilon > 0$ and a corresponding K_ε satisfying the conditions in (7.14). Let

$$G_z = \left(1 - \frac{\varepsilon}{2}\right) F + \frac{\varepsilon}{2}\Delta_z;$$

then $d_p(F, G_z) < \varepsilon$. Because the dimension is one, there exists a closed interval, $[-b, b]$ such that $K_\varepsilon \subset [-b, b]$. It is easy to see that by choosing z large enough one can make $T(G_z) > b$, and it is not true that $G(\{T_n \in K_\varepsilon\}) \to 1$. Thus, the conditions in (7.14) are violated and ε^* cannot be greater than 0. Intuitively, one needs only one extreme contamination point to make the maximum bias arbitrarily large. The finite-sample breakdown point is $\varepsilon_n^* = 1/n$ and note that $\lim_{n\to\infty} \varepsilon_n^* = \varepsilon^* = 0$.

Example 7.5
Consider the sample median: $T_n = x_{(m+1)}$ if $n = 2m + 1$ and $T_n = \frac{1}{2}(x_{(m)} + x_{(m+1)})$ if $n = 2m$, where $x_{(k)}$ is the kth-order statistic and m is a positive integer. The asymptotic breakdown point is $\frac{1}{2}$ (Hampel *et al.*, 1986, p. 99; Huber, 1981, p. 60). The finite-sample breakdown point is $([(n+1)/2])/n$, where $[\cdot]$ is the greatest integer function. Here the finite-sample breakdown point tends to the asymptotic breakdown point as the sample size tends to infinity.

A useful intuitive picture of the roles of the influence function and the breakdown point was given by Yohai and Zamar (1990). Suppose we consider contamination of the type given by (7.1). Given $0 \leq \alpha < 0.5$, an α-contamination neighbourhood about the 'true' distribution, $F_{\boldsymbol{\theta}}$, is the set

$$\mathcal{F}_{\boldsymbol{\theta},\alpha} \equiv \{F : F = (1-\alpha)F_{\boldsymbol{\theta}} + \alpha H\}, \tag{7.15}$$

where H is an arbitrary distribution on $\Re^p$. Let $\mathbf{T}$ be Fisher consistent; then the asymptotic bias at F of the estimator $\mathbf{T}(F_n)$ is defined to be

$$B(\mathbf{T}, F, \boldsymbol{\theta}) = \|\mathbf{T}(F) - \mathbf{T}(F_{\boldsymbol{\theta}})\| = \|\mathbf{T}(F) - \boldsymbol{\theta}\|.$$

The maximum asymptotic bias of T at $\mathcal{F}_{\boldsymbol{\theta},\alpha}$ is defined as

$$\tilde{B}_\alpha(T, \boldsymbol{\theta}) \equiv \sup_{F \in \mathcal{F}_{\boldsymbol{\theta},\alpha}} B(\mathbf{T}, F, \boldsymbol{\theta}).$$

T is called bias robust if $\tilde{B}_\alpha(T, \boldsymbol{\theta})$ is 'small'.

Definition 7.3 The **gross-error breakdown point** is

$$\varepsilon_g^* = \inf\{\alpha : \tilde{B}_\alpha(T, \boldsymbol{\theta}) = \infty\}.$$

A plot of $\tilde{B}_\alpha(T, \boldsymbol{\theta})$ versus α is called the maximum bias curve (Figure 7.2). Figure 7.2 shows the roles of the breakdown point and the influence function (gross-error sensitivity). The gross-error sensitivity corresponds to the slope of the maximum asymptotic bias at the origin, $\alpha = 0$, and the breakdown point is the asymptote of the gross-error sensitivity.

7.1.3 Qualitative robustness

The influence function and the breakdown point are examples of **quantitative robustness**. They describe quantitatively how much a small change in the underlying distribution affects the estimate. Another concept introduced by Hampel (1971) is **qualitative robustness**, sometimes simply called **robustness**. It is essentially equivalent to the continuity of an estimator, T, with respect to the weak topology on the domain of the T. By definition, the weak topology is the weakest topology on the space of all probability measures such that $F \rightarrow \int \psi \mathrm{d}F$ is continuous whenever ψ is bounded and continuous; see Huber (1981, Chapter 2) and Hampel *et al.* (1986, Chapter 2) for details. Note that the influence function is related to differentiability and qualitative robustness is related to continuity. Logically, continuity comes first. The problem is, however, that qualitative robustness is of little help in the actual selection of a robust procedure for a particular application. Suppose there are several (qualitatively) robust estimators; then the qualitative robust criterion will not help in selecting among them. We will not pursue qualitative robustness in this book.

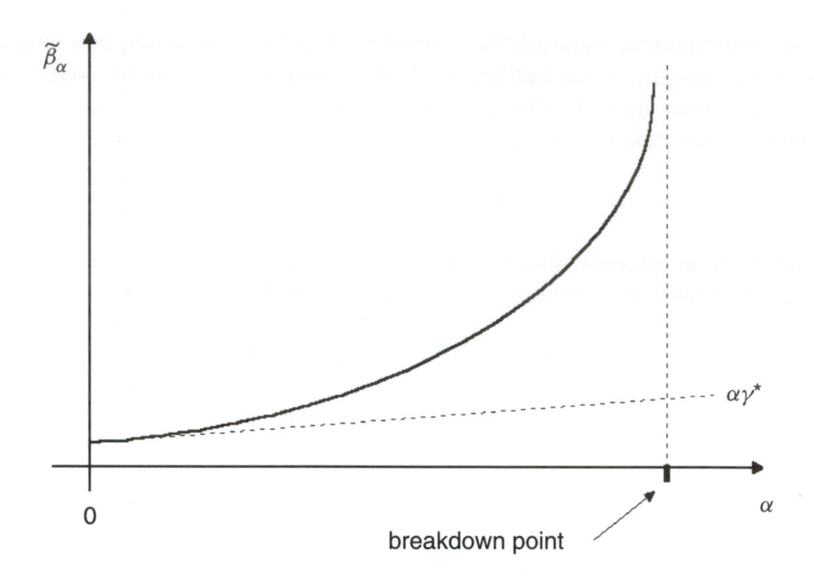

Figure 7.2: Typical maximum-bias curve.

7.1.4 M-estimators

Perhaps the most important class of estimators in robust statistical theory is M-estimators. The ML estimator is the value $\mathbf{T}_n = \mathbf{T}(F_n)$ minimizing the log-likelihood function

$$\sum\{-\log f(\mathbf{x}_i, \mathbf{T}_n)\} = -n \int \log f(\mathbf{x}, \mathbf{T}_n)\, \mathrm{d}F_n(\mathbf{x}). \tag{7.16}$$

Huber (1964) generalized the above minimization to

$$\min_{\mathbf{T}_n} \sum \rho(\mathbf{x}_i, \mathbf{T}_n), \tag{7.17}$$

where ρ is an arbitrary loss function. Basically, the idea is to choose ρ in such a way that it is not as sensitive to outliers as (7.16) is. A typical choice for ρ is given by Huber's function in Example 7.9. We can assume, without loss of generality, that $\rho(0) = 0$. If ρ has a derivative, then the **score function** is $\boldsymbol{\psi}(\mathbf{x}, \mathbf{t}) = (\frac{\partial}{\partial t_1}\rho(\mathbf{x}, \mathbf{t}), \ldots, \frac{\partial}{\partial t_p}\rho(\mathbf{x}, \mathbf{t}))'$ and one obtains the unbiased estimating equation,

$$\sum \boldsymbol{\psi}(\mathbf{x}_i, \mathbf{T}_n) = \mathbf{0} \tag{7.18}$$

(see, for example, Example 7.9). The class of estimators defined by either (7.17) or (7.18) are called M-estimators. The name M-estimator comes from 'generalized maximum likelihood' estimator (Huber, 1964). Note that (7.17) and (7.18) are not necessarily equivalent. For example, (7.18) can have more solutions than (7.17) and the solution of (7.17) might not occur at a 'critical point' given by (7.18). This is analogous to the difference between ML estimates and LE estimates discussed in Section 1.3. In the literature (cf. Sections 3.5 and 6.4), the unbiased estimating equation approach, (7.18), seems more popular than the loss function approach, (7.17).

The functional form of (7.18) is $\int \boldsymbol{\psi}(\mathbf{x}, \mathbf{T}(F_n))\,\mathrm{d}F_n(\mathbf{x}) = \mathbf{0}$, and in general we write

$$\int \boldsymbol{\psi}(\mathbf{x}, \mathbf{T}(F))\,\mathrm{d}F\,(\mathbf{x}) = \mathbf{0}. \tag{7.19}$$

The influence function of $\mathbf{T}(F)$ defined by (7.19) can be shown (see, for example, Huber, 1981, p. 45) to be

$$\mathbf{IF}(\mathbf{x}; \mathbf{T}, F) = \mathbf{M}(\boldsymbol{\psi}, F)^{-1} \cdot \boldsymbol{\psi}(\mathbf{x}, \mathbf{T}(F)) \tag{7.20}$$

where $\mathbf{M}^{(p\times p)}$ is given by

$$\mathbf{M}(\boldsymbol{\psi}, F) = -\int \frac{\partial}{\partial \boldsymbol{\theta}} \boldsymbol{\psi}(\mathbf{x}, \boldsymbol{\theta})|_{\boldsymbol{\theta}=\mathbf{T}(F)}\,\mathrm{d}F\,(\mathbf{x}). \tag{7.21}$$

The asymptotic covariance matrix is then

$$\mathbf{V}^{(p\times p)}(\mathbf{T}, F) = \mathbf{M}(\boldsymbol{\psi}, F)^{-1} \cdot \mathbf{Q}(\boldsymbol{\psi}, F) \cdot \mathbf{M}'(\boldsymbol{\psi}, F)^{-1},$$

where

$$\mathbf{Q}^{(p\times p)}(\boldsymbol{\psi}, F) = \int \boldsymbol{\psi}(\mathbf{x}, \mathbf{T}(F))\boldsymbol{\psi}'(\mathbf{x}, \mathbf{T}(F))\,\mathrm{d}F\,(\mathbf{x}).$$

Example 7.6 (*Example 7.3 continued*)
The estimator, (7.2), is the ML estimator so (7.16) and (7.17) yield

$$\min_{T_n} \sum \rho(x_i, T_n) = \min_{T_n} \sum (x_i - T_n)^2,$$

which, of course, implies $T_n = \bar{x}$. Furthermore, $\psi\,(x, t) = -2(x - t)$ and (7.21) becomes

$$M(\psi, F) = -\int \left[\frac{\partial}{\partial \theta}\psi(x, \theta)\right]_{T(F)} \mathrm{d}F\,(x) = -2$$

so that (7.20) is

$$IF(x; T, F) = M(\psi, F)^{-1} \cdot \psi(x, T(F)) = x - T(F),$$

which is the same as (7.12).

If we set $\rho(x_i, T_n) = |x_i - T_n|$, which is less sensitive to outliers, then $T_n =$ median $(x_1, \ldots, x_n)$ and

$$\psi(x, t) = \begin{cases} 1 & |x - t| > 0 \\ -1 & |x - t| < 0. \end{cases}$$

Suppose F_θ has a density function, f_θ, with respect to a certain measure (not necessarily Lebesgue measure) and that the score function of the log-likelihood,

$$\mathbf{s}(\mathbf{x}, \boldsymbol{\theta}) = \left(\frac{\partial}{\partial \theta_1} \log\, f_\theta(\mathbf{x}), \ldots, \frac{\partial}{\partial \theta_p} \log\, f_\theta(\mathbf{x})\right)'$$

exists. Furthermore, assume the Fisher information,

$$J(\boldsymbol{\theta}) = \int \mathbf{s}(\mathbf{x}, \boldsymbol{\theta})\, \mathbf{s}'(\mathbf{x}, \boldsymbol{\theta})\, \mathrm{d}F_{\boldsymbol{\theta}}(\mathbf{x}),$$

exists. For a Fisher consistent estimator, one can write (7.19) as

$$\int \boldsymbol{\psi}(\mathbf{x}, \boldsymbol{\theta})\, \mathrm{d}F_{\boldsymbol{\theta}}(\mathbf{x}) = \mathbf{0},$$

for all $\boldsymbol{\theta}$ in the parameter space. An important and useful formula can be derived for $\mathbf{M}$ under Fisher consistency:

$$\mathbf{M}(\boldsymbol{\psi}, F_{\boldsymbol{\theta}}) = \int \boldsymbol{\psi}(\mathbf{x}, \boldsymbol{\theta})\, \mathbf{s}'(\mathbf{x}, \boldsymbol{\theta})\, \mathrm{d}F_{\boldsymbol{\theta}}(\mathbf{x}). \tag{7.22}$$

All these results can be found in standard literature on robustness theory, in particular, Hampel *et al.* (1986, pp. 230 ff.).

For an M-estimator defined by the unbiased estimating equation (7.19), the influence function is proportional to defining function, $\boldsymbol{\psi}$. Moreover, under Fisher consistency and the model $F_{\boldsymbol{\theta}}$, the influence function is determined by the unbiased estimating function $\boldsymbol{\psi}(\cdot, \boldsymbol{\theta})$ alone, without any reference to the other $\boldsymbol{\theta}$s and without using derivatives of $\boldsymbol{\psi}$. Note that the asymptotic covariance matrix agrees with the corresponding formula provided in Section 3.5.

7.2 Robust Orthogonal Regression

In this section, robust ME model estimators are derived that have bounded influence function. An estimator called the bias robust estimator is also discussed.

Consider the no-equation-error model

$$\mathbf{x}_i = \boldsymbol{\xi}_i + \boldsymbol{\delta}_i, \quad y_i = \eta_i + \varepsilon_i, \quad \eta_i = \beta_0 + \boldsymbol{\beta}'\boldsymbol{\xi}_i, \qquad i = 1, \ldots, n,$$

where $\boldsymbol{\beta}$, $\boldsymbol{\xi}_i$ and $\boldsymbol{\delta}_i$ are p-dimensional vectors. Furthermore, assume that the error covariance matrix is $\sigma_\varepsilon^2 \mathbf{I}$, where $\mathbf{I}^{((p+1)\times(p+1))}$ is the identity matrix and σ_ε^2 is either known or unknown. As we have seen in Chapters 3 and 5, the orthogonal regression estimators of β_0 and $\boldsymbol{\beta}$ satisfy

$$\min_{\beta_0, \boldsymbol{\beta}} \sum \left(\frac{y_i - \beta_0 - \boldsymbol{\beta}'\mathbf{x}_i}{K(\boldsymbol{\beta})} \right)^2, \tag{7.23}$$

where $K(\boldsymbol{\beta}) = (1 + \boldsymbol{\beta}'\boldsymbol{\beta})^{1/2}$ is the weight that converts to orthogonal residuals, $(y_i - \beta_0 - \boldsymbol{\beta}'\mathbf{x}_i)/K(\boldsymbol{\beta})$. Following an idea of Huber (1973) for robust estimation in ordinary regression; Zamar (1989) and Cheng and Van Ness (1990; 1992) independently proposed replacing the L_2 loss function in (7.23) by a more general function ρ,

$$\min_{\beta_0, \boldsymbol{\beta}} \sum \rho \left(\frac{y_i - \beta_0 - \boldsymbol{\beta}'\mathbf{x}_i}{K(\boldsymbol{\beta})} \right). \tag{7.24}$$

Because (7.23) squares the orthogonal residuals, points with large deviations from the regression plane tend to drive the choice of $\hat{\beta}_0$ and $\hat{\boldsymbol{\beta}}$, as is seen in Figure 7.1. The idea is to downweight the influence of such large orthogonal residuals by choosing ρ so that it places less weight on such residuals.

Example 7.7
One common loss function is the L_1 loss function, $\rho(d) = |d|$. This at least avoids squaring the residuals as in (7.23), but, as we shall see, this still does not lead to very robust estimates. In Section 7.4 we will give an algorithm for converting an L_1 ordinary regression algorithm into an L_1 orthogonal regression algorithm.

The functional form of (7.24) is

$$\min_{\beta_0,\boldsymbol{\beta}} \int \rho\left(\frac{y - \beta_0 - \boldsymbol{\beta}'\mathbf{x}}{K(\boldsymbol{\beta})}\right) \mathrm{d}F(x, y). \tag{7.25}$$

One needs to take the scale, σ_ε, into account in choosing ρ. If σ_ε were known, one could modify (7.25) above to something like

$$\min_{\beta_0,\boldsymbol{\beta}} \int \rho\left(\frac{y - \beta_0 - \boldsymbol{\beta}'\mathbf{x}}{K(\boldsymbol{\beta})\sigma_\varepsilon}\right) \mathrm{d}F(x, y).$$

If σ_ε is not known, then some estimate of the scale, σ_ε, is needed. One suggestion is to estimate σ_ε along with the βs as has been proposed for robust ordinary regression; see Hampel *et al.* (1986, p. 312). More precisely, one might attempt to solve

$$\sum \psi\left(\frac{y_i - \beta_0 - \boldsymbol{\beta}'\mathbf{x_i}}{K(\boldsymbol{\beta})\sigma_\varepsilon}\right)\mathbf{x}_i = 0$$

and

$$\sum \chi\left(\frac{y_i - \beta_0 - \boldsymbol{\beta}'\mathbf{x_i}}{K(\boldsymbol{\beta})\sigma_\varepsilon}\right) = 0$$

simultaneously, where χ is some suitable score function for scale. This appears to be mathematically difficult. Henceforth, we will, for simplicity, assume that the scale is either known or estimated independently and taken to be unity.

Zamar (1985; 1989) showed that the estimates defined by (7.25) are consistent if the distribution of the error $(\boldsymbol{\delta}, \varepsilon)$ is spherically symmetric and the $\boldsymbol{\xi}_i$ are independent and identically distributed. The asymptotic normality and the large-sample results of the functional model are also proven under certain regularity conditions in Zamar (1985).

7.2.1 M-estimators in orthogonal regression

From now on, we consider only the structural model. Differentiating (7.25) with respect to β_0 gives

$$\int \psi\left(\frac{y - \beta_0 - \boldsymbol{\beta}'\mathbf{x}}{K(\boldsymbol{\beta})}\right) \mathrm{d}F(\mathbf{x}, y) = 0; \tag{7.26}$$

and differentiating (7.25) with respect to $\beta_1, \beta_2, \ldots, \beta_p$ respectively and using (7.26) gives

$$\int \psi \left(\frac{y - \beta_0 - \boldsymbol{\beta}'\mathbf{x}}{K(\boldsymbol{\beta})} \right) (\mathbf{x} + \boldsymbol{\beta} y) \, \mathrm{d}F(\mathbf{x}, y) = \mathbf{0}, \tag{7.27}$$

where $\mathrm{d}\rho/\mathrm{d}z = \psi(z)$ and that the regularity conditions hold that allow interchange of differentiation and integration. One can treat the slope parameters, $\boldsymbol{\beta}$, separately because of the following argument. Define the centred random variables, $\dot{\mathbf{x}} = \mathbf{x} - \boldsymbol{\mu}_x$ and $\dot{y} = y - \mu_y$; then (7.26) and (7.27) reduce to

$$\int \psi \left(\frac{\dot{y} - \boldsymbol{\beta}'\dot{\mathbf{x}}}{K(\boldsymbol{\beta})} \right) (\dot{\mathbf{x}} + \boldsymbol{\beta}'\dot{y}) \, \mathrm{d}G(\dot{\mathbf{x}}, \dot{y}) = \mathbf{0}, \tag{7.28}$$

where G is the joint distribution of $\dot{\mathbf{x}}$ and $\dot{y}$. If one writes the estimates in functional form, $\hat{\boldsymbol{\beta}} = \mathbf{T}(G)$, then $\mathbf{T}(G) = \mathbf{T}(F)$. In practice, $\boldsymbol{\mu}_\mathbf{x}$ and μ_y are not known; however, one can centre the data on robust estimates of the means in the following sense. Suppose that one desires bounded influence estimates, $S(F)$ of β_0 and $\mathbf{T}(F)$ of $\boldsymbol{\beta}$; then choose any bounded influence and consistent estimates, $\boldsymbol{\mu}_\mathbf{x}(F)$ of $\boldsymbol{\mu}_\mathbf{x}$ and $\mu_y(F)$ of μ_y (see, for example, Hampel *et al.*, 1986, Section 2.4) and use the following proposition from Cheng and Van Ness (1992).

Proposition 7.1 *Let $\tilde{\boldsymbol{\mu}}_\mathbf{x}(F)$, $\tilde{\mu}_y(F)$, and $\mathbf{T}(F)$ be bounded influence estimators of $\boldsymbol{\mu}_\mathbf{x}$, μ_y, and $\boldsymbol{\beta}$ respectively, and assume that either $\tilde{\boldsymbol{\mu}}_\mathbf{x}(F)$ or $\mathbf{T}(F)$ are continuous functionals in the sense that $\tilde{\boldsymbol{\mu}}_\mathbf{x}(H) \to \tilde{\boldsymbol{\mu}}_\mathbf{x}(F)$ or $\mathbf{T}(H) \to \mathbf{T}(F)$ whenever $H \to F$ weakly (see Section 7.1.1). Then $S(F) \equiv \tilde{\mu}_y(F) - \mathbf{T}'(F)\tilde{\mu}_\mathbf{x}(F)$ is a bounded influence estimator of β_0. If $\tilde{\boldsymbol{\mu}}_\mathbf{x}(F_n)$, $\tilde{\mu}_y(F_n)$, and $\mathbf{T}(F_n)$ are consistent then so is $S(F_n)$.*

Proof. Let $F_{t,\mathbf{z}}(\mathbf{x}) = (1-t)F(\mathbf{x}) + t\Delta_\mathbf{z}$, where $0 \le t \le 1$ and $\Delta_\mathbf{z}$ is the distribution function of a unit mass at $\mathbf{x}$. Suppose that $\mathbf{T}$ is continuous; then, by definition, the influence function for S at $\mathbf{z}$ is defined to be

$$\begin{aligned}
\mathbf{IF}(\mathbf{z}; S, F) &= \lim_{t \to 0^+} \frac{S(F_{t,\mathbf{z}}) - S(F)}{t} \\
&= \lim_{t \to 0^+} \left\{ \frac{\tilde{\mu}_y(F_{t,\mathbf{z}}) - \tilde{\mu}_y(F)}{t} - \mathbf{T}'(F_{t,\mathbf{x}}) \frac{\tilde{\boldsymbol{\mu}}_\mathbf{x}(F_{t,\mathbf{z}}) - \tilde{\mu}_\mathbf{x}(F)}{t} \right. \\
&\quad \left. - \tilde{\boldsymbol{\mu}}_\mathbf{x}(F) \frac{\mathbf{T}(F_{t,\mathbf{z}}) - \mathbf{T}(F)}{t} \right\}
\end{aligned}$$

and therefore

$$\mathbf{IF}(\mathbf{z}; S, F) = \mathbf{IF}(\mathbf{z}; \tilde{\mu}_y, F) - \mathbf{T}'(F)\mathbf{IF}(\mathbf{z}; \tilde{\boldsymbol{\mu}}_\mathbf{x}, F) - \tilde{\boldsymbol{\mu}}_\mathbf{x}(F)\mathbf{IF}(\mathbf{z}; \mathbf{T}, F). \tag{7.29}$$

Because all the terms on the right-hand side of this equation are bounded, $\mathbf{IF}(\mathbf{z}; S, F)$ is bounded. The consistency is obvious. □

The influence function for $\hat{\boldsymbol{\beta}} = \mathbf{T}(G)$ can be obtained from (7.20) as

$$\mathbf{IF}(\dot{\mathbf{x}}, \dot{y}, \mathbf{T}, G_{\boldsymbol{\beta}}) = \mathbf{M}(\psi, G_{\boldsymbol{\beta}})^{-1} \psi \left(\frac{\dot{y} - \boldsymbol{\beta}'\dot{\mathbf{x}}}{K(\boldsymbol{\beta})} \right) (\dot{\mathbf{x}} + \boldsymbol{\beta}\dot{y}), \tag{7.30}$$

where

$$\mathbf{M}(\psi, G_{\boldsymbol{\beta}}) = -\int \frac{\partial}{\partial \boldsymbol{\zeta}} \left[\psi \left(\frac{\dot{y} - \boldsymbol{\zeta}'\dot{\mathbf{x}}}{K(\boldsymbol{\zeta})} \right) (\dot{\mathbf{x}} + \boldsymbol{\zeta}\dot{y}) \right] \Bigg|_{\boldsymbol{\zeta}=\boldsymbol{\beta}} \mathrm{d}G(\dot{\mathbf{x}}, \dot{y}). \tag{7.31}$$

Note that for any nontrivial ψ, the influence function (7.30) is unbounded, and therefore, this form of M-estimation cannot lead to bounded influence estimators of $\boldsymbol{\beta}$. This also happens in ordinary regression; see Hampel *et al.* (1986, Chapter 6). One can, therefore, turn to generalized M-estimators; see Section 7.2.2.

Example 7.8 (*Cheng and Van Ness, 1990*)
Kelly (1984) obtained the influence function for the orthogonal regression estimator (1.30) and (1.31a) under L_2 loss, (7.23), by using the definition of influence function directly. It is much simpler to use (7.30) to find this influence function. Suppose $p = 1$. For the L_2 loss function, one can take $\psi(z) = z$ and the estimating equation becomes

$$\int (\dot{y} - \beta_1 \dot{x})(\dot{x} + \beta_1 \dot{y})\, \mathrm{d}G(\dot{x}, \dot{y}) = 0.$$

Using this, (7.31) becomes

$$\begin{aligned} M(\psi, G_{\beta_1}) &= -\int \frac{\partial}{\partial \zeta} \left[\left(\frac{\dot{y} - \zeta \dot{x}}{\sqrt{1+\zeta^2}} \right) (\dot{x} + \zeta \dot{y}) \right] \Bigg|_{\zeta=\beta_1} \mathrm{d}G_{\beta_1}(\dot{x}, \dot{y}) \\ &= -\frac{1}{K(\beta_1)} - \int \frac{\partial}{\partial \zeta} [(\dot{y} - \zeta \dot{x})(\dot{x} + \zeta \dot{y})] \Bigg|_{\zeta=\beta_1} \mathrm{d}G_{\beta_1}(\dot{x}, \dot{y}) \end{aligned}$$

and the influence function is

$$\begin{aligned} IF(x, y; T, F) &= \frac{(\dot{y} - \beta_1 \dot{x})(\dot{x} + \beta_1 \dot{y})}{-\int \frac{\partial}{\partial \zeta} [(\dot{y} - \zeta \dot{x})(\dot{x} + \zeta \dot{y})] \Big|_{\zeta=\beta_1} \mathrm{d}G_{\beta_1}(\dot{x}, \dot{y})} \\ &= \frac{(\dot{y} - \beta_1 \dot{x})(\dot{x} + \beta_1 \dot{y})}{\sigma_x^2 - \sigma_y^2 + 2\beta_1 \sigma_{xy}} \\ &= \frac{(y - \mu_y - \beta_1(x - \mu_x))(x - \mu_x + \beta_1(y - \mu_y))}{\sigma_x^2 - \sigma_y^2 + 2\beta_1 \sigma_{xy}}. \end{aligned}$$

If we define the estimator of β_0 as

$$S(F) \equiv \tilde{\mu}_y(F) - T(F)\tilde{\mu}_{\mathbf{x}}(F),$$

then by (7.29) and Fisher consistency

$$\begin{aligned} IF(x, y; S, F) &= y - \mu_y - \beta_1(x - \mu_x) - \mu_x IF(x, y; T, F) \\ &= y - \beta_0 - \beta_1 x - \mu_x IF(x, y; T, F). \end{aligned}$$

Note that this influence function is quadratic in x and y, which means that the estimate is very nonrobust.

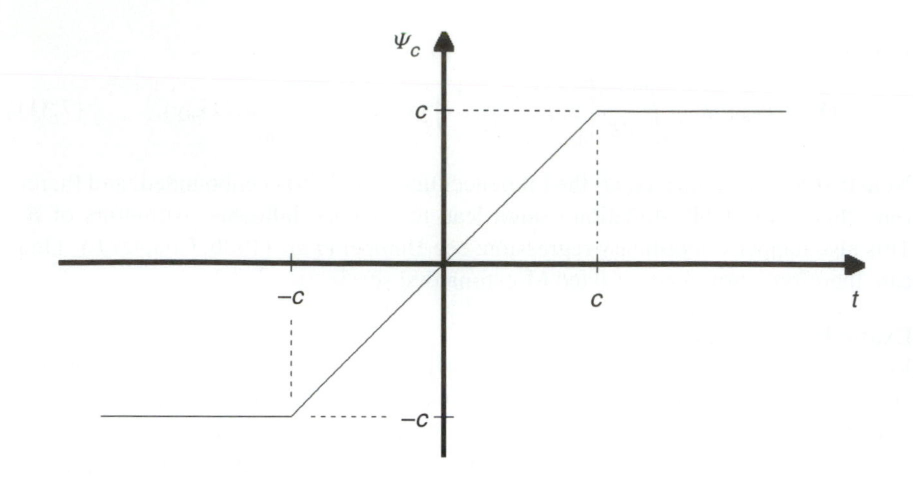

Figure 7.3: Huber's score function.

Example 7.9 (*Cheng and Van Ness, 1990*)
A very commonly used score function is Huber's function (Figure 7.3): for any $c > 0$,

$$\psi_c(t) = \begin{cases} t & |t| \leq c, \\ c\,\mathrm{sign}(t) & |t| > c. \end{cases} \tag{7.32}$$

This is equivalent to a loss function, ρ, which is quadratic on the interval $[-c, c]$ and linear outside that interval. Assume the no-intercept model, $\beta_0 = 0$, and that $p = 1$; then the influence function for β_1 can be calculated via (7.30), to be

$$IF(x, y; T, F_{\beta_1}) = \begin{cases} \dfrac{(y - \beta_1 x)(x + \beta_1 y)}{B(\beta_1)K(\beta_1)} & |y - \beta_1 x| \leq c, \\ c(x + \beta_1 y)\dfrac{\mathrm{sign}(y - \beta_1 x)}{B(\beta_1)} & \text{otherwise}, \end{cases}$$

where

$$B(\beta_1) = -\int \frac{\partial}{\partial \zeta}\psi_c\left(\frac{y - \zeta x}{K(\zeta)}\right)(x + \zeta y)|_{\zeta=\beta_1}\, \mathrm{d}F_{\beta_1}.$$

Note that $IF(x, y; T, F_{\beta_1})$ is linear in x and y. Thus, by this measure of robustness, the Huber estimator is considerably more robust than the L_2 (orthogonal regression) estimator. However, this estimator still does not have a bounded influence function.

In any case, the influence function of the estimator defined by (7.30) cannot be bounded for any ψ function. One must take a more general approach if one wishes a bounded influence function estimator for β_1, as we shall see in the next subsection.

7.2.2 Bounded influence estimators and Hampel's optimality problem

The first goal of this subsection is to define a class of M-estimators for the ME model, possessing finite gross-error sensitivities, γ^*. The second goal is to find the lower bound, γ_L^*, on the gross-error sensitivity over this class of estimators and to specify the estimator attaining this lower bound. Finally, the third goal is to find the solution to Hampel's optimality problem, which is to obtain the most efficient estimator whose gross-error sensitivity is bounded above by γ_B^*, where γ_B^* is some chosen bound, $\gamma_B^* > \gamma_L^*$. There is a trade-off here: by permitting a larger value for γ_B^*, one can obtain a more efficient estimator. Note that the bounded influence estimator is calculated at the assumed model, for example, the normal model.

Let us first return to the general estimation problem. If the parameter under investigation is a scalar, then the term 'most efficient estimator' is simply defined as the estimator with smallest asymptotic variance. For the multiparameter case, efficiency is a little more complicated. Generally speaking, there are three ways to define multivariate efficiency (see Staudte and Sheather, 1990, p. 246). Suppose there are two consistent estimators, $\hat{\boldsymbol{\theta}}_1$ and $\hat{\boldsymbol{\theta}}_2$, of the p-vector parameter, $\boldsymbol{\theta}$, having asymptotic covariance matrices $\boldsymbol{\Sigma}_1$ and $\boldsymbol{\Sigma}_2$, respectively. Then $\hat{\boldsymbol{\theta}}_1$ is commonly defined to be more efficient than $\hat{\boldsymbol{\theta}}_2$ in one of the following ways:

1. $\boldsymbol{\Sigma}_2 - \boldsymbol{\Sigma}_1$ is positive definite, or
2. $\mathrm{tr}(\boldsymbol{\Sigma}_2) > \mathrm{tr}(\boldsymbol{\Sigma}_1)$, or
3. $\det(\boldsymbol{\Sigma}_2) > \det(\boldsymbol{\Sigma}_1)$;

where tr indicates the trace and det indicates the determinant of a matrix. The first criterion is the most desirable because it says that for the linear combination $\mathbf{a}'\boldsymbol{\theta}$, the estimator $\mathbf{a}'\hat{\boldsymbol{\theta}}_1$ has smaller variance than $\mathbf{a}'\boldsymbol{\theta}_2$ for any choice of the p-vector $\mathbf{a}$. Also it is obvious that (1) implies both (2) and (3). If one does not restrict the gross-error sensitivity, the ML estimator is the most efficient estimator. If the restriction of bounded gross-error sensitivity is imposed, however, the most efficient estimator according to the first criterion does not exist; see Hampel *et al.* (1986, p. 238) and Ruppert (1985). The second criterion is motivated by the asymptotic mean squared error because if $\hat{\boldsymbol{\theta}}$ is a Fisher consistent estimator of $\boldsymbol{\theta}$ with asymptotic covariance $\boldsymbol{\Sigma}$, then

$$nE\|\hat{\boldsymbol{\theta}} - \boldsymbol{\theta}\|^2 \to \int \|IF(x; T(F_{\boldsymbol{\theta}}), F_{\boldsymbol{\theta}})\|^2 \,\mathrm{d}F_{\boldsymbol{\theta}}\,(x) = \mathrm{tr}\,\boldsymbol{\Sigma}.$$

Finally, the third criterion is motivated by confidence ellipsoid considerations when the estimators have a multivariate normal distribution. The ellipsoids have the form

$$\{\boldsymbol{\theta} : (\hat{\boldsymbol{\theta}} - \boldsymbol{\theta})' \boldsymbol{\Sigma}^{-1} (\hat{\boldsymbol{\theta}} - \boldsymbol{\theta}) \le c\}.$$

The volume of the confidence ellipsoid is proportional to the square root of determinant of $\boldsymbol{\Sigma}$. Therefore, the estimator that permits the smaller volume of confidence ellipsoid is to be preferred.

If an estimator has finite gross-error sensitivity, γ^* defined by (7.8), then it is called B-robust (B means 'bias' here). If there exists a finite minimum, γ_L^*, of γ^*, then an estimator that attains this minimum is called a **most B-robust** estimator. Given a bound, $\gamma_B^* > \gamma_L^*$, the **optimal bounded influence estimator** is the one that is the most efficient, among all estimators whose gross-error sensitivity is less than or equal to γ_B^*. This estimator is the solution to Hampel's optimality problem, so named because this problem was first proposed and solved by Hampel (1968) for the general one-dimensional parameter case. Hampel *et al.* (1986, Chapter 4) extended this to multidimensional M-estimators and showed that, if such estimators exist, they must take a certain form; however, it is not clear how this existence can be determined when applied to a specific parametric model. The efficiency is taken in the sense of (2) above, that is, minimizing the trace of the asymptotic covariance of the estimator. Hampel *et al.* (1986, Chapter 4) also discuss the optimal estimators with respect to a variety of other types of sensitivities, such as the self-standardized sensitivity, but these will not be discussed here.

Generalized M-estimators.

We now return to orthogonal regression for the structural model. We concentrate on the estimation of the slope parameters, $\boldsymbol{\beta}$, and assume that the data are centred about their means (thus β_0 is to be taken care of separately – see Proposition 7.1). It is clear that the M-estimators defined by (7.28) do not possess bounded influence because while one can bound ψ in (7.30), the term $\mathbf{x}+\boldsymbol{\beta}y$ is not bounded. Any score function defined as a derivative of some loss function, ρ, will have this shortcoming. Following the terminology used in robust ordinary regression (Maronna and Yohai, 1981), we call an M-estimator defined by (7.28) a **classical M-estimator**. Again, in analogy to robust ordinary regression, **generalized M-estimators** (GM-estimators) for the ME model are defined by

$$\int \tau[\mathbf{w}(\mathbf{T}(F)),\, v(\mathbf{T}(F))]\,\mathbf{w}(\mathbf{T}(F))\,\mathrm{d}F\,(\mathbf{x},y)=\mathbf{0}, \tag{7.33}$$

where

$$\mathbf{w}(\boldsymbol{\beta})=\frac{\mathbf{x}+\boldsymbol{\beta}y}{K(\boldsymbol{\beta})} \qquad \text{and} \qquad v(\boldsymbol{\beta})=\frac{y-\boldsymbol{\beta}'\mathbf{x}}{K(\boldsymbol{\beta})}. \tag{7.34}$$

Thus GM-estimators are obtained by replacing $\psi(v(\boldsymbol{\beta}))$ in (7.28) by $\tau(\mathbf{w}(\boldsymbol{\beta}),v(\boldsymbol{\beta}))$. The second variable in the τ function enables one to obtain bounded influence estimators, as we shall see shortly. Intuitively, one can think of $\psi(v(\boldsymbol{\beta}))$ as only accounting for high orthogonal residuals, while the second variable in τ also accounts for high-leverage data points. Under the structural model,

$$\operatorname{cov}(\mathbf{w}(\boldsymbol{\beta}),\, v(\boldsymbol{\beta}))=\mathbf{0}. \tag{7.35}$$

Note that both the classical M-estimator and the GM-estimator are defined via unbiased estimating equations and hence are in the class of M-estimators defined by (7.19). Thus 'GM-estimator' is a slight abuse of terminology.

All known proposals for the score function, τ, for ordinary regression GM-estimators, are of the form

$$\tau(\mathbf{x},r)=\zeta(\mathbf{x})\,\psi(r\,\varphi(\mathbf{x})),$$

for appropriate functions ζ and φ, where r is the residual. The classical M-estimator uses $\zeta = \varphi = 1$. The optimal bounded influence estimator turns out to be of the Schweppe type (see below and Hampel *et al.*, 1986, p. 316), which uses $\varphi = 1/\zeta$.

Some regularity conditions on τ are needed; in particular, some continuity and integrability conditions and the assumption that $\tau(w, \cdot)$ is odd in the second argument (see Hampel *et al.*, 1986, p. 315; Maronna and Yohai, 1981; Ronchetti and Rousseeuw, 1985). The following result establishes Fisher consistency (Cheng and Van Ness, 1992, Theorem 3.1).

Theorem 7.1 *In the normal structural ME model, if a unique GM-estimator is defined by (7.33) with certain regularity conditions on* τ, *it is Fisher consistent and* $E_v\tau(\mathbf{w}, v) = 0$ *for all* $\mathbf{w}$.

The proof of this theorem requires the strong assumption of joint normality of $(\mathbf{x}, y)$, and it unfortunately has not been generalized to other model distributions. The technical reason is that the known proofs require the independence of $\mathbf{w}$ and v and not just (7.35).

Now we are ready to solve Hampel's optimality problem for orthogonal regression. Unfortunately, due to the complexity of the algebra, we can only solve the simple ME model with univariate x. Recall that the intercept is eliminated so the only structural parameter is the slope β_1. All the results below can be found in Cheng and Van Ness (1992). The influence function of the GM-estimator defined by (7.33) is

$$IF(x, y; T, F) = M^{-1}(\tau, F)\tau[w(T(F), v(T(F)]\, w(T(F)),$$

where

$$\begin{aligned} M(\tau, F) &= -\int \frac{\partial}{\partial t}[\tau[w(t), v(t)]w(t)]_{t=T(F)}\, \mathrm{d}F(x, y), \\ Q(\tau, F) &= \int \tau^2[w(T(F)), v(T(F))]\, w^2(T(F))\, \mathrm{d}F(x, y), \end{aligned}$$

and

$$\gamma^* = \sup_{x,y} |IF(x, y; T, F)| = \sup_{x,y} |M^{-1}(\tau, F)\tau[w(T(F)), v(T(F))]\, w(T(F))|.$$

The asymptotic variance is

$$V(T, F) = \int IF^2(x, y; T, F)\, \mathrm{d}F(x, y) = M^{-2}(\tau, F)Q(\tau, F). \tag{7.36}$$

Note that M can be expressed as

$$M(\beta_1) = \int \tau[(w(\beta_1)), (v(\beta_1))]\, w(\beta_1)\, \frac{\partial}{\partial \beta_1} \log f_{\beta_1}(x, y)\, \mathrm{d}F_{\beta_1}(x, y),$$

by the Fisher consistency (cf. (7.22)). If one rotates (x, y) into the principal components (w, v) (see Cheng and Van Ness, 1992, p. 390), it can be shown that

$$M(\beta_1) = a(\beta_1) \int \tau(w, v)\, w^2 v\, \mathrm{d}H(w, v), \tag{7.37}$$

where $H(w, v)$ is the distribution of (w, v) and

$$a(\beta_1) = \frac{1}{1+\beta_1^2} \frac{[(\sigma_x^2 - \sigma_y^2)^2 + 4\sigma_{xy}^2]^{1/2}}{\sigma_x^2\sigma_y^2 - \sigma_{xy}^2}. \tag{7.38}$$

The following theorem gives the lower bound of gross-error sensitivity.

Theorem 7.2 *It is always true that*

$$\gamma^* \geq \gamma_L^*(\beta_1) \equiv \frac{(\sigma_x^2\sigma_y^2 - \sigma_{xy}^2)(1+\beta_1^2)\pi}{2[(\sigma_x^2 - \sigma_y^2)^2 + 4\sigma_{xy}^2]^{1/2}} \tag{7.39}$$

and $\tau(w, v) = \text{sign}(v)/|w|$ *attains this lower bound.*

Proof. By (7.37), $M(\beta_1) = a(\beta_1)E[\tau(w, v)\, w^2 v]$, and

$$\begin{aligned} 1 &= M^{-1}(\beta_1)a(\beta_1)E\left[\tau(w, v)\, w^2 v\right] \leq a(\beta_1)E[|\tau(w, v)||M^{-1}(\beta_1)w||wv|] \\ &\leq a(\beta_1)\,\gamma^*\, E|wv| = a(\beta_1)\,\gamma^*\, E|w|E|v| \end{aligned}$$

because w and v are independent. Moreover, $E|w| = 2\sigma_w/\sqrt{2\pi}$ and $E|v| = 2\sigma_v/\sqrt{2\pi}$ because w and v are normal with mean 0. A simple calculation yields (7.39). The score function $\tau(w, v) = \text{sign}(v)/\mid w \mid$ is Fisher consistent because τ is odd in v, $M(\beta_1) = a(\beta_1)\, E[w^2 v\ \text{sign}(v)/|w|] = a(\beta_1)\, E|wv|$, and

$$\gamma^* = \sup_{w,v} |\tau(w, v)||M^{-1}w| = \sup_{w,v} \frac{1}{\|w|}|M^{-1}w| = \gamma_L^*(\beta_1).$$

□

The estimator corresponding to $\tau(w, v) = \text{sign}(v)/|w|$ attains the lower bound, γ_L^*, is the most B-robust estimator, and for a given data set is the solution of

$$\sum \frac{\text{sign}(y_i - \beta_1 x_i)}{\mid x_i + \beta_1 y_i \mid}(x_i + \beta_1 y_i) = 0. \tag{7.40}$$

Note that the corresponding most B-robust estimator for ordinary simple linear regression is

$$\sum \frac{\text{sign}(y_i - \beta_1 x_i)}{\mid x_i \mid} x_i = 0. \tag{7.41}$$

This has solution $\hat{\beta}_1 =$ median $\{y_i/x_i\}$ (Krasker, 1980). The solution of (7.40) is similar. Rotate coordinates so that the regression line is the new x-axis; then the solution is $\hat{\beta}_1 =$ median $\{y_i^*/x_i^*\}$, where (x_i^*, y_i^*) are the rotated observations. In practice, this would require an iterative procedure similar to that in Section 7.4. Martin *et al.* (1989) showed that the estimator (7.41) has the property that it is optimal in the sense that it minimizes the maximum bias among all regression equivariant estimators over an α-contaminated neighbourhood, $\mathcal{F}_{0,\alpha}$, $0 \leq \alpha \leq 0.5$ (see (7.15)) where F_0

is the underlying distribution (H in (7.15) is sometimes taken to be symmetric). We conjecture that if one uses the minimax approach in the orthogonal ME model, a similar result is true for (7.40).

Finally, we conclude this subsection with the solution to Hampel's optimality problem for the univariate normal structural model.

Theorem 7.3 (*Cheng and Van Ness, 1992, Theorem 5.1*) *Consider the class of score functions, τ, that satisfy the regularity conditions given in Cheng and Van Ness (1992, p. 359) (see also Hampel et al., 1986, p. 315). The estimator corresponding to the score function, $\tau_{\gamma_B^*(\beta_1)}$, minimizing the asymptotic variance subject to a bound, $\gamma_B^*(\beta_1)$, on the gross-error sensitivity exists and is unique if and only if $\gamma_B^*(\beta_1) > \gamma_L^*(\beta_1)$. This score function is defined by*

$$\tau_{\gamma_B^*(\beta_1)}(w, v) = \frac{\psi_{\gamma_B^*(\beta_1)}(a(\beta_1)b(\beta_1)\,|w|\,v)}{b(\beta_1)\,|w|},$$

where ψ_c is Huber's score function, (7.32), $a(\beta_1)$ is defined by (7.38), and $b(\beta_1)$ is defined implicitly by

$$a(\beta_1)E\psi_{\gamma_B^*(\beta_1)}(a(\beta_1)b(\beta_1)\,|w|\,v) = 1.$$

7.2.3 Bias robust orthogonal regression

We now discuss the bias robust estimation method for orthogonal regression given in Zamar (1992). The model is equivalent to that used in Section 5.4.1, but in a slightly different form. Assume there are n independent observations,

$$\mathbf{z}_i = \boldsymbol{\zeta}_i + \boldsymbol{\varepsilon}_i, \quad \boldsymbol{\varphi}_0'\boldsymbol{\zeta}_i = \vartheta_0, \qquad i = 1, \ldots, n, \tag{7.42}$$

where $\mathbf{z}_i$, $\boldsymbol{\zeta}_i$, $\boldsymbol{\varepsilon}_i$ and $\boldsymbol{\varphi}_0$ are $(p+1)$-dimensional vectors while ϑ_0 is a scalar. Usually the intercept, ϑ_0, is considered a nuisance parameter. In order to make this model uniquely defined, let $\|\boldsymbol{\varphi}_0\| = 1$. Moreover, the normal structural relationship is assumed in this subsection; that is, $\boldsymbol{\zeta}$ and $\boldsymbol{\varepsilon}$ are independent Gaussian vectors with $E(\boldsymbol{\zeta}) = \boldsymbol{\mu}$, $E(\boldsymbol{\varepsilon}) = \mathbf{0}$, $\mathrm{var}(\boldsymbol{\zeta}) = \boldsymbol{\Sigma}_{\zeta\zeta}$ and $\mathrm{var}(\boldsymbol{\varepsilon}) = \sigma_\delta^2\mathbf{I}_{p+1}$, where $\mathbf{I}_{p+1}$ is the identity matrix. The eigenvalues of $\boldsymbol{\Sigma}_{\zeta\zeta}$, in increasing order, are denoted by λ_k, $k = 0, \ldots, p$. The corresponding eigenvectors are denoted by $\boldsymbol{\varphi}_k$ for $k = 0, \ldots, p$. Note that $\lambda_0 = 0$ because the $\boldsymbol{\zeta}$s are linearly related by (7.42), and therefore, the parameter vector $\boldsymbol{\varphi}_0$ is the eigenvector corresponding to the eigenvalue 0. We assume that λ_1 is positive. It is also assumed that σ_δ^2 is known. If it is not known, then it should be replaced by a robust estimate. The signal to noise ratio, defined as

$$\Delta = \frac{\sqrt{\lambda_1}}{\sigma_\delta}, \tag{7.43}$$

plays an important role in the minimax bias theory discussed below.

Let the assumed model be F_θ and the α-contamination neighbourhood $\mathcal{F}_{\theta,\alpha}$ be as defined in (7.15). Zamar (1992) defined the asymptotic bias at a distribution F of an estimate $\hat{\boldsymbol{\varphi}}_n$ of $\boldsymbol{\varphi}_0$ as

$$B(F) = 1 - |\hat{\boldsymbol{\varphi}}'\boldsymbol{\varphi}_0|,$$

where $\hat{\boldsymbol{\varphi}}$ is the limiting value of $\hat{\boldsymbol{\varphi}}_n$ under F. The maximum asymptotic bias of $\hat{\boldsymbol{\varphi}}_n$ over $\mathcal{F}_{\theta,\alpha}$ and its breakdown point are defined as

$$\bar{B}(\alpha) = \sup_{F \in \mathcal{F}_{\theta,\alpha}} B(F) \qquad \text{and} \qquad BP = \sup\{\alpha : \bar{B}(\alpha) < 1\}, \tag{7.44}$$

respectively. These are similar to, but not exactly the same as, the definitions in Section 7.1.2.

Huber (1964) showed that, in the simple location model, the median minimizes the maximum asymptotic bias among all translation equivariant location estimates. In a general parametric model, an estimate that minimizes the maximum asymptotic bias over a certain class of distributions will be called **bias robust**.

Now consider the class of M-estimators defined by

$$\arg \min_{\|\boldsymbol{\varphi}\|=1, \vartheta \in \Re} \sum \rho(r_i(\vartheta, \boldsymbol{\varphi})), \tag{7.45}$$

where ρ is a general loss function and $r_i(\vartheta, \boldsymbol{\varphi}) = \boldsymbol{\varphi}'\mathbf{z}_i - \vartheta$ is the **orthogonal residual** and it is assumed that these estimates exist and are unique. That is, the M-estimates, $\hat{\boldsymbol{\varphi}}$ and $\hat{\vartheta}$ minimize (7.45). Using the functional notation common in M-estimation, the notation would be $\hat{\boldsymbol{\varphi}}(F_n)$ and $\hat{\vartheta}(F_n)$, where F_n is the sample distribution function of the data. Note that (7.45) is equivalent to (7.24) and ρ must be even in order to ensure Fisher consistency. Let $\bar{B}_\rho(\alpha)$ be the maximum asymptotic bias of the M-estimate defined by (7.45) with loss function ρ. It can be shown that if ρ is unbounded then $\bar{B}_\rho(\alpha) = 1$ for all positive α. Therefore, we restrict our attention to even and bounded ρ and assume, without loss of generality, that $\rho(\infty) = 1$.

The goal is to find bias robust M-estimates for α-contaminated neighbourhoods when the loss function, ρ, is still further restricted to be monotone increasing on $[0, \infty)$ and to have a finite number of discontinuities. We denote by $\mathcal{L}$ the class of all such ρ:

$$\mathcal{L} = \{\rho : \rho(0) = 0,\ \rho(\infty) = 1,\ \rho \text{ is even, } \rho \text{ is monotone increasing on } [0, \infty), \\ \text{and } \rho \text{ has a finite number of discontinuities}\}.$$

Because the estimates of the $\boldsymbol{\varphi}$s are defined implicitly using the loss function ρ, it is ρ that one wishes to specify, that is, if one specifies ρ, one specifies the estimates. For example, $\bar{B}_\rho(\alpha)$ is thought of as the asymptotic bias of the M-estimate defined by ρ.

One interesting connection between $\bar{B}_\rho(\alpha)$ when α is near zero and the gross-error sensitivity defined by Hampel (1968) (see Section 7.1.1.) is also described by Zamar (1992). He defined $\frac{\mathrm{d}}{\mathrm{d}\alpha}\bar{B}_\rho(\alpha)|_{\alpha=0}$ to be the sensitivity of the estimate. It is intuitively clear that, these two sensitivities are the same under certain regularity conditions, because both of them are derivatives of the estimate. Zamar (1992), however, gave a formal proof of this equivalence provided that the supremum and derivative can be interchanged.

Zamar (1992) showed that the bias robust orthogonal regression M-estimate is implicitly defined by the loss function

$$\rho(t) = \begin{cases} 0 & \text{if } |t| \leq t^* \\ 1 & \text{otherwise.} \end{cases} \tag{7.46}$$

The bias robust fit has a simple geometric interpretation. It is the 'centre' of a strip of width $2t^*\sigma_\delta$, which includes the maximum possible number of data points. The crucial number t^* is related to α in (7.15).

In order to state Zamar's results, let γ be a number between 0 and 1 and define

$$\boldsymbol{\varphi}(\gamma) = (1-\gamma)\boldsymbol{\varphi}_0 + \{1-(1-\gamma)^2\}^{1/2}\boldsymbol{\varphi}_1, \qquad \vartheta(\gamma) = \boldsymbol{\mu}'\boldsymbol{\varphi}(\gamma),$$

and the function

$$\begin{aligned} g_\rho(\gamma) &= E_{F_\theta}[\rho\{\boldsymbol{\varphi}(\gamma)'\mathbf{z} - \vartheta(\gamma)\}] - E_{F_\theta}\{\rho(\boldsymbol{\varphi}_0'\mathbf{z} - \vartheta_0)\} \\ &= E[\rho\{r(\gamma)N\}] - E\{\rho(\sigma_\delta N)\}, \end{aligned}$$

where N is the standard normal random variable and $r^2(\gamma) = \sigma_\delta^2 + \lambda_1\{1-(1-\gamma)^2\}$.

For any given α with $0.5 < \alpha < 1$, Zamar (1992) considered the following problem:

PROBLEM P. Find ρ^* such that $\bar{B}_{\rho^*}(\alpha) \leq \bar{B}_\rho(\alpha)$ for all ρ in $\mathcal{L}$.

In order to solve Problem P, the following auxiliary problem is considered for each $0 < \gamma < 1$:

PROBLEM P$_\gamma$. Find ρ_γ such that $g_{\rho_\gamma}(\gamma) \geq g_\rho(\gamma)$ for all ρ in $\mathcal{L}$.

The main result is the following theorem.

Theorem 7.4 (*Zamar, 1992, Theorem 2*) *Assume* $\rho \in \mathcal{L}$.

(*a*) *For each* $0 < \gamma < 1$, *Problem* P_γ *has a solution denoted by* ρ_γ.
(*b*) *The function* $G(\gamma) = g_{\rho_\gamma}(\gamma)$ *is continuous and strictly increasing in* $(0, 1)$.
(*c*) *Let* $0 < \alpha < 0.5$ *be given. If* $\lim_{\gamma\to 1} G(\gamma) > \alpha/(1-\alpha)$ *then there exists a unique* $\gamma^* = \gamma^*(\alpha)$ *such that* $G(\gamma^*) = \alpha/(1-\alpha)$ *and the bias robust estimate has loss function (7.46) with* $t^* = \gamma^*$. *On the other hand, if* $\lim_{\gamma\to 1} G(\gamma) \leq \alpha/(1-\alpha)$, *then all the orthogonal regression M-estimates break down, that is, either* $\hat{\vartheta}_n$ *breaks down (its maximum asymptotic bias is unbounded) or* $\hat{\varphi}_n$ *breaks down (its asymptotic bias is equal to 1).*

In his proof, Zamar (1992) showed that the solution ρ_γ is the loss function (7.46) with $t^* = t_{\gamma^*}$, which is defined by

$$t_{\gamma^*} = \sigma_\delta \left\{ \frac{1+\Delta^2 b^2}{\Delta^2 b^2} \log(1+\Delta^2 b^2) \right\}^{1/2}, \tag{7.47}$$

where Δ is defined by (7.43) and $b^2 = 1 - (1-\gamma)^2$. He also showed that

$$G(\gamma) = 2\Phi\left\{\left(\frac{1+\Delta^2 b^2}{\Delta^2 b^2}\log(1+\Delta^2 b^2)\right)^{1/2}\right\} - 2\Phi\{(\log(1+\Delta^2 b^2))^{1/2}\}, \quad (7.48)$$

where Φ is the standard normal distribution function.

The bias robust estimate therefore has loss function (7.46) with $t^* = t_{\gamma^*}$, which is given by (7.47). For a given fraction of contamination α, the next step is to find t^*, which means finding γ^* by solving the nonlinear equation $G(\gamma) = \alpha/(1-\alpha)$.

It appears that the bias robust estimate depends on Δ because of (7.48), but Zamar (1992) showed that this is not so. While the bias robust estimate does not depend on Δ, the corresponding minimax bias (and hence the breakdown point) of the estimate does. Zamar (1992) also gave an upper bound of the breakdown point of general orthogonal M-estimates. The upper bound also depends on the signal to noise ratio Δ, and it is sharp and attained by the most bias robust estimate. The upper bound increases with Δ, which is intuitively reasonable.

7.3 Robust Measurement Error Model Estimation via Robust Covariance Matrices

Another way to find robust estimators in ME models is to use robust covariance matrices. The idea is to use some robust mean estimation procedure to obtain robust estimates, $(\hat{\boldsymbol{\mu}}_x', \hat{\mu}_y)$, of the mean vector and some robust covariance matrix estimation procedure to obtain a robust estimate, $\hat{\boldsymbol{\Sigma}}$, of the covariance matrix of $(\mathbf{x}', y)$; then use a method of moments approach to solve for robust estimates of $(\beta_0, \boldsymbol{\beta}')$. That is, replace the sample means and sample covariance matrices in (5.5) or (5.17) and (5.18) by the robust mean and robust covariance matrix estimates, respectively. One could use existing robust estimates, or construct new ones designed for the particular kind of contamination one wishes to guard against.

In the equation error model (5.4) with $\boldsymbol{\Sigma}_{\delta\delta}$ and $\boldsymbol{\Sigma}_{\delta\varepsilon}$ known, the robust estimate of $\boldsymbol{\beta}$ is

$$\hat{\boldsymbol{\beta}} = (\mathbf{R_{xx}} - \boldsymbol{\Sigma}_{\delta\delta})^{-1}(\mathbf{R_{xy}} - \boldsymbol{\Sigma}_{\delta e}) \quad (7.49)$$

(cf. (5.5)), where $\mathbf{R_{xx}}$ and $\mathbf{R_{xy}}$ are the robust estimates of the covariance matrix of $\mathbf{x}$ and the covariance between $\mathbf{x}$ and y, respectively.

For the no-equation-error functional model, in order to adapt (5.18), one needs to modify the notation. Let $\mathbf{z_i} = (y_i, x_{1i}, \ldots, x_{pi})'$ and

$$\operatorname{cov}\left(\begin{pmatrix}\varepsilon\\ \delta\end{pmatrix}\right) = \begin{bmatrix}\sigma_\varepsilon^2 & \boldsymbol{\Sigma}_{\varepsilon\delta}\\ \boldsymbol{\Sigma}_{\delta\varepsilon} & \boldsymbol{\Sigma}_{\delta\delta}\end{bmatrix} = \sigma_\delta^2\boldsymbol{\Omega}_\mathbf{0} = \sigma_\delta^2\begin{bmatrix}\lambda & \boldsymbol{\Lambda}_{\varepsilon\delta}\\ \boldsymbol{\Lambda}_{\delta\varepsilon} & \boldsymbol{\Lambda}_{\delta\delta}\end{bmatrix}.$$

Formula (5.19) needs a slight modification as well, and becomes

$$\hat{\boldsymbol{\beta}} = (\mathbf{R_{xx}} - \hat{\tau}\boldsymbol{\Lambda}_{\delta\delta})^{-1}(\mathbf{R_{xy}} - \hat{\tau}\boldsymbol{\Lambda}_{\delta\varepsilon}) \quad (7.50)$$

where $\hat{\tau}$ is the smallest root of $|\mathbf{R_{zz}} - \tau\boldsymbol{\Omega}_\mathbf{0}| = \mathbf{0}$, and $\mathbf{R_{zz}}$ is the robust covariance matrix estimate of $\mathbf{z}$.

The rest of the parameters, in the equation error model, can be estimated by

$$\hat{\beta}_0 = \hat{\mu}_y - \hat{\boldsymbol{\beta}}'\hat{\boldsymbol{\mu}}_{\mathbf{x}}, \qquad \hat{\boldsymbol{\Sigma}}_{\xi\xi} = \mathbf{R}_{\mathbf{xx}} - \boldsymbol{\Sigma}_{\delta\delta}, \qquad \hat{\sigma}_e^2 = s_{yy} - \hat{\boldsymbol{\beta}}'\mathbf{R}_{\mathbf{xx}}\hat{\boldsymbol{\beta}} + \hat{\boldsymbol{\beta}}'\boldsymbol{\Sigma}_{\delta\delta}\hat{\boldsymbol{\beta}}, \tag{7.51}$$

where $\hat{\boldsymbol{\beta}}$ is given by (7.49) and $\hat{\mu}_y$, $\hat{\boldsymbol{\mu}}_{\mathbf{x}}$ are the robust estimates of the means of y and $\mathbf{x}$, respectively. For the no-equation-error model, $\hat{\sigma}_\delta^2 = \hat{\tau}$ and β_0 is estimated by the first equation in (7.51) with $\hat{\boldsymbol{\beta}}$ given by (7.50).

There are a number of articles on robust covariance estimation. Much of this material is summarized in Huber (1981, Chapter 8) and Hampel *et al.* (1986, Chapter 5). Let $\mathbf{x_1}, \ldots, \mathbf{x_n}$ be a sample from a p-variate density f of the form

$$f(\mathbf{x}) = |(\det\ \mathbf{V})| f(\|\mathbf{V}(\mathbf{x} - \mathbf{t})\|),$$

where $f(\mathbf{x}) = \tilde{f}(\|\mathbf{x}\|)$ is a spherically symmetric density in $\mathbb{R}^p$. The problem here is to estimate the location vector $\mathbf{t}$ and the covariance matrix, $\mathbf{V}$. Maronna (1976) proposed solutions $(\mathbf{t}, \mathbf{V})$ of the system of unbiased estimating equations

$$\int u_1(d(\mathbf{x}, \mathbf{t}; \mathbf{V}^{-1}))(\mathbf{x} - \mathbf{t}) = \mathbf{0}, \tag{7.52}$$

$$\int u_2(d^2(\mathbf{x}, \mathbf{t}; \mathbf{V}^{-1}))(\mathbf{x} - \mathbf{t})(\mathbf{x} - \mathbf{t})' = \mathbf{V}, \tag{7.53}$$

where the functions $u_1(s)$ and $u_2(s)$ are defined for nonnegative s and $d^2(\mathbf{x}, \mathbf{y}; \mathbf{M}) = (\mathbf{x} - \mathbf{y})'\mathbf{M}(\mathbf{x} - \mathbf{y})$ for p-vectors $\mathbf{x}$ and $\mathbf{y}$ and positive semidefinite matrix, $\mathbf{M}$. He proved the existence and uniqueness of the solution to (7.53) and (7.53) under general conditions on u_1 and u_2. Moreover, he proved large-sample results for the resulting estimators and derived the influence function and breakdown point for the α-contamination neighbourhood (7.15). The breakdown point is not very satisfactory because it is bounded above by $1/(p + 1)$. This means that the breakdown point decreases with the dimension. A computation procedure for the estimator is described in Huber (1981, p. 238).

Stahel (1987) (see also Hampel *et al.*, 1986, Chapter 5) studied the influence function and asymptotic covariance of M-estimators of the covariance matrix. He also obtained the optimal B-robust estimators.

Ammann (1993) studied robust singular value decompositions (RSVDs). Consider the **combined data matrix,** $\dot{\mathbf{Z}}^{(n\times(p+1))} = [\dot{\mathbf{X}}|\dot{\mathbf{y}}]$ whose rows are $(\mathbf{x}_1', y_1), \ldots, (\mathbf{x}_n', y_n)$ centred on the column means of this array. A singular value decomposition (SVD) of $\dot{\mathbf{Z}}$ exists: $\dot{\mathbf{Z}} = \mathbf{UDV}'$, where $\mathbf{U}$ is an $n \times n$ orthogonal matrix, $\mathbf{V}$ is a $(p + 1) \times (p + 1)$ orthogonal matrix, and $\mathbf{D}$ is a diagonal matrix with diagonal elements, $d_1 \geq \ldots, \geq d_{p+1} \geq 0$. Then $\dot{\mathbf{Z}}'\dot{\mathbf{Z}} = \mathbf{V\Lambda V}'$, where $\boldsymbol{\Lambda}$ is the eigenvalue matrix, $\text{diag}[d_1^2, \ldots, d_{p+1}^2]$. The first column of $\mathbf{V}$ gives the direction along which the data have the highest variability, the second column gives the direction along which the data have the second highest variability, and so on. The unit vector orthogonal to the orthogonal regression hyperplane is the last column of $\mathbf{V}$, and thus, this is one way of doing orthogonal regression. Ammann noted that the SVD can be represented as an iteration of two steps: a least-squares regression fit of the data matrix, followed by a rotation to the regression hyperplanes. A robust version of SVD can be obtained by centring $\dot{\mathbf{Z}}$ on robust estimates of the column means and replacing the least-squares regression by a robust regression. Because there is vast literature on robust regression (Hampel *et al.*, 1986, Chapter 6), this approach provides quite a few robust covariance matrix estimates. Ammann's approach is particularly well adapted to obtaining robust estimates of the parameters in the orthogonal regression model, that is, the no-equation-error ME model with error covariance matrix known up to a scalar multiple, because the robust eigenvalues and eigenvectors can be obtained directly from the RSVD (cf. Section 5.4). A simpler direct method of converting regression algorithms to orthogonal regression algorithms is given in the next section. Ammann (1993, p. 508) gives an algorithm for performing RSVD (see Example 7.10).

The exact robust properties such as the breakdown point and the bounded influence of estimators of $\boldsymbol{\beta}$ via robust covariance matrices are, however, not known. For example, if one chooses the bounded influence estimator of the covariance matrix described in Hampel *et al.* (1986, Chapter 5), it is not clear the associated robust estimator of $\boldsymbol{\beta}$ will also be of bounded influence. One property, namely consistency, can be carried over. In other words, if the robust covariance matrix estimator is consistent then so is the resulting estimator of $\boldsymbol{\beta}$.

7.4 Computational Methods for Robust Orthogonal Regression

In this section, we give a general algorithm (Ammann and Van Ness, 1988; 1989) for converting any regression algorithm into a corresponding orthogonal regression algorithm. There is a rich class of regression algorithms available. The algorithm given here converts these algorithms into orthogonal regression algorithms, and thereby provides a corresponding rich class of orthogonal algorithms. Thus, one can perform L_1, robust, weighted, etc. orthogonal regression. This conversion algorithm is an iterative algorithm, but it converges in very few iterations.

The idea is quite simple. For one explanatory variable, if the regression line is parallel to the x-axis (i.e. $\beta_1 = 0$), then the ordinary regression line and the orthogonal regression line are the same. They both measure the distance from a point to the regression line along a line parallel to the y-axis. For a p-dimensional explanatory variable, if the regression hyperplane is parallel to the hyperplane of the x-axes (i.e. $\boldsymbol{\beta} = 0$), then ordinary regression and orthogonal regression again give the same answer. Thus, if one could rotate the data so that the regression hyperplane is parallel to the x-axes, one could then use an ordinary regression procedure and obtain a hyperplane which is both the regression hyperplane and the orthogonal regression hyperplane. Rotating this hyperplane back to the original coordinate would give the orthogonal regression hyperplane in the original variables. The problem is, of course, that one does not know the hyperplane in the first place in order to obtain the first rotation. This suggests the following iterative procedure.

Suppose one has some ordinary regression routine, let us call it REG, that one wishes to convert to a corresponding orthogonal regression routine. For example, REG might be L_1 regression (mean absolute error regression) or some form of robust ordinary regression. For simplicity, first consider the case $p = 1$. The conversion algorithm proceeds as follows (Ammann and Van Ness, 1988):

1. Perform REG on the data to obtain a regression line.
2. Rotate the data so that the latest regression line is horizontal.
3. Perform REG on the rotated data, where the new x variable is the variable parallel to the regression line and the new y variable is the variable perpendicular to that line.
4. Check to see if the latest $\hat{\beta}_1$ is within a predetermined accuracy tolerance of zero. If so, stop and rotate the data and the latest line back to the original coordinates. If not, go to step 2.

The tolerance in step 4 will depend on the accuracy required. Typically, one might take a tolerance giving three or four significant figures. Experience indicates that this algorithm converges very fast and that it can easily accommodate small tolerances. If $p > 1$, the algorithm is the same except that the rotation takes the p-dimensional regression hyperplane parallel to plane of the p x-variable axes.

In more detail, the algorithm for p-dimensional $\mathbf{x}$ can be described as follows. Suppose that REG expects the data in the form of the data matrix

$$\mathbf{D}^{(n\times(p+1))} = \left\{ \begin{array}{cccccc} x_{1,1} & x_{2,1} & \cdots & x_{p,1} & y_1 \\ x_{1,2} & x_{2,2} & \cdots & x_{p,2} & y_2 \\ \vdots & \vdots & \ddots & \vdots & \vdots \\ x_{1,n} & x_{2,n} & \cdots & x_{p,n} & y_n \end{array} \right\}$$

Centre the data on some bounded influence estimates of the means, $(\hat{x}_1,\hat{x}_2, \ldots ,\hat{x}_p, \hat{y}) = (\hat{\mathbf{x}}', \hat{y})$; for example, use the column medians. Thus, define

$$\dot{\mathbf{D}} = \mathbf{D} - \{\hat{x}_1\mathbf{1}, \hat{x}_2\mathbf{1}, \ldots ,\hat{x}_p\mathbf{1}, \hat{y}\mathbf{1}\}, \tag{7.54}$$

where $\mathbf{1}$ is a column vector of 1s. Amman and Van Ness (1988; 1989) found that centring the data first greatly improved the performance of the algorithm.

The algorithm for the conversion of REG to a corresponding orthogonal regression routine is as follows:

Initialize $\mathbf{R}$ to the identity matrix, $\mathbf{I}^{((p+1)\times(p+1))}$;
Get the initial fit $\mathbf{b}^{(p+1)} := REG\left(\dot{\mathbf{D}}\right)$;
while $\|\mathbf{b}\| > tolerance$ **do**
  begin get rotation matrix $\mathbf{P}^{((p+1)\times(p+1))}$;
    rotate data, $\dot{\mathbf{D}} := \dot{\mathbf{D}} * \mathbf{P}$;
    update $\mathbf{R}$, $\mathbf{R} := \mathbf{R} * \mathbf{P}$;
    get new fit, $\mathbf{b} := REG\left(\dot{\mathbf{D}}\right)$;
  end;
$\mathbf{b}_0$:= last column of $\mathbf{R}$;
$\hat{\boldsymbol{\beta}} = \left(-b_{0,1}/b_{0,p+1}, \ldots, -b_{0,p}/b_{0,p+1}\right)'$;

$\beta_0 = \hat{y} - \hat{\boldsymbol{\beta}}'\hat{\mathbf{x}};$

If one wishes to do bounded influence orthogonal regression, then REG should be a robust ordinary regression routine that suitably downweights data points having either high leverage or high residual (or both). In other words, REG could be chosen to be an algorithm implementing a GM-estimate of the form (7.33). Many robust regression routines in computer packages do *not* do this, they downweight only high-residual data points.

7.4.1 The roreg algorithm

In Ammann and Van Ness (1988; 1989) this conversion algorithm is implemented using the computer statistics package, S (see, for example, Becker *et al.*, 1988), and a robust regression algorithm called 'rreg' that is contained in S and S-PLUS (see, for example, Venables and Ripley, 1994, p. 215). A series of Monte Carlo tests of the resulting algorithm were done. The routine, rreg, in its standard form only downweights high residual data points, but it has a feature for implementing 'intrinsic weights', which enables one to also downweight high-leverage data points. Denote by 'roreg' (robust orthogonal regression) the routine that implements the conversion routine with REG = rreg, where intrinsic weights have been used to make roreg a GM-estimating routine.

Roreg calculates the intrinsic weights as follows. Assume the data have been centred as in (7.54), the ith iteration has yielded $\hat{\boldsymbol{\beta}}_i$ and some weight function, w, has been chosen. The routine, rreg, provides a selection of weight functions that can be used: andrews, bisquare, cauchy, fair, hampel, huber, logistic, talworth, and welsch. Now, to calculate the intrinsic weights for rreg in the $(i+1)$th iteration first calculate the leverage values,

$$v_j(\hat{\boldsymbol{\beta}}_i) = \frac{y_j - \hat{\boldsymbol{\beta}}_i \mathbf{x}_j}{K(\hat{\boldsymbol{\beta}}_i)}, \qquad j = 1, 2, \ldots, n$$

(recall (7.34)), and standardize these by dividing them by their standard deviation:

$$v_j^{(s)}(\hat{\boldsymbol{\beta}}_i) = \frac{v_j(\hat{\boldsymbol{\beta}}_i)}{\sqrt{\operatorname{var}(v_1, \ldots, v_n)}}.$$

The intrinsic weights for the $(i+1)$th iteration are then $w(v_1^{(s)}(\hat{\boldsymbol{\beta}}_i)), \ldots, w(v_n^{(s)}(\hat{\boldsymbol{\beta}}_i))$. Ammann and Van Ness choose the starting value of $\hat{\boldsymbol{\beta}}$ to be

$$\begin{aligned}\hat{\boldsymbol{\beta}}_0 \;=\; & \left(\underset{j=1,2,\ldots,n}{\text{median}} \left(\frac{y_j - \text{median}_{k=1,2,\ldots,n}\, y_k}{x_{1,j} - \text{median}_{k=1,2,\ldots,n}\, x_{1,k}} \right), \right. \\ & \left. \ldots, \underset{j=1,2,\ldots,n}{\text{median}} \left(\frac{y_j - \text{median}_{k=1,2,\ldots,n}\, y_k}{x_{p,j} - \text{median}_{k=1,2,\ldots,n}\, x_{p,k}} \right) \right)',\end{aligned}$$

which is an extremely robust estimator of $\hat{\boldsymbol{\beta}}$.

7.4.2 Monte Carlo study of roreg

Ammann and Van Ness (1988; 1989) presented the results of several Monte Carlo studies of roreg on the structural model. These results show that, for the ME model data studied,

1. orthogonal regression can be quite unstable when the data are contaminated;
2. roreg provides dramatic estimator improvement on contaminated data;
3. roreg is very efficient compared to orthogonal regression when applied to uncontaminated data;
4. roreg requires very few iterations (between two and seven);
5. the performance of both orthogonal regression and roreg deteriorates as the slope parameters, $\boldsymbol{\beta}$, increase;
6. both orthogonal regression and roreg improve dramatically as σ^2 increases; and
7. roreg performs better than orthogonal regression on structural model with heavy-tailed error distributions.

The uncontaminated model studied was the normal structural model with $\lambda = 1$ known. The contaminated data were created by simulating misrecorded data or by generating data from a structural model with extremely heavy tails. In Ammann and Van Ness (1989), the model characteristics were varied as follows:

- two dimensions for $\boldsymbol{\beta}$, $p = 1,\ 3$;
- two sample sizes, $n = 20,\ 40$ for $p = 1$ and $n = 40,\ 80$ for $p = 3$;
- two slope parameters, $\beta = 1,\ 5$ when $p = 1$ and $\boldsymbol{\beta} = (1,1,1)',\ (1,5,5)'$ when $p = 3$;
- two values of σ, $\sigma = 2,\ 4$ when $p = 1$ and $\boldsymbol{\sigma} = (2,2,2)',\ (4,4,2)'$ when $p = 3$;
- three error distributions, normal, double exponential, and t with three degrees of freedom; and
- contaminated and uncontaminated data.

Ammann and Van Ness (1988) give some results for $p = 4$, $n = 30$ and 80, and $\beta = 10$; and also study the relative performance of three of the weight functions available in rreg – fair, welsch and cauchy.

The contaminated data contained 10% 'outliers' simulating misrecorded data. The outliers were created by taking 10% of the generated data and either multiplying the x or y component of each chosen data point by 10 (misrecording the data by slipping the decimal point one place to the right) or by multiplying it by -1 (recording the wrong sign). The details are given with the summary table for each Monte Carlo run.

We give here a few typical results. These results are based on 400 replications and the weight function, welsch, used in rreg. Tables 7.1–7.4 give the sample mean squared error and the sample standard deviation of the mean squared error so that the reader can get some idea of the scale of the differences between the estimators.

First consider the efficiency of roreg. Because the efficiency appears good, one can confidently use roreg even when one is not certain whether the data are contaminated. For example, for the structural model with $p = 1,\ n = 40,\ \beta_0 = 0,\ \beta_1 = 5, \xi \sim N(0, 4),\ \delta \sim N(0, 1)$, and $\varepsilon \sim N(0, 1)$, the results from Ammann and Van Ness (1989) are given in Table 7.1.

Next consider the same model as in the previous paragraph but now with four points contaminated in each of the 40 data point samples. The contamination is created by multiplying the x value in the first data vector by 10, multiplying the y value in the second data vector by -1, multiplying the x value in the third data vector by -1, and multiplying the y value in the fourth data vector by 10. The results are given in Table 7.2. The improvement in using roreg is dramatic.

Again consider the same model as in Table 7.1 only now assume that δ and ε have a double exponential (Laplace) distribution, $f(x) = \frac{1}{2}e^{-|x|}$. There are no misrecorded data in the sense of the previous paragraph. The results are given in Table 7.3. The improvement in using roreg is again dramatic.

Table 7.1 Performance of roreg on uncontaminated data: 400 replications, $n = 40,\ \beta_0 = 0,\ \beta_1 = 5$

Method	Average mean squared error (standard deviation)		Average (s.d.)
	$\hat{\beta}_0$	$\hat{\beta}_1$	No. of iterations
orthog. regress.	0.686 (0.048)	0.194 (0.018)	
roreg	0.704 (0.049)	0.191 (0.016)	2.21 (0.41)

Now we consider a $p = 3$ dimensional model. Assume the structural model with $n = 80;\ \beta_0 = 0;\ \beta_1 = \beta_2 = \beta_3 = 1;\ \xi_1, \xi_2, \xi_3 \sim N(0, 4);\ \delta \sim N(0, 1)$, and $\varepsilon \sim N(0, 1)$. The contamination was on the first eight points. The first data point had its first x value multiplied by -1, the second data point had its second x value multiplied by 10, the third data point had its third x value multiplied by -1, the fourth data point had its y value multiplied by 10, the fifth data point had its first x value multiplied by 10, the sixth data point had its second x value multiplied by -1, the seventh data point had its third x value multiplied by 10, and the eighth data point had its y value multiplied by -1. The results from Ammann and Van Ness (1989) are given in Table 7.4.

Table 7.2 Performance of roreg on contaminated data: 400 replications, $n = 40$, $\beta_0 = 0$, $\beta_1 = 5$

Method	Average mean squared error (standard deviation)		Average (s.d.)
	$\hat{\beta}_0$	$\hat{\beta}_1$	No. of iterations
orthog. regress.	54.313 (7.123)	126.816 (12.462)	
roreg	0.872 (0.063)	0.345 (0.027)	2.72 (0.47)

Table 7.3 Performance of roreg on data with double exponential errors: 400 replications, $n = 40$, $\beta_0 = 0$, $\beta_1 = 5$

Method	Average mean squared error (standard deviation)		Average (s.d.)
	$\hat{\beta}_0$	$\hat{\beta}_1$	No. of iterations
orthog. regress.	54.064 (6.415)	136.012 (11.742)	
roreg	0.703 (0.048)	0.344 (0.036)	2.58 (0.50)

Table 7.4 Performance of roreg on three-dimensional contaminated data: 400 replications, $n = 80$, $\beta_0 = 0$, $\beta_1 = \beta_2 = \beta_3 = 1$

Method	Average mean squared error (standard deviation)				Average (stan. dev.)
	$\hat{\beta}_0$	$\hat{\beta}_1$	$\hat{\beta}_2$	$\hat{\beta}_3$	No. iterations
orthog. reg.	8.530 (3.991)	30.216 (21.118)	199.502 (174.613	426.154 (415.588)	
roreg	0.075 (0.006)	0.335 (0.003)	0.034 (0.003)	0.040 (0.004)	4.60 (1.23)

Example 7.10 (*Cheng and Van Ness, 1997b*)
We now compare five different regression methods on the Hertzsprung–Russell data for the star cluster CYG OB1 (see Rousseeuw and Leroy, 1987, p. 27). Here x is the logarithm of the effective temperature at the surface of the star and y is the logarithm of the star's light intensity. The values for the 47 stars are plotted in Figure 7.4; the 43 stars grouped more or less together are considered main sequence stars and the four stars in the upper left-hand corner are red giants. For our purposes we can imagine

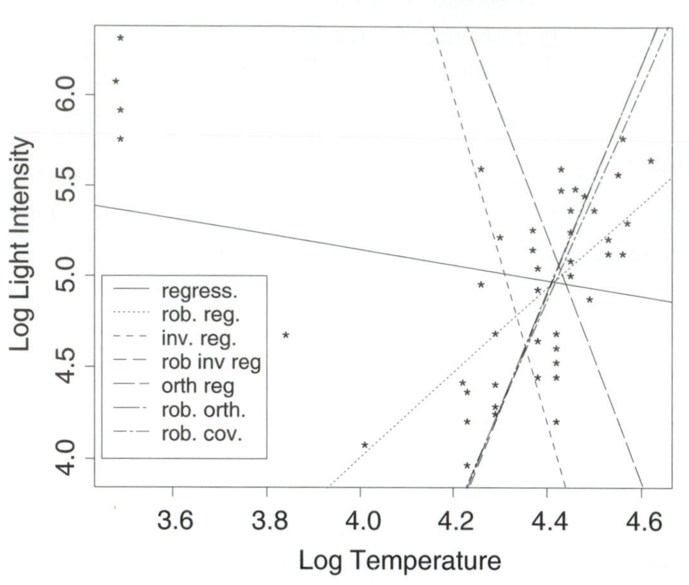

Figure 7.4: Regression comparison on the Hertzsprung–Russell data.

that we are trying to fit a model for the main sequence stars but have four clear outliers (the red giants) contaminating our data. There are some main sequence stars that are between the bulk of the main sequence stars and the red giants. These might be on their way to becoming red giants.

We fitted these data using ordinary regression, robust ordinary regression (rreg), roreg, and a robust covariance method, rsvd (see Ammann, 1993). Not having information on the relative accuracy of the two measurements, we took $\lambda = 1$ for the ME model methods. We found that the results were relatively insensitive to the choice of λ. The bisquare weight function was used in rreg for both the residual weights and the leverage weights. The default tuning constant, 4.685, was used for the residual weights and the tuning constant, 4, was used for the leverage weights. These were values the authors had found worked well in other examples (see Cheng and Van Ness, 1997b). The bisquare weight function with the default tuning constant was also used for roreg. The results are shown in Figure 7.4. The differences are again dramatic. Regression is dominated by the four red giants. Robust regression is much less influenced by the 'outliers', but still sensitive to them. Orthogonal regression is also heavily influenced by the four red giants. Roreg and the robust covariance matrix methods give very similar results and fit the main body of the data well.

7.4.3 Robust calibration

The calibration problem was introduced in Section 2.5.2. Calibration estimates have the same robustness problems as do regression and ME regression estimates. Thus, robust procedures are called for in calibration. Cheng and Van Ness (1997b) investigated robustness for the linear random calibration problem. In that paper seven different calibration methods were compared theoretically, in Monte Carlo studies, and on real data. We now summarize some of these results.

As was pointed out in Section 2.5.2, if the goal is to predict ξ given y in the normal structural model, then inverse regression is the best mean squared error solution. This is not necessarily true when the data are contaminated and no longer normally distributed. It is also not true if the primary goal is to estimate β_0 and β_1 rather than to predict ξ.

The seven methods studied are regression, robust regression (rreg), inverse regression, robust inverse regression (inverse rreg), orthogonal regression, roreg, and Ammann's (1993) rsvd. Two issues were addressed in the Monte Carlo study – efficiency and robustness. The weight functions and tuning constants used were the same as those in Example 7.10.

As in the previous section, two forms of 'contamination' were considered. The first was to generate a normal structural data set and contaminate 10% of it with outliers simulating misrecorded data, and the second was to generate structural data with normal ξs but heavy-tailed errors, ε and δ.

For the structural model with $p = 1$, $n = 20$, $\beta_0 = 0$, $\beta_1 = 1, \xi \sim N(0, 4)$, $\delta \sim N(0, 1)$, and $\varepsilon \sim N(0, 1)$, and contamination consisting of multiplying x_1 by 10 and y_2 by -1, the results from Cheng and Van Ness (1997b) are given in Table 7.5. In this table the sample mean absolute deviations, $\frac{1}{m}\sum_{i=1}^{m}|\hat{\theta}_i - \theta|$, are given, where m is the number of replications. The MAE is followed by its standard deviation in parentheses. These results indicate that robust inverse regression is best in estimating ξ, with roreg and rsvd a close second. However, roreg and rsvd are better at estimating β_1.

Table 7.5 Comparison of seven methods of calibration on 10% contaminated data: 100 replications, $n = 20$, $\beta_0 = 0$, $\beta_1 = 1$

	Mean absolute deviation of estimate from true value		
Method	$\hat{\beta}_0$ (s.d.)	$\hat{\beta}_1$ (s.d.)	$\hat{\xi}$ (s.d.)
regression	0.305 (0.025)	0.661 (0.024)	8.043 (20.3)
robust regression	0.252 (0.018)	0.248 (0.015)	1.332 (0.068)
inverse regression	0.840 (0.061)	0.782 (0.343)	1.422 (0.159)
robust inv. regress.	0.394 (0.032)	0.390 (0.037)	0.844 (0.027)
orthog. regress.	0.331 (0.026)	0.686 (0.121)	6.482 (0.703)
roreg	0.290 (0.019)	0.174 (0.014)	1.118 (0.166)
rsvd	0.281 (0.021)	0.171 (0.013)	0.917 (0.032)

Table 7.6 Comparison of seven methods of calibration on data with t_3 errors: 100 replications, $n = 20$, $\beta_0 = 0$, $\beta_1 = 1$

	Mean absolute deviation of estimate from true value		
Method	$\hat{\beta}_0$ (s.d.)	$\hat{\beta}_1$ (s.d.)	$\hat{\xi}$ (s.d.)
regression	0.263 (0.021)	0.191 (0.015)	1.109 (0.094)
robust regression	0.226 (0.019)	0.166 (0.012)	0.958 (0.054)
inverse regression	0.305 (0.026)	0.244 (0.041)	0.678 (0.029)
robust inv. regress.	0.261 (0.021)	0.204 (0.021)	0.682 (0.034)
orthog. regress.	0.271 (0.024)	0.169 (0.030)	0.859 (0.073)
roreg	0.230 (0.019)	0.132 (0.012)	0.810 (0.053)
rsvd	0.237 (0.020)	0.139 (0.012)	0.798 (0.049)

For the same model but with the outliers removed and the error distributions changed from normal to t with three degrees of freedom, the results are given in Table 7.6. These results are much more uniform. Again, robust inverse regression performs well in estimating ξ, but so does inverse regression. Both roreg and rsvd also perform reasonably well, but significantly worse than inverse regression. Again roreg and rsvd were best at estimating β_1.

7.5 Bibliographic Notes and Discussion

Robust statistics has existed implicitly for a long time. The term 'robust' was coined by Box (1953). Because of a lack of systematic development, it was as late as the 1960s before the subject was recognized as a legitimate separate statistical topic. The major pioneering paper was Huber (1964). Another major early contribution was by F. Hampel, in his Ph.D. dissertation (Hampel, 1968). Since then, robust statistics has developed rapidly in many directions. Most of the results are summarized in Huber (1981) and Hampel *et al.* (1986). In this chapter, we briefly mention the concepts of breakdown point and influence function and focus on the use of M-estimators in ME models.

The maximum likelihood estimators and method of moments estimators for the ME model are in some sense even more sensitive to contamination – particularly outliers – than are the least squares estimates for regression. For example, the expression for the influence function for the ME model estimates, (7.30), has the term $(\dot{\mathbf{x}} + \boldsymbol{\beta}\dot{y})$ in it; while in regression, this term is replaced by $\dot{\mathbf{x}}$ (see, for example, Hampel *et al.*, 1986, p. 316). Thus, robust ME model methods will be needed in many applications. Because of the difficulty of the problem, robust estimation for ME models remained unsolved for a long time after Huber (1972) commented on its importance. Carroll and Gallo (1982) seem to have been the first to propose an robust estimator for the slope in the simple ME model. This was for the case when the explanatory variable has replications. Kelly (1984) derived the influence function of the estimator (1.30) but no robust estimator was proposed.

Zamar (1989) studied (classical) M-estimators for the ME model when the error covariance matrix is known up to a proportionality factor.

Cheng and Van Ness (1990) and Yohai and Zamar (1990) independently proposed a bounded influence M-estimator in the orthogonal regression assumption. Cheng and Van Ness (1992) solved Hampel's optimality problem (see Section 7.2) for simple orthogonal regression.

Croos and Fuller (1991) proposed a robust estimator based on the robust regression estimator of y on and x and the robust regression estimator of x on y. The resulting estimator is consistent and asymptotically normal when the error is normally distributed with known covariance matrix.

Carroll *et al.* (1993) derived a bounded influence robust estimator when some of the explanatory variables are measured without error. They also address the important special case when the explanatory variables are replicated and when δ and ε are correlated. The resulting estimators are variant of the Mallows-type estimator (see Hampel *et al.*, 1986, p. 321).

There are several approaches to robustness, including M-estimators, L-estimators and R-estimators (see Huber, 1981; Hampel *et al.*, 1986), which correspond respectively to ML-type estimates, linear combinations of order statistics, and estimates derived from rank tests. These are competing approaches – Huber (1981, Chapter 3) has a general discussion. Because of the difficulty in obtaining the robustness properties of L- and R-estimates, especially in the multidimensional parametric problem, M-estimates has received more attention in the literature. Even in the class of M-estimates, there are several conceptually different approaches to robustness; see Huber (1983, p. 67) for a summary. Of course, there are still other approaches, such as the bias robustness approach (cf. Section 7.2). Unfortunately, it seems inevitable that (except in certain very simple cases such as the location problem) any single robust estimator cannot fulfil all or even more than one criterion for robustness. For example, the bounded influence estimators, described in Section 7.2, generally have poor breakdown points. In robust ordinary regression, the breakdown point of GM-estimators is a decreasing function of the dimension, $p+1$, of the explanatory variable vector (Maronna *et al.*, 1979; Maronna and Yohai, 1981). Although the corresponding results are not yet known for ME models, we suspect that they are not any better. The practitioner needs to determine the kind of robustness properties that are most important for his or her specific problem and then choose the appropriate estimator.

There are still many unanswered questions in robust ME models, as there are in most areas of robustness. Some are mentioned in this chapter. For example, the robust ME model estimators utilizing robust covariance matrices are attractive because this method provides robust estimators for all kinds of ME models, but their exact robustness properties are not yet known. What robust properties of the estimators are induced by each different robust covariance matrix methods? There are many general research areas open in robust ME model estimation: for example, scaling problems, higher-dimensional models, and nonlinear models.

7.6 Exercises

7.1 *Give the details of the derivation of (7.13).*

7.2 *Derive (7.27).*

7.3 *Let $p = 1$ and T be a functional (estimator) with influence function $IF(x; T, F)$. The self-standardized sensitivity is defined (Hampel et al., 1986, p. 228) as*

$$\gamma_s^* = \sup_x \{IF'(x; T, F)V(T, F)^{-1}IF(x; T, F)\}^{1/2},$$

where $V(T, F)$ is the asymptotic covariance of T defined in (7.36). Find the optimal B-robust estimator with respect to γ_s^* for the ME model with $\lambda = 1$ known.

7.4 *The α-trimmed mean ($0 < \alpha < 1$) is given by*

$$T(F) = \frac{\int_\alpha^{1-\alpha} F^{-1}(t)\,\mathrm{d}t}{1 - 2\alpha}$$

(Hampel et al., 1986, p. 99). The finite-sample version is

$$T_n = \bar{x}_{n,\alpha} = \frac{1}{n - 2r} \sum_{k=r+1}^{n-r} x_{(k)},$$

where $x_{(k)}$ is the kth-order statistic and $r = [n\alpha]$. Find the breakdown point of the α-trimmed mean.

7.5 *Find the breakdown point of the least-squares estimator for ordinary regression.*

7.6 *Find the breakdown point of the orthogonal regression estimator for the ME model ($\lambda = 1$).*

7.7 Research Problem

7.7 *The Mallows-type estimators for orthogonal regression are defined, similar to those in robust ordinary regression (Hampel et al., 1986, p. 321), as*

$$\tau(\mathbf{w}, v) = \zeta(\mathbf{w})\psi(v),$$

for some appropriate functions ζ and ψ, with $\mathbf{w}$ and v as defined in (7.34). Find the optimal B-robust estimator within this class.

8

Additional Topics

In this chapter we take up some additional topics not previously discussed.

8.1 Estimation of the True Variables

The estimates of the true values of the variables can be useful in a number of ways. For example, Fuller (1987, p. 25) suggested that they can be used in model checking (see also Exercises 8.3 and 8.4). In particular, the analogy of the residual versus independent variables plot in classical regression theory is the residual $\hat{v}_i$ versus estimated true values $\hat{\boldsymbol{\xi}}_i$ plot in the ME model (cf. Exercise 8.3 and 8.4). They might also be useful in diagnostics analysis for the ME model, which is an area that is still largely unexplored (cf. Cheng and Tsai, 1993; Wellman and Gunst, 1991).

The general idea is that one first estimates the unknown structural parameters and then treats these estimates as if they were known for the purpose of estimating the true values. As we have seen in Chapter 2, the estimates of the βs generally possess no first moment. Consequently, any formula estimating the moments of $\hat{\boldsymbol{\xi}}_i$ will be only estimating the asymptotic moments of $\hat{\boldsymbol{\xi}}_i$ (see Fuller, 1987 p. 21).

Generally speaking, the ME model parameters of major point and interval estimation interest are the βs. The true values $\boldsymbol{\xi}_i$ and η_i are nuisance parameters, which one can, hopefully, by-pass in the estimation procedure. Some properties of the estimates of the true variables are discussed in the literature, however (see, for example, Fuller 1987; 1990).

8.1.1 The linear structural model without equation error

Because the given data are $\mathbf{z}_i' = (y_i, \mathbf{x}_i')$, the natural estimate of $(\eta_i, \boldsymbol{\xi}_i')'$ is the conditional expectation of $(\eta_i, \boldsymbol{\xi}_i')'$ given $\mathbf{z}_i$. Under the joint normality assumption on $\boldsymbol{\xi}_i$, $\boldsymbol{\delta}_i$, and ε_i, it can be shown (see, for example, Fuller, 1990) that

$$E((\eta_i, \boldsymbol{\xi}_i')' \mid \mathbf{z}_i) = \mathbf{z}_i - \boldsymbol{\Omega}\boldsymbol{\Sigma}_{\mathbf{zz}}^{-1}(\mathbf{z}_i - \boldsymbol{\mu}_{\mathbf{z}}), \tag{8.1}$$

where $\mathbf{\Omega}$ is the covariance matrix of the measurement error $(\varepsilon_i, \boldsymbol{\delta}_i')'$, that is,

$$\mathbf{\Omega} = \begin{bmatrix} \sigma_\varepsilon^2 & \mathbf{\Sigma}_{\varepsilon\delta} \\ \mathbf{\Sigma}_{\delta\varepsilon} & \mathbf{\Sigma}_{\delta\delta} \end{bmatrix}.$$

Therefore, the estimate of $(\eta_i, \boldsymbol{\xi}_i')'$ is (when the parameters are identifiable)

$$(\hat{\eta}_i, \hat{\boldsymbol{\xi}}_i')' = \mathbf{z}_i - \hat{\mathbf{\Omega}}\mathbf{S}_{\mathbf{zz}}^{-1}(\mathbf{z}_i - \bar{\mathbf{z}}), \tag{8.2}$$

where $\bar{\mathbf{z}}$ is the sample mean, $\mathbf{S}_{\mathbf{zz}}$ is the sample covariance matrix of $\mathbf{z}_i$, and $\hat{\mathbf{\Omega}}$ is an estimate of $\mathbf{\Omega}$.

If the normality assumption does not hold, the estimate (8.2) is the best linear predictor of $(\eta_i, \boldsymbol{\xi}_i')'$ in the sense of minimum mean squared error.

8.1.2 The linear functional model without equation error

We begin with the simple ME model. In Section 1.3.2, when the errors are jointly normal and the ratio of the error variances, λ, is known, the ML estimate of ξ_i is given by (1.51). For the normal vector explanatory variables model, when the error covariance matrix is known up to a scalar multiple, the ML estimates of the true values $\boldsymbol{\xi}_i$ and η_i are given in Section 5.4.2.

Fuller (1987, p. 20) suggested using the following approach. First, suppose that the structural parameters are known; then for one explanatory variable,

$$\begin{bmatrix} y_i - \beta_0 \\ x_i \end{bmatrix} = \begin{bmatrix} \beta_1 \\ 1 \end{bmatrix} \xi_i + \begin{bmatrix} \varepsilon_i \\ \delta_i \end{bmatrix}. \tag{8.3}$$

Equation (8.3) can be treated as a classical regression model with ξ_i being the unknown parameters. Therefore, the best linear unbiased estimator of ξ_i is

$$\tilde{\xi}_i = [(\beta_1, 1)\mathbf{\Omega}^{-1}(\beta_1, 1)']^{-1}(\beta_1, 1)\mathbf{\Omega}^{-1}(y_i - \beta_0, x_i)',$$

where $\mathbf{\Omega}$ is the covariance matrix of the measurement error $(\varepsilon_i, \delta_i)$.

In practice, one can replace structural parameters by their estimates. Fuller (1987, pp. 20 ff.) also derived some properties, such as the variance, of the estimate of ξ_i. Another useful formula for estimating ξ_i, which is equivalent to $\tilde{\xi}_i$, is (Fuller, 1987, p. 21 (1.2.27))

$$\hat{\xi}_i = x_i - \hat{\sigma}_v^{-2}\hat{\sigma}_{\delta v}\hat{v}_i, \tag{8.4}$$

where $v_i = y_i - \beta_0 - \beta_1 x_i = \varepsilon_i - \beta_1\delta_i$ is the residual defined in Section 3.5, $\hat{v}_i = y_i - \hat{\beta}_0 - \hat{\beta}_1 x_i$ is its estimate, $\sigma_{\delta v} = \sigma_{\delta\varepsilon} - \beta_1\sigma_\delta^2$, and $\hat{\sigma}_{\delta v}$ is its estimate.

If $\sigma_\varepsilon^2/\sigma_\delta^2 = \lambda$ is known and ε and δ are uncorrelated, then (8.4) coincides with the ML estimate (1.51).

For the vector explanatory variables model, the same argument holds; one need only treat x_i, ξ_i, and δ_i as vectors in the above formulae. Therefore the estimate of $\boldsymbol{\xi}_i$ becomes

$$\hat{\boldsymbol{\xi}}_i = \mathbf{x}_i - \hat{\sigma}_v^{-2}\hat{\mathbf{\Sigma}}_{\delta v}\hat{v}_i, \tag{8.5}$$

where $\hat{v}_i = y_i - \hat{\beta}_0 - \hat{\boldsymbol{\beta}}'\mathbf{x}_i$ and $\hat{\mathbf{\Sigma}}_{\delta v} = \hat{\mathbf{\Sigma}}_{\delta\varepsilon} - \hat{\mathbf{\Sigma}}_{\delta\delta}\hat{\boldsymbol{\beta}}$.

One need not assume normality of the errors in the preceding estimation procedure.

8.1.3 The equation error model

In the functional model, there is no problem estimating η_i when the model does not have an equation error because it is obvious that

$$\hat{\eta}_i = \hat{\beta}_0 + \hat{\boldsymbol{\beta}}'\hat{\boldsymbol{\xi}}_i, \tag{8.6}$$

where the right-hand side of this equation employs the appropriate estimates of the parameters. On the other hand, if the functional model does have equation error, Fuller (1987, p. 114) suggested that one first estimate the measurement errors $(\varepsilon_i, \boldsymbol{\delta}_i')$ and then estimate the true variables $(\eta_i, \boldsymbol{\xi}_i')$ by

$$(\hat{\eta}_i, \hat{\boldsymbol{\xi}}_i') = (y_i, \mathbf{x}_i') - (\hat{\varepsilon}_i, \hat{\boldsymbol{\delta}}_i').$$

The estimate of the measurement error $(\varepsilon_i, \boldsymbol{\delta}_i')'$, when the errors are normally distributed, is $\hat{v}_i\hat{\sigma}_v^2\widehat{\text{cov}}(v, (\varepsilon, \boldsymbol{\delta}')')$, where $v_i = y_i - \beta_0 - \boldsymbol{\beta}'\mathbf{x}_i$ is the residual and

$$\text{cov}(v, (\varepsilon, \boldsymbol{\delta}')') = (1, -\boldsymbol{\beta}')\boldsymbol{\Omega}.$$

Therefore, the estimate of the true value $(\eta_i, \boldsymbol{\xi}_i')$ is

$$(\hat{\eta}_i, \hat{\boldsymbol{\xi}}_i') = (y_i, \mathbf{x}_i') - \hat{\sigma}_v^2\hat{v}_i(1, -\hat{\boldsymbol{\beta}}')\hat{\boldsymbol{\Omega}}. \tag{8.7}$$

As for the structural model, one still can use (8.1) with the $\boldsymbol{\Omega}$ replaced by $\boldsymbol{\Omega}_\mathbf{e}$, where

$$\boldsymbol{\Omega}_\mathbf{e} = \begin{bmatrix} \sigma_e^2 & \boldsymbol{\Sigma}_{\varepsilon\delta} \\ \boldsymbol{\Sigma}_{\delta\varepsilon} & \boldsymbol{\Sigma}_{\delta\delta} \end{bmatrix}$$

and $\sigma_e^2 = \sigma_\varepsilon^2 + \sigma_q^2$. Thus the estimate of the true values is similar to (8.2) with $\hat{\boldsymbol{\Omega}}$ replaced by $\hat{\boldsymbol{\Omega}}_\mathbf{e}$.

8.1.4 The nonlinear model

In the nonlinear model,

$$\mathbf{x}_i = \boldsymbol{\xi}_i + \boldsymbol{\delta}_i, \qquad y_i = g(\boldsymbol{\xi}_i, \boldsymbol{\beta}) + e_i,$$

where $e_i = \varepsilon_i + q_i$ (q_i could be identically zero), the procedures in the previous subsection are not valid. For the functional model, equation (8.3) is no longer true. For the structural model, the joint distribution of $(\mathbf{x}_i', y_i)$ is not known even if $(\boldsymbol{\xi}_i, \boldsymbol{\delta}_i, e_i)$ are jointly normal. For the structural model with $\boldsymbol{\delta}_i$ and ε_i independent and the $\mathbf{x}_i$ normal,

$$E(\boldsymbol{\xi}_i \mid \mathbf{x}_i) = \mathbf{K}'\mathbf{x}_i + (\mathbf{I} - \mathbf{K}')\boldsymbol{\mu},$$

where $\boldsymbol{\mu}$ is the mean of the $\boldsymbol{\xi}_i$, $\mathbf{K} = \boldsymbol{\Sigma}_{\mathbf{xx}}^{-1}\boldsymbol{\Sigma}_{\xi\xi}$ is the reliability matrix defined in Section 5.3.2, and $\mathbf{I}$ is the identity matrix. Then, if the parameters are identifiable, the estimate of $\boldsymbol{\xi}_i$ becomes

$$\hat{\boldsymbol{\xi}}_i = \hat{\mathbf{K}}'\mathbf{x}_i + (\mathbf{I} - \hat{\mathbf{K}}')\bar{\mathbf{x}}, \tag{8.8}$$

where $\bar{\mathbf{x}}$ is the sample mean of $\mathbf{x}_i$ and $\hat{\mathbf{K}}$ is the estimate of the reliability matrix. Even if $\mathbf{x}_i$ is not normal, one still can view (8.8) as the best linear (in $\mathbf{x}_i$) predictor of $\boldsymbol{\xi}_i$ under mean squared error.

An important use of the estimate (8.8) is that one can use it to replace $\boldsymbol{\xi}_i$ in the original relation. After this approximation, one can use classical nonlinear regression on the model

$$y_i = g(\hat{\boldsymbol{\xi}}_i; \boldsymbol{\beta}) + e_i.$$

This is called **regression calibration** and has been suggested by many authors, for example, Gleser (1990) and Carroll and Stefanski (1990). Regression calibration is simple and applicable to any ME model, but remains only an approximation in most cases. For more details, see Carroll *et al.*, (1995, Chapter 3). For the functional model, the meaning of $E(\boldsymbol{\xi}_i|\mathbf{x}_i)$ is not well defined; many authors still use (8.8) as an approximation. There is not much difference from a point estimation point of view, however; see the comment of Gleser (1990, p. 101).

8.2 Obtaining Identifiability Assumption Information

Almost all the estimation procedures for ME models require some form of additional knowledge about the measurement errors. We now briefly discuss some possible ways of obtaining such knowledge.

We begin with the classical univariate no-equation-error ME model with $\lambda = \sigma_\varepsilon^2/\sigma_\delta^2$ known. In many situations, the experimenter might have some knowledge of how ξ_i and η_i are measured and hence have some information on the relative scale of the measurement error. The most common situation is that the two variables are measured in a similar way and it is reasonable to assume that $\lambda = 1$. This does not require explicit knowledge of either σ_δ^2 or σ_ε^2. Some authors implicitly assume $\lambda = 1$. This is particularly common in articles using orthogonal regression. In the general (multivariate) ME model, this assumption becomes that the error covariance matrix is equal to $\sigma_e^2\mathbf{I}_p$, where $\mathbf{I}_p$ is the $p \times p$ identity matrix and σ_e^2 is unknown. This assumption only works in the no-equation-error model. If the model contains equation error, knowing the ratio of the measurement error variances does not make consistent estimation of $\boldsymbol{\beta}$ possible under the normality assumption; see Section 3.3.

8.2.1 Equal number of replications with uncorrelated errors

We now examine the use of replications in determining side conditions.

First consider the case when the measurement errors $\boldsymbol{\delta}_i$ and ε_i are uncorrelated. In practice, the experimenter might be able to obtain knowledge about an error variance via replications. Suppose that y_i, $\mathbf{x}_{ij}$, $j = 1, \ldots, m$ $(m \geq 2)$, $i = 1, \ldots, n$, are observed, where

$$y_i = \beta_0 + \boldsymbol{\beta}'\boldsymbol{\xi}_i + e_i, \qquad \mathbf{x}_{ij} = \boldsymbol{\xi}_i + \boldsymbol{\delta}_{ij}, \tag{8.9}$$

the $\boldsymbol{\delta}_{ij}$ are independent and identically distributed measurement errors, and the assumptions on $\boldsymbol{\xi}_i$, e_i, and $\boldsymbol{\delta}_i$ are those appropriate for the corresponding structural, functional or ultrastructural models. Note that e_i can contain equation error. Let $\bar{\mathbf{x}}_{i\cdot}$

be the mean of the m replicate measurements of $\boldsymbol{\xi}_i$, that is, $\bar{\mathbf{x}}_{i\cdot} = \sum_{j=1}^{m} \mathbf{x}_{ij}/m$; then the usual estimate of the error covariance, $\boldsymbol{\Sigma}_{\delta\delta}$, is

$$\hat{\boldsymbol{\Sigma}}_{\delta\delta} = \frac{1}{n(m-1)} \sum_{i=1}^{n} \sum_{j=1}^{m} (\mathbf{x}_{ij} - \bar{\mathbf{x}}_{i\cdot})(\mathbf{x}_{ij} - \bar{\mathbf{x}}_{i\cdot})'. \tag{8.10}$$

In order to use the formulae from the previous chapters for estimating $\boldsymbol{\beta}$, one can replace the data $(y_i, \mathbf{x}_i')$ with $(y_i, \bar{\mathbf{x}}_{i\cdot}')$, remembering that $\bar{\mathbf{x}}_{i\cdot}$ has mean $\boldsymbol{\mu} = E\mathbf{x}_{ij}$ and variance

$$\boldsymbol{\Sigma}_{\xi\xi} + \frac{\boldsymbol{\Sigma}_{\delta\delta}}{m},$$

where $\boldsymbol{\Sigma}_{\xi\xi}$ is the variance of $\boldsymbol{\xi}$ and the sample variance of $\bar{\mathbf{x}}_{i\cdot}$ is

$$\frac{1}{n} \sum_{i=1}^{n} (\bar{\mathbf{x}}_{i\cdot} - \hat{\boldsymbol{\mu}})(\bar{\mathbf{x}}_{i\cdot} - \hat{\boldsymbol{\mu}})', \tag{8.11}$$

where $\hat{\boldsymbol{\mu}} = \sum_{i=1}^{n} \bar{\mathbf{x}}_{i\cdot}/n$. One must adjust for this change in covariance when using the formulae in the previous chapters. For example, (5.5) becomes

$$\hat{\boldsymbol{\beta}} = \left[\sum_{i=1}^{n} ((\bar{\mathbf{x}}_{i\cdot} - \hat{\boldsymbol{\mu}})(\bar{\mathbf{x}}_{i\cdot} - \hat{\boldsymbol{\mu}})' - \hat{\boldsymbol{\Sigma}}_{\delta\delta}/m) \right]^{-1} \sum_{i=1}^{n} (\bar{\mathbf{x}}_{i\cdot} - \hat{\boldsymbol{\mu}}) y_i.$$

If the $\mathbf{x}_{ij}$ are normally distributed, then $\hat{\boldsymbol{\Sigma}}_{\delta\delta}$ is independent of $(y_i, \bar{\mathbf{x}}_i')$. Moreover, $\hat{\boldsymbol{\Sigma}}_{\delta\delta}$ is the ML estimate of $\boldsymbol{\Sigma}_{\delta\delta}$ provided the associated estimate of $\boldsymbol{\Sigma}_{\xi\xi}$ is positive semidefinite (Gleser, 1992, p. 699).

8.2.2 Unequal number of replications

If the replicate numbers are not the same, that is, (8.9) holds with $j = 1, \ldots, m_i$ $(m_i \geq 2)$; then the error variance $\boldsymbol{\Sigma}_{\delta\delta}$ can be estimated by

$$\hat{\boldsymbol{\Sigma}}_{\delta\delta} = \frac{1}{\sum_{i=1}^{n}(m_i - 1)} \sum_{i=1}^{n} \sum_{j=1}^{m_i} (\mathbf{x}_{ij} - \bar{\mathbf{x}}_{i\cdot})(\mathbf{x}_{ij} - \bar{\mathbf{x}}_{i\cdot})'.$$

The other parameters are estimated by

$$\hat{\boldsymbol{\mu}} = \sum_{i=1}^{n} m_i \bar{\mathbf{x}}_{i\cdot} \Big/ \sum_{i=1}^{n} m_i \quad \text{and} \quad \hat{\boldsymbol{\Sigma}}_{\xi\xi} = \frac{1}{\gamma} \left[\sum_{i=1}^{n} m_i (\bar{\mathbf{x}}_{i\cdot} - \hat{\boldsymbol{\mu}})(\bar{\mathbf{x}}_{i\cdot} - \hat{\boldsymbol{\mu}})' - (n-1)\hat{\boldsymbol{\Sigma}}_{\delta\delta} \right],$$

where $\boldsymbol{\mu}$ is the mean of $\boldsymbol{\xi}$ (and hence of $\mathbf{x}$) and $\gamma = \sum_{i=1}^{n} m_i - \sum_{i=1}^{n} m_i^2 / \sum_{i=1}^{n} m_i$.

Gleser (1992) distinguished between replicated models and **reliability studies**. In a reliability study, the experimenter often gathers the replicated data in *different* environments and circumstances from the ME model data $(y_i, \mathbf{x}_i')$. The purpose is to find identifiability information, such as an estimate of $\boldsymbol{\Sigma}_{\delta\delta}$ (or reliability matrix $\mathbf{K}$)

that can be estimated via (8.10) using the replicated reliability study data. Because of the differences in environments and circumstances, one needs to be concerned about whether the assumption obtained from estimates over the reliability study data (for example, estimates of the error variance $\mathbf{\Sigma}_{\delta\delta}$) are good in the environment in which the ME model data are obtained.

The difficulty with unequal numbers of replications is that each $\bar{\mathbf{x}}_{i\cdot}$ has a different error variance $\mathbf{\Sigma}_{\delta\delta}/m_i$. One needs to be careful in adopting formulae from the previous chapters.

On the other hand, the reliability matrix, **K**, which is defined in Section 5.3.2, will be different for each i and hence the procedures described in Section 5.3.2. are not directly applicable. Fuller (1987) and Hasabelnaby *et al.* (1989) suggested methods for dealing with lack of balance in number of replications.

If a reliability study is conducted independently, then the ME regression data $(y_i, \mathbf{x}_i)$ are, of course, independent of the estimate of the error variance, $\hat{\mathbf{\Sigma}}_{\delta\delta}$, obtained from the reliability study. In this situation, one can treat the estimated error variance as a fixed known constant, and then all the procedures and formulae in the previous chapters are directly applicable without change.

8.2.3 Correlated measurement errors

If the measurement errors $\boldsymbol{\delta}_i$ and ε_i are correlated, then one needs a full replication of the model: $y_{ij}, \mathbf{x}_{ij}, j = 1, \ldots, m\ (m \geq 2), i = 1, \ldots, n$, are observed, where

$$\mathbf{x}_{ij} = \boldsymbol{\xi}_i + \boldsymbol{\delta}_{ij}, \qquad y_{ij} = \eta_i + \varepsilon_{ij}.$$

Let $\mathbf{z}_i' = (y_i, \mathbf{x}_i')$ and $\mathbf{\Omega}$ be the covariance matrix of the measurement error $(\varepsilon_i, \boldsymbol{\delta}_i')'$. Then the estimate of $\mathbf{\Omega}$ is

$$\hat{\mathbf{\Omega}} = \frac{1}{\sum_{i=1}^{n}(m_i - 1)} \sum_{i=1}^{n} \sum_{j=1}^{m_i} (\mathbf{z}_{ij} - \bar{\mathbf{z}}_{i\cdot})(\mathbf{z}_{ij} - \bar{\mathbf{z}}_{i\cdot})',$$

where $\bar{\mathbf{z}}_{i\cdot}$ is the mean of the m_i replicate measurements of $(\eta_i, \boldsymbol{\xi}_i')'$.

Again, if one has equal numbers of replications, then the data $\bar{\mathbf{z}}_{i\cdot}$ have error variance $\mathbf{\Omega}/m$. On the other hand, if the numbers of replications are unequal, then $\bar{\mathbf{z}}_{i\cdot}$ has variance $\mathbf{\Omega}/m_i$. Therefore, some adjustments, similar to those mentioned before, are necessary when adopting formulae in the previous chapters.

8.2.4 Validation data

In many applications, particularly in the physical sciences, it is possible to obtain measurements on objects for which $\boldsymbol{\xi}$ is known. For example, one can arrange laboratory experiments where the true values of the variables are set and therefore known. It is then quite simple to estimate $\mathbf{\Sigma}_{\delta\delta}$. If one has m validation data, $\binom{\boldsymbol{\xi}_1'}{\mathbf{x}_1'}, \ldots, \binom{\boldsymbol{\xi}_m'}{\mathbf{x}_m'}$, then

$$\hat{\mathbf{\Sigma}}_{\delta\delta} = \frac{1}{m} \sum (\mathbf{x}_i - \boldsymbol{\xi}_i)(\mathbf{x}_i - \boldsymbol{\xi}_i)'.$$

8.3 Conclusions

Replications can only be used to estimate the covariance matrix of the measurement errors and not the variance of the equation error q_i directly. If one has full replications of the equation error model, then σ_q^2 can be estimated once $\boldsymbol{\beta}$ is estimated. Thus, prior information about σ_q^2 is not available from replications.

The major complaint concerning the use of ME models is the need for further information, not needed in ordinary regression, in order to be able to estimate the parameters. Finding such information in many practical situations can be difficult. Replications can provide the solution when they are possible. An alternative is to ignore the measurement error and, say, use least squares, but this can result in a large bias and one might be estimating parameters that are not even identifiable under the 'true' model. There might be cases where one simply has no choice but to use some such alternative. Another warning concerning obtaining identifiability information is that such knowledge has to be from the 'outside'. It cannot be found within the (unreplicated) ME model data $(y_i, \mathbf{x}_i')$ themselves. For example, this is why the grouping method does not work when the grouping is based on **x**. For another example, for the simple ME model, one might try to regress y on x and regress x on y to find the information about the 'error variances' and/or their ratio λ. This is not legitimate because the resulting (estimated) 'error variances' are not σ_δ^2 and σ_ε^2.

8.4 Relations to Other Latent Variables Models

In this section, we briefly discuss the relation between the linear ME model and other latent variable models, particularly the simultaneous equations ME model and the factor analysis model. The purpose is not to give details, but only to show that such a relation exists for a general class of latent variables models, which are widely discussed and used in many disciplines, such as econometrics, psychology, and social sciences. The latent variables models discussed below all involve multivariate linear ME models, that is, linear ME models with both **y** and **x** multidimensional, which have not been previously discussed in this book.

Before proceeding, we define the multivariate model as follows

$$\boldsymbol{\eta}_i = \boldsymbol{\alpha} + \mathbf{B}\boldsymbol{\xi}_i + \mathbf{q}_i, \qquad \mathbf{x}_i = \boldsymbol{\xi}_i + \boldsymbol{\delta}_i, \qquad \mathbf{y}_i = \boldsymbol{\eta}_i + \boldsymbol{\varepsilon}_i$$

where **y**, $\boldsymbol{\eta}$, $\boldsymbol{\varepsilon}$, **q**, and $\boldsymbol{\alpha}$ are m-vectors; **x**, $\boldsymbol{\xi}$, and $\boldsymbol{\delta}$ are p-vectors; while **B** is an $m \times p$ matrix. This is a direct extension of the vector explanatory variables model discussed in Chapter 5. We also allow the equation error **q** to be identically zero – the no-equation-error model. Again, some authors do not separate out $\boldsymbol{\alpha}$ explicitly.

8.4.1 Simultaneous equations ME model

The simultaneous equations model, commonly used in econometrics, is defined as follows (see Schneeweiss and Mittag, 1986). Suppose that there are unknown true values, $\boldsymbol{\xi}_i$ and $\boldsymbol{\eta}_i$, related by

$$\boldsymbol{\Gamma}\boldsymbol{\eta}_i = \mathbf{B}\boldsymbol{\xi}_i + \mathbf{q}_i, \tag{8.12}$$

$$\mathbf{y}_i = \boldsymbol{\eta}_i + \boldsymbol{\delta}_i, \qquad \mathbf{x}_i = \boldsymbol{\xi}_i + \boldsymbol{\varepsilon}_i, \tag{8.13}$$

for $i = 1, \ldots, n$, where $\mathbf{\Gamma}$ is an $m \times m$ matrix and $\mathbf{B}$ is a $m \times p$ matrix; and $\boldsymbol{\eta}_i$ is an m-vector and $\boldsymbol{\xi}_i$ is a p-vector. One needs some prior restrictions on the parameter matrices $\mathbf{\Gamma}$ and $\mathbf{B}$, whose elements are not all unknown. These prior restrictions are part of the model. If $\mathbf{\Gamma}$ is nonsingular, then (8.12) can be written as

$$\boldsymbol{\eta}_i = \mathbf{A}\boldsymbol{\xi}_i + \mathbf{f}_i,$$

where $\mathbf{A} = \mathbf{\Gamma}^{-1}$ and $\mathbf{f} = \mathbf{\Gamma}^{-1}\mathbf{q}_i$. The model (8.12)–(8.13), therefore, has the same form as the usual multivariate ME model but with some prior restrictions on the parameters.

The simultaneous equations model can be written in a symmetric form as follows (see Aigner *et al.*, 1984, p. 1362). Let $\mathbf{Z}$ be an $n \times l$ matrix of observed data, which is generated by the unobservable true values $n \times l$ matrix, Ξ, plus errors matrix, $\mathbf{\Delta}$, that is,

$$\mathbf{Z} = \mathbf{\Xi} + \mathbf{\Delta}. \tag{8.14}$$

The true (latent) variables matrix, $\mathbf{\Xi}$, is subject to r $(r \leq l)$ linear constraints

$$\mathbf{\Xi}\mathbf{\Pi} = \mathbf{0}, \tag{8.15}$$

where $\mathbf{\Pi}$ is an $l \times r$ parameter matrix. If $r = 1$, then the simultaneous equations model reduces to the familiar ME model. If some row of $\mathbf{\Delta}$ is independent of the other rows of $\mathbf{\Delta}$, it may be interpreted as an equation error. If $r > 1$, model (8.14)–(8.15) defines a simultaneous equations model with errors in the variables.

8.5 The factor analysis model

Factor analysis has a long history in psychometrics and other social sciences. Because it is closely related to the use of instrumental variables, we will look at factor analysis model from this point of view, thereby relating it to previous chapters in this book. We use the following simple example as an illustration:

$$y_{i1} = \beta_{01} + \beta_{11}\xi_i + e_{i1}, \tag{8.16}$$

$$y_{i2} = \beta_{02} + \beta_{12}\xi_i + e_{i2}, \tag{8.17}$$

$$x_i = \xi_i + \delta_i. \tag{8.18}$$

If we assume that ξ_i, e_{i1}, e_{i2}, and δ_i are independent and identically distributed with diagonal covariance matrix, then all parameters are identifiable. One can treat (8.16) and (8.18) as a usual ME model and (8.17) as an instrumental variable used to help estimate β_{01} and β_{11}. Alternatively, one could consider (8.17) and (8.18) as the ME model and (8.16) as the instrumental variable in estimating β_{02} and β_{12}.

Although there is a close analogy between instrumental variable analysis and factor analysis, these two models are conceptually different. In factor analysis, the latent variables are fully conceptual variables and are not treated as variables observed with errors.

The general form for factor analysis (see Schneeweiss and Mittag, 1986) can be written as

$$\mathbf{y}_i = \boldsymbol{\mu} + \mathbf{B}\boldsymbol{\xi}_i + \mathbf{e}_i, \qquad \mathbf{x}_i = \boldsymbol{\xi}_i + \boldsymbol{\delta}_i,$$

where $\mathbf{y}$ is an m-vector, called the **indicator** variables, $\boldsymbol{\xi}$ is the p-vector of latent variables, called **common** or **latent factors**, the m-vector error $\mathbf{e}$ is called the **specific** or **unique factors** and the $m \times p$ matrix, $\mathbf{B}$, is called the matrix of **loadings**. It is assumed that $m > p$, usually $m \gg p$, and the rank of $\mathbf{B}$ is p. Identifiability assumptions need to be imposed on the matrix, $\mathbf{B}$, and/or on the error, $\mathbf{e}$. Typically, the error covariance matrix is assumed to be diagonal as in the example above. Note that the factor analysis model is not exactly the same as the usual linear ME model unless some $p \times p$ submatrix of the loading matrix, $\mathbf{B}$, is the identity matrix, which is not usually true in factor analysis.

8.5.1 Other latent variable models

There are many other latent variables models. A very general model is the **linear structural relationship** (LISREL), which, unfortunately, bears the same name as one of the ME models discussed in depth throughout this book. The LISREL model is general because many latent variables models, such as the factor analysis and simultaneous equations models, can be viewed as special cases. We will not discuss the LISREL, but refer readers to Schneeweiss and Mittag (1986).

For references on latent variables models and their relations to the linear ME model, see Anderson (1976), Gleser (1981), Aigner *et al.* (1984), Schneeweiss and Mittag (1986), Fuller (1987), and Bartholomew and Knott (1999).

8.6 Terminology

The measurement error model bears many different names, such as errors-in-variables model, functional relation, structural relation, and ultrastructural relation. We also mentioned, in Section 3.6, the alternative uses of 'functional modelling' and 'structural modelling' introduced by Carroll *et al.* (1995). Many articles concerning generalized linear models, particularly those in biostatistics and biomedical science, use the term *covariate measurement error*. Covariate refers to the independent variable in those articles.

There is another important term concerning measurement error that has not been mentioned heretofore. If the distribution of y_i does not depend on $\boldsymbol{\delta}_i$, then $\mathbf{x}_i$ carries no information about y_i other than what is available in $\boldsymbol{\xi}_i$. Carroll *et al.* (1995, p. 16) called such measurement error **nondifferential**. In other words, measurement error is nondifferential if the distribution of y_i given $(\boldsymbol{\xi}_i, \mathbf{x}_i)$ depends only on $\boldsymbol{\xi}_i$. This corresponds to the case when the measurement error $\boldsymbol{\delta}_i$ is independent of ε_i for the models of this book. The measurement error is called **differential** otherwise. In the linear ME model, differential measurement errors do not create serious problems in estimation. Problems do occur with differential measurement errors in problem formulation and parameter estimation for the nonlinear ME model. Although differential measurement error for the polynomial model was successfully dealt with in

Chapter 6, it remains a difficult task to treat the general nonlinear ME model with differential measurement error. (cf. Carroll *et al.*, 1995, p. 245).

In the vector explanatory variables model, some of the independent variables can bc measured without error, but such variables were not given a separate notation in this book. If one wishes to distinguish the independent variables measured with and without error, then the model becomes something like

$$y_i = \eta_i + e_i = \beta_0 + \boldsymbol{\beta}_1' \mathbf{w}_i + \boldsymbol{\beta}_2' \boldsymbol{\xi}_i + e_i, \qquad \mathbf{x}_i = \boldsymbol{\xi}_i + \boldsymbol{\delta}_i, \tag{8.19}$$

where y_i, $\mathbf{w}_i$, and $\mathbf{x}_i$ are observable. In our opinion, there is no significant difference between these two formulations. The option seems to be circumstantial. If the experimenter wishes to emphasize that some variables are error-free, then the choice of (8.19) is natural.

Another term frequently used in the literature is **surrogate**. Unfortunately, this terminology is not universally defined. Usually it means an (unbiased) estimate of the true value. For example, for the model, (8.19), $\mathbf{x}_i$ is a surrogate for $\boldsymbol{\xi}_i$ and y_i is a surrogate for η_i,

$$E\mathbf{x}_i = \boldsymbol{\xi}_i \qquad \text{and} \qquad Ey_i = \eta_i.$$

Carroll *et al.* (1995, p. 16) called $\mathbf{x}_i$ a surrogate of $\boldsymbol{\xi}_i$ if it is conditionally independent of the response given the true independent variables; for example, in (8.19) the conditional probability densities must satisfy

$$f(y|x, \mathbf{w}, \boldsymbol{\xi}) = f(y|\mathbf{w}, \boldsymbol{\xi}).$$

Therefore, under this criterion, $\mathbf{x}_i$ is a surrogate for $\boldsymbol{\xi}_i$ and y_i is a surrogate for η_i only when the measurement error is nondifferential.

8.7 Exercises

8.1 *In the linear functional model with an equation error, the estimates of η_i and $\boldsymbol{\xi}_i$ are defined in (8.7). Show that if the equation error vanishes, then the estimate of η_i coincides with (8.6), while with or without equation error, the estimate of $\boldsymbol{\xi}_i$ is the same for (8.5) and (8.7). If the equation error does not vanish, what is the difference between the estimates of η using (8.6) and (8.7)?*

8.2 *In the normal linear structural model, if we use (8.8) and (8.6) to estimate $\boldsymbol{\xi}_i$ and η_i respectively, does this agree with (8.2)? Does the presence of equation error make any difference?*

8.3 *Show that for the linear functional equation error model in which $\boldsymbol{\Sigma}_{\delta\delta}$ and $\Sigma_{\delta\varepsilon}$ are known,*

$$\sum \hat{\boldsymbol{\xi}}_i \hat{v}_i = 0,$$

where $\hat{\boldsymbol{\xi}}_i$ is defined in (8.5). (Hint: Use the modified least-squares method described in (5.23).)

8.4 *Show that for the linear univariate functional no-equation-error model with the error covariance matrix known up to a scalar multiple,*

$$\sum \hat{\xi}_i \hat{v}_i = 0,$$

where $\hat{\xi}_i$ *is defined in (8.5). Is the above equation still true for the linear vector explanatory variables model? (Hint: Use the generalized least-squares method described in (5.22).)*

8.5 *Show that neither the conclusion of Exercise 8.3 nor that of Exercise 8.4 is true if the model is structural where* $\hat{\xi}_i$ *is defined by (8.2). On the other hand, the conclusions of Exercises 8.3 and 8.4 hold true in the functional model regardless of the error distribution.*

Appendix A

Identification in Measurement Error Models

A.1 Overview

This appendix discusses some basic results on identification in linear measurement error models. We shall adopt a notation close to that used in Reiersol (1950). We say that we have specified the **structure** of a model when the distribution of the data generated by the model is completely specified. A structure is thus a particular realization of a model. For example, the mean and covariance completely specify a normal distribution; then the structure of the model is specified if we specify that the model is normal and give specific values for the mean and covariance. For an arbitrary distribution, we usually use characteristic functions to specify the structure since the characteristic function uniquely defines the distribution. A structure, therefore, generates a unique distribution. There may be several structures generating the same distribution of the observations. In this case, we call these structures equivalent. A parameter is identifiable if it has the same value in all equivalent structures.

A general result obtained by Rothenberg (1971) and Bowden (1973) states that a (vector) parameter is identified if the expected Fisher information matrix is nonsingular. Of course, we need to know the joint distribution of the data in order to check the identifiability of the parameter in this way.

A.2 Structural Model

In this section, we give an important result due to Reiersol (1950). Consider the structural model (1.2)–(1.3) where we allow δ_i and ε_i to be correlated but assume ξ_i to be independent of δ_i and ε_i. Denote the characteristic function of the random vector z by $\phi_z(t)$ and denote $\ln \phi_z(t)$ by $\psi_z(t)$. If $(\delta_i, \varepsilon_i)$ are normally distributed, then

$$\phi_{(\delta,\varepsilon)}(t_1, t_2) = \exp\{-\tfrac{1}{2}(\sigma_\delta^2 t_1^2 + 2\sigma_{\delta\varepsilon} t_1 t_2 + \sigma_\varepsilon^2 t_2^2)\}.$$

Because of (1.2),

$$\phi_{(\xi,\eta)}(t_1, t_2) = \exp\{\beta_0 \mathrm{i} t_2\}\phi_\xi(t_1 + \beta_1 t_2)$$

where $\mathrm{i} = \sqrt{-1}$. Therefore

$$\phi_{(x,y)}(t_1, t_2) = \exp\{\beta_0 \mathrm{i} t_2 - \tfrac{1}{2}(\sigma_\delta^2 t_1^2 + 2\sigma_{\delta\varepsilon} t_1 t_2 + \sigma_\varepsilon^2 t_2^2)\}\phi_\xi(t_1 + \beta_1 t_2).$$

Our aim to study the identifiability of the key parameter, β_1. Suppose that there exist two different structures for the structural model with normal errors,

$$S = \{\beta_0, \beta_1, \sigma_\delta^2, \sigma_{\delta\varepsilon}, \sigma_\varepsilon^2, \psi_\xi(t)\} \quad \text{and} \quad \tilde{S} = \{\tilde{\beta}_0, \tilde{\beta}_1, \tilde{\sigma}_\delta^2, \tilde{\sigma}_{\delta\varepsilon}, \tilde{\sigma}_\varepsilon^2, \tilde{\psi}_\xi(t)\},$$

which generate the same probability distribution of the observed variables (x_i, y_i). Then we have

$$\begin{aligned} &\exp\{\beta_0 \mathrm{i} t_2 - \tfrac{1}{2}(\sigma_\delta^2 t_1^2 + 2\sigma_{\delta\varepsilon} t_1 t_2 + \sigma_\varepsilon^2 t_2^2)\}\phi_\xi(t_1 + \beta_1 t_2) \\ &\quad = \exp\{\tilde{\beta}_0 \mathrm{i} t_2 - \tfrac{1}{2}(\tilde{\sigma}_\delta^2 t_1^2 + 2\tilde{\sigma}_{\delta\varepsilon} t_1 t_2 + \tilde{\sigma}_\varepsilon^2 t_2^2)\}\tilde{\phi}_\xi(t_1 + \tilde{\beta}_1 t_2). \end{aligned} \tag{A.1}$$

If $\beta_1 \neq \tilde{\beta}_1$, then we can always find values of t_1 and t_2 such that

$$t_1 + \beta_1 t_2 = z, \qquad t_1 + \tilde{\beta}_1 t_2 = 0, \tag{A.2}$$

where z is arbitrary. Solving the equations above, we obtain

$$t_1 = -\frac{\tilde{\beta}_1}{\beta_1 - \tilde{\beta}_1}, \qquad t_2 = \frac{z}{\beta_1 - \tilde{\beta}_1}.$$

Substituting these expressions in (A.1), we obtain

$$\begin{aligned} \phi_\xi(z) &= \exp\left\{\mathrm{i}\frac{\tilde{\beta}_0 - \beta_0}{\beta_1 - \tilde{\beta}_1} z - \frac{z^2}{2(\beta_1 - \tilde{\beta}_1)^2}\right. \\ &\qquad \left. \times[(\tilde{\sigma}_\delta^2 - \sigma_\delta^2)\tilde{\beta}_1^2 + 2(\tilde{\sigma}_{\delta\varepsilon} - \sigma_{\delta\varepsilon})\tilde{\beta}_1 + \tilde{\sigma}_\varepsilon^2 - \sigma_\varepsilon^2]\right\}. \end{aligned}$$

Therefore, ξ_i is either normally distributed or a constant. The latter case happens if $\sigma_\delta^2 = \tilde{\sigma}_\delta^2$, $\sigma_{\delta\varepsilon} = \tilde{\sigma}_{\delta\varepsilon}$, and $\sigma_\varepsilon^2 = \tilde{\sigma}_\varepsilon^2$. Reiersol (1950) also treated the constant case as a normal random variable. It is also clear that if ξ_i is normally distributed so is η_i. Recall that we have shown in Chapter 1 that, if (x_i, y_i) is normal, then β_1 is not identifiable. Hence the conclusion is:

Theorem A.1 *In the structural model with normally distributed errors, the slope β_1 is identifiable if and only if ξ_i is not normally distributed.*

There is a pitfall in the above result when ξ_i is constant. Then β_1 is not identifiable so ξ_i should be considered as a normal structural model with ξ_i having a degenerate normal distribution, $\xi_i \sim N(\mu, 0)$. The model might also be interpreted as a normal functional model with all the $\xi_i = \mu$. In either case, the model becomes

$$x_i = \mu + \delta_i, \qquad y_i = \beta_0 + \beta_1 \mu + \varepsilon_i. \tag{A.3}$$

Although the definitions are not very clear in this situation, we argue as follows. The most important characteristic of the structural model is that the ξ_i are independent and identically distributed while the most important characteristic of functional model is the presence of incidental parameters which increase with sample size. Therefore, model (A.3) should be considered a structural model and not a functional model according to these characteristics of structural and functional models. This interpretation has the further advantage that we can still say the normal functional model is always identifiable (recall Section 1.2).

Reiersol (1950) went on to characterize the identification of the rest of the parameters when the slope β_1 is identifiable and he proved the following:

Theorem A.2 *When β_1 is identifiable in the structural model with normally distributed errors, then β_0 is also identifiable and the other components of S are identifiable if and only if both of the following conditions are satisfied:*

(a) *neither the distribution of ξ nor the distribution of η is divisible by a normal distribution, and*
(b) *either $\delta \equiv 0$ or $\varepsilon \equiv 0$.*

Reiersol's proof goes as follows. Suppose β_1 is identifiable; for any two equivalent models, equation (A.1) holds with $\beta_1 = \tilde{\beta}_1$. Substituting (A.2), namely $t_1 = z - \beta_1 t_2$, into (A.1) and taking logarithms, we have

$$\begin{aligned} \psi_\xi(z) - \tilde{\psi}_\xi(z) &= (\tilde{\beta}_0 - \beta_0)\mathrm{i}(z - \beta_1 t_2) + \tfrac{1}{2}(\sigma_\delta^2 - \tilde{\sigma}_\delta^2)(z - \beta_1 t_2)^2 \\ &\quad + (\sigma_{\delta\varepsilon} - \tilde{\sigma}_{\delta\varepsilon})(z - \beta_1 t_2)t_2 + \tfrac{1}{2}(\sigma_\varepsilon^2 - \tilde{\sigma}_\varepsilon^2)t_2^2. \end{aligned} \qquad \text{(A.4)}$$

Since this is an identity in z and t_2, the coefficients of t_2, zt_2, and t_2^2 must be zero:

$$\beta_0 = \tilde{\beta}_0, \qquad \sigma_{\delta\varepsilon} - \tilde{\sigma}_{\delta\varepsilon} = \beta_1(\sigma_\delta^2 - \tilde{\sigma}_\delta^2), \qquad \sigma_\varepsilon^2 - \tilde{\sigma}_\varepsilon^2 = \beta_1(\sigma_{\delta\varepsilon} - \tilde{\sigma}_{\delta\varepsilon}). \quad \text{(A.5)}$$

Note that β_0 is always identifiable when β_1 is. Moreover, inserting (A.5) in (A.4), we obtain

$$\tilde{\phi}_\xi(z) = \exp(-\tfrac{1}{2}(\sigma_\delta^2 - \tilde{\sigma}_\delta^2)z^2)\phi_\xi(z). \qquad \text{(A.6)}$$

Now suppose the model is not identifiable and that (a) and (b) hold, say $\delta \equiv 0$. Then $\tilde{\phi}_\xi(z)$ is a legitimate characteristic function, different from $\phi_\xi(z)$. We have from (A.6)

$$\phi_\xi(z) = \exp(-\tfrac{1}{2}\tilde{\sigma}_\delta^2 z^2)\tilde{\phi}_\xi(z),$$

which says that $\phi_\xi(z)$ is divisible by a normal distribution, contradicting the assumption that (a) holds. The case $\varepsilon \equiv 0$ can be treated using the equation for $\phi_\eta(z)$.

Conversely, suppose that one of the conditions is not satisfied. If condition (a) is not satisfied then the distribution of ξ_i is divisible by some normal distribution and can be written as, say,

$$\phi_\xi(z) = \exp(\mathrm{i}\mu_0 z - \tfrac{1}{2}\sigma_0^2 z^2)\phi_0(z),$$

where μ_0, $\sigma_0^2 > 0$, and ϕ_0 are some mean, variance and characteristic function respectively. Substituting into (A.6) gives

$$\tilde{\phi}_\xi(z) = \exp[i\mu_0 z - \tfrac{1}{2}(\sigma_\delta^2 - \tilde{\sigma}_\delta^2 + \sigma_0^2 z^2)]\phi_0(z) \tag{A.7}$$

which will be a characteristic function provided

$$\sigma_\delta^2 - \tilde{\sigma}_\delta^2 + \sigma_0^2 \geq 0. \tag{A.8}$$

We also need to know that there is a solution of (A.8) with $\tilde{\sigma}_\delta^2 > 0$ and $\tilde{\sigma}_\varepsilon^2 > 0$. From (A.5) we see that we must have

$$\tilde{\sigma}_\varepsilon^2 = \sigma_\varepsilon^2 + \beta_1^2(\tilde{\sigma}_\delta^2 - \sigma_\delta^2) > 0. \tag{A.9}$$

Thus we can choose $\tilde{\sigma}_\delta^2 > \sigma_\delta^2$ so that (A.9) is satisfied and also have $\tilde{\sigma}_\delta^2 - \sigma_\delta^2$ small enough so that (A.8) is also satisfied. Determining $\tilde{\phi}_\xi(z)$, $\tilde{\sigma}_{\delta\varepsilon}$, and $\tilde{\sigma}_\varepsilon^2$ from equations (A.5) and (A.6), we can obtain a structure $\tilde{S}$ equivalent to S in which the value of β_1 is the same as in S.

If condition (b) is violated, then both σ_δ^2 and σ_ε^2 are positive, and we can choose $\tilde{\sigma}_\delta^2$ such that $\sigma_\delta^2 > \tilde{\sigma}_\delta^2 > 0$ and (A.9) holds. Then from (A.6), $\tilde{\phi}_\xi(z)$ is a characteristic function and also the equations (A.5) have a solution. Thus, we can again find a structure $\tilde{S}$ equivalent to S such that $\beta_1 = \tilde{\beta}_1$ because (A.5) and (A.6) are satisfied. This proves the theorem. □

The above theorem might be interpreted incorrectly because condition (b) makes the structural model an ordinary regression model (with random regressor). This may suggest that if we need to identify the ME structural model we have to require the measurement error (in either x or y) to be identically zero. First of all, the intercept β_0 is identifiable if the slope β_1 is identifiable and the converse is true. In fact, we can set

$$\bar{y} = \hat{\beta}_0 + \hat{\beta}_1 \bar{x}; \tag{A.10}$$

then if either of the βs is identifiable and admits a consistent estimator denoted by $\hat{\beta}$ (see Section A.4 below), then the other β can be consistently estimated by (A.10). Therefore, if our major interest is to estimate the intercept and slope there is no need for any further conditions.

Secondly, if any of the components of S are known in addition to the identifiability of β_1, then the rest of the structure is identifiable without condition (a) or (b) above. For example, when $\sigma_{\delta\varepsilon} = 0$, a case which we have discussed extensively in the main text, it is obvious from (A.5) and (A.6) that we have $\sigma_\delta^2 = \tilde{\sigma}_\delta^2$, $\sigma_\varepsilon^2 = \tilde{\sigma}_\varepsilon^2$, and $\psi_\xi(z) = \psi_\xi^*(z)$. In other words, the preceding statement is strictly restricted to lack of any knowledge about the structure S other than the identifiability of β_1.

A.3 Functional Model

As we stated in Section 1.2, the normal functional model is identifiable without any additional assumption but lacks consistent estimators. A simple example illuminates the situation.

Example A.1 (*Neyman and Scott, 1948, and Hsiao, 1989*)
Consider the data set $(x_1, \ldots, x_n)$ with

$$x_i = \theta_i + \varepsilon_i,$$

where ε_i are independent and identically distributed with mean zero and variance σ^2 and the common distribution is normal. Then the expected Fisher information matrix is an $(n+1) \times (n+1)$ diagonal matrix, with the first n diagonal elements equal to $1/\sigma^2$ and the last diagonal element equal to $n/2\sigma^4$, which is nonsingular. By the theorem of Rothenberg (1971), the model is identifiable. Yet it is clear that with one observed x_i for θ_i, we cannot obtain two separate estimates for θ_i and σ^2. Furthermore, there are no consistent estimates as $n \to \infty$.

It is also clear intuitively that if two structures

$$S = \{\theta_1, \ldots, \theta_n, \sigma^2\}, \qquad \tilde{S} = \{\tilde{\theta}_1, \ldots, \tilde{\theta}_n, \tilde{\sigma}^2\}$$

have any pair of elements different, then they will generate samples with two different distributions.

This example introduced the concept of incidental parameters θ_i which increase with sample size. Neyman and Scott (1948) showed that, in the presence of incidental parameters, ML estimation may have difficulties. The functional model also exemplifies this point, as we saw earlier.

A.4 Identifiability and Consistent Estimation

In this section we clarify the connection between the identifiability and the consistent estimation of a parameter. First of all, if a parameter is not identifiable, then there does not exist a consistent (weak or strong) estimator of this parameter. This is because, if a parameter is not identifiable, then at least two different parameters can generate data with the same distribution. Consistent estimation means that the estimate converges to the true parameter. Since the limit of any convergent sequence has to be unique the parameter cannot be identifiable. An equivalent statement is that if a consistent estimate exists for a parameter, then that parameter must be identifiable.

Oddly enough, identifiability is not sufficient for consistent estimation. Consistency is a property of convergence of infinite sequences of observations. If the number of parameters increases with sample size then we have to deal with an infinite sequence of parameters. This is the situation in the functional model. In practice the parameters are always identifiable in the functional model but such identifiability is not helpful in seeking a consistent estimator because such an estimator might not exist. The following important theorem describing consistent estimation for the functional model has been obtained independently by Gleser (1983) and Höschel (see Bunke and Bunke, 1989, p. 239):

Theorem A.3 *In ME models, if a structural parameter is not identifiable in the structural model, then it cannot be consistently estimated in the corresponding functional model.*

Or equivalently:

Theorem A.4 *In ME models, if a structural parameter has a consistent estimate in the functional model, then it is also identifiable in the corresponding structural model.*

From this result and Reiersol's results it is clear that, in a normal functional model, the slope cannot be consistently estimated. This is the reason why some authors (for example, Malinvaud, 1970, p. 401) define identifiability of a parameter to mean that the parameter permits a consistent estimator.

If the number of parameters is fixed with respect to sample size, then consistent estimation is equivalent to identifiability. This was shown by Deistler and Seifert (1978, p. 978). However, this is only an existence theorem. It is not very helpful in the sense of telling us how to search for a consistent estimate when a parameter is identifiable. We commonly use maximum likelihood, the method of moments, etc., to search for consistent estimates. For example, in the structural model with normally distributed errors, for any given nonnormal ξ_i we know that β_1 is identifiable thus a consistent estimate exists for β_1. The existing results on how to find a consistent estimate are limited. Usually, higher-order moments are used (see Chapter 4). On the other hand, the ML estimate has not been obtained. The difficulty is that we need to maximize the likelihood of the (x_i, y_i), $i = 1, \ldots, n$, under a specific distributional assumption on the ξ_i.

Bibliography

Adcock, R. J. (1877). Note on the method of least squares. *Analyst*, **4**, 183–184.

Adcock, R. J. (1878). A problem in least squares. *Analyst*, **5**, 53–54.

Aigner, D. J., Hsiao, C., Kapteyn, A., and Wansbeek, T. J. (1984). Latent variables models in econometrics. In *Handbook of Econometrics*, Vol. **II** (eds Z. Griliches and M. D. Intriligator), pp. 1323–1393. Elsevier Science, Amsterdam.

Amemiya, Y. (1997). Generalization of the TLS approach in the errors-in-variables problem. In *Recent Advances in Total Least Squares Techniques and Errors-in-Variables Modeling* (ed. S. Van Huffel), pp. 77–86. SIAM, Philadelphia.

Amemiya, Y. and Fuller, W. A. (1984). Estimation for the multivariate errors-in-variables model with estimated error covariance. *Ann. Statist.*, **12**, 497– 509.

Amemiya, Y. and Fuller, W. A. (1988). Estimation for the nonlinear functional relationship. *Ann. Statist.*, **16**, 147–160.

Ammann, L. P. (1993). Robust singular value decompositions: a new approach to projection pursuit. *J. Amer. Statist. Assoc.*, **88**, 505–514.

Ammann, L. P. and Van Ness, J. W. (1988). A routine for converting regression algorithms into corresponding orthogonal regression algorithms. *ACM Trans. Math. Software*, **14**, 76–87.

Ammann, L. P. and Van Ness, J. W. (1989). Standard and robust orthogonal regression. *Comm. Statist. Simulation Comput.*, **18**, 145–162.

Anderson, T. W. (1951). Estimating linear restrictions on regression coefficients for multivariate normal distributions. *Ann. Math. Statist.*, **22**, 327–351.

Anderson, T. W. (1958). *Introduction to Multivariate Analysis*. Wiley, New York.

Anderson, T. W. (1976). Estimation of linear functional relationships: approximate distributions and connections with simultaneous equations in econometrics (with discussion). *J. R. Statist. Soc. B*, **38**, 1–36.

Anderson, T. W. (1984). Estimating linear statistical relationships. *Ann. Statist.*, **12**, 1–45.

Anderson, T. W. and Rubin, H. (1956). Statistical inference in factor analysis. In *Proceedings of the Third Berkeley Symposium on Mathematical Statistics and Probability* (ed. J. Neyman), **Vol. 5**, pp. 111–150. University of California Press,

Berkeley.

Anderson, T. W. and Sawa, T. (1982). Exact and approximate distributions of the maximum likelihood estimator of a slope coefficient. *J. R. Statist. Soc. B*, **44**, 52–62.

Bal, S. and Ojha, T. P. (1975). Determination of biological maturity and effect of harvesting and drying conditions on milling quality of paddy. *J. Agric. Engng. Res.*, **20**, 353–361.

Barker, F., Soh, Y. C., and Evans, R. J. (1988). Properties of the geometric mean functional relationship. *Biometrics*, **44**, 279–281.

Bartholomew, D. J. and Knott, M. (1999). *Latent Variable Models and Factor Analysis*, 2nd Edition. Arnold, London.

Bartlett, M. S. (1936). The information available in small samples. *Proc. Camb. Phil. Soc. A*, **32**, 560–566.

Bartlett, M. S. (1949). Fitting a straight line when both variables are subject to error. *Biometrics*, **5**, 207–212.

Becker, R. A., Chambers, J. M., and Wilks, A. R. (1988). *The NEW S Language*. Chapman & Hall, New York.

Bekker, P. A. (1986). Comment on identification in the linear errors in variables model. *Econometrica*, **54**, 215–217.

Berkson, J. (1950). Are there two regressions? *J. Amer. Statist. Assoc.*, **45**, 164–180.

Bickel, P. J. and Doksum, A. D. (1977). *Mathematical Statistics: Basic Ideas and Selected Topics*. Holden-Day, San Francisco.

Birch, M. W. (1964). A note on the maximum likelihood estimation of a linear structural relationship. *J. Amer. Statist. Assoc.*, **59**, 1175–1178.

Boos, D. D. and Serfling, R. J. (1980). A note on differentials and the CLT and the LIL for statistical functions, with applications to M-estimates. *Ann. Statist.*, **8**, 618– 624.

Bowden, R. J. (1973). The theory of parametric identification. *Econometrica*, **41**, 1069–1174.

Bowden, R. J. and Turkington, D. A. (1984). *Instrumental Variables*. Cambridge University Press, Cambridge.

Box, G. E. P. (1953). Non-normality and tests on variances. *Biometrika*, **40**, 318–335.

Britt, H. I. and Luecke, R. H. (1973), The estimation of parameters in nonlinear implicit models. *Technometrics*, **15**, 233–247.

Brown, P. J. (1982). Multivariate calibration. *J. R. Statist. Soc. B*, **44**, 287–321.

Brown, R. L. (1957). Bivariate structural relation. *Biometrika*, **44**, 84–96.

Bunke, H. and Bunke, O. (eds) (1989). *Nonlinear Regression, Functional Relations and Robust Methods*. Wiley, New York.

Buonaccorsi, J. P. (1995). Prediction in the presence of measurement error: general discussion and an example predicting defoliation. *Biometrics*, **51**, 1562–1569.

Burr, D. (1988). On errors-in-variables in binary regression – Berkson case. *J. Amer. Statist. Assoc.*, **83**, 739–743.

Carroll, R. J. and Gallo, P. P. (1982). Some aspects of robustness in functional errors-in-variables regression models. *Commun. Statist. A*, **11**, 2573–2585.

Carroll, R. J. and Ruppert, D. (1988). *Transformation and Weighting in Regression*. Chapman & Hall, London.

Carroll, R. J. and Ruppert, D. (1996). The use and misuse of orthogonal regression in linear errors-in-variables models. *Amer. Statist.*, **50**, 1–6.

Carroll, R. J. and Stefanski, L. A. (1990). Approximate quasilikelihood estimation in models with surrogate predictors. *J. Amer. Statist. Assoc.*, **85**, 652–663.

Carroll, R. J., Eltinge, J. L., and Ruppert, D. (1993). Robust linear regression in replicated measurement error models; regions for errors-in-variables model. *Statist. Probab. Lett.*, **16**, 169–175.

Carroll, R. J., Ruppert, D., and Stefanski, L. A. (1995). *Measurement Error in Nonlinear Models*. Chapman & Hall, London.

Carter, R. L. and Fuller, W. A. (1980). Instrumental variable estimation of the simple errors in variables model. *J. Amer. Statist. Assoc.*, **75**, 687–692.

Casella, G. and Berger, R. L. (1990). *Statistical Inference*. Wadsworth & Brooks/Cole, Pacific Grove, CA.

Chan, L. K. and Mak, T. K. (1984). Maximum likelihood estimation in multivariate structural relationships. *Scand. J. Statist.*, **11**, 45–50.

Chan, L. K. and Mak, T. K. (1985). On the polynomial functional relationship. *J. R. Statist. Soc. B*, **47**, 510–518.

Chang, Y. P. and Huang, W. T. (1997). Inferences for the linear errors-in-variables with changepoint model. *J. Amer. Statist. Assoc.*, **92**, 171–178.

Cheng, C.-L. (1994). On generalized least squares and least squares. Unpublished manuscript. Institute of Statistical Science, Academia Sinica, Taiwan, Rep. of China.

Cheng, C.-L. (1997). On the polynomial Berkson model. Unpublished manuscript. Institute of Statistical Science, Academia Sinica, Taiwan, Rep. of China.

Cheng, C.-L. and Schneeweiss, H. (1996). The polynomial regression with errors in the variables. Discussion paper 42 of Sonderforschungsbereich 386. Institute of Statistics, University of Munich, Germany.

Cheng, C.-L. and Schneeweiss, H. (1998a). Polynomial regression with errors in the variables. *J. R. Statist. Soc. B*, **60**, 189–199.

Cheng, C.-L. and Schneeweiss, H. (1998b). Note on two estimators for polynomial regression with errors in the variables. In *Proceedings of the 13th International Workshop on Statistical Modeling* (eds B. Marx and H. Friedl), pp. 141–147. New Orleans, LA.

Cheng, C.-L. and Tsai, C.-L. (1993). Diagnostics in measurement error model. Unpublished manuscript. Institute of Statistical Science, Academia Sinica, Taiwan, Rep. of China.

Cheng, C.-L. and Tsai, C.-L. (1994). Comparison of three different linear calibration estimators in measurement error models. Unpublished manuscript. Institute of Statistical Science, Academia Sinica, Taiwan, Rep. of China.

Cheng, C.-L. and Tsai, C.-L. (1995). Estimating linear measurement error models via M-estimators. In *Symposia Gaussiana: Proceedings of Second Gauss Symposium, Conference B: Statistical Sciences* (eds V. Mammitzsch and H. Schneeweiss), pp. 247–259. Walter de Gruyter, Berlin.

Cheng, C.-L. and Tsai, C.-L. (1996). Score tests in measurement error models: heteroscedasticity, autocorrelation and transformation. Unpublished manuscript. Institute of Statistical Science, Academia Sinica, Taiwan, Rep. of China.

Cheng, C.-L. and Van Ness, J. W. (1990). Bounded influence errors-in-variables regression. In *Statistical Analysis of Measurement Error Models and Application* (eds P. Brown and W. Fuller), *Contemporary Mathematics*, **112**, pp. 227–241. American Mathematical Society, Providence, RI.

Cheng, C.-L. and Van Ness, J. W. (1991). On the unreplicated ultrastructural model. *Biometrika*, **78**, 442–445.

Cheng, C.-L. and Van Ness, J. W. (1992). Generalized M-estimators for errors-in-variables regression. *Ann. Statist.*, **20**, 385–397.

Cheng, C.-L. and Van Ness, J. W. (1994). On estimating linear relationships when both variables are subject to errors. *J. R. Statist. Soc. B*, **56**, 167–183.

Cheng, C.-L. and Van Ness, J. W. (1997a). Structural and functional models revisited. *Recent Advances in Total Least Squares Techniques and Errors-in-Variables Modeling* (ed S. Van Huffel), pp. 37–50. SIAM, Philadelphia.

Cheng, C.-L. and Van Ness, J. W. (1997b). Robust calibration. *Technometrics*, **39**, 401–411.

Cheng, C.-L., Schneeweiss, H. and Thamerus, M. (1998). A small sample estimator for a polynomial regression with errors in the variables. Discussion paper 113 of Sonderforschungsbereich 386. Institute of Statistics, University of Munich, Germany.

Cox, D. R. (1993). Unbiased estimating equations derived from statistics that are functions of a parameter. *Biometrika*, **80**, 905–909.

Cox, D. R. and Hinkley, D. V. (1974). *Theoretical Statistics*. Chapman & Hall, London.

Cox, N. R. (1976). The linear structural relation for several groups of data. *Biometrika*, **63**, 231– 237.

Cramér, H. (1946). *Mathematical Methods of Statistics*. Princeton University Press, Princeton, NJ.

Creasy, M. A. (1956). Confidence limits for the gradient in the linear functional relationship. *J. R. Statist. Soc. B*, **18**, 65–69.

Croos, J. and Fuller, W.A. (1991). Robust estimation in measurement error models. In *American Statistical Association Proceedings of the Business and Economic Statistics Section*, pp. 283–288. American Statistical Association, Alexandria, VA.

Deistler, M. and Seifert, H. G. (1978). Identifiability and consistent estimability in econometrics models. *Econometrica*, **46**, 969–980.

Deming, W. E. (1931). The application of least squares, *Philos. Mag. Ser. 7*, **11**, 146–158.

Dent, B. M. (1935). On observation of points connected by linear relation. *Proc. Phys. Soc. London*, **47**, 92–108.

Dolby, G. R. (1976). The ultrastructural relation: a synthesis of the functional and structural relations. *Biometrika*, **63**, 39–50.

Donoho, D. L. and Huber, P. J. (1983). The notion of breakdown point. In *A Festschrift for Erich Lehmann* (eds P. J. Bickel, K. Doksum, and J. L. Hodges, Jr), pp. 157–184. Wadsworth, Belmont, CA.

Dorff, M. and Gurland, J. (1961). Small sample behavior of slope estimates in a linear functional relation. *Biometrics,* **17**, 283–298.

Durbin, J. (1954). Errors in variables. *Int. Statist. Rev.,* **22**, 23–32.

Edwards, A. F. W. (1992). *Likelihood* (expanded edition). Johns Hopkins University Press, Baltimore, MD.

Ezekiel, M. and Fox, K. A.(1954). *Methods of Correlation and Regression Analysis.* Wiley, New York.

Fedorov V. V. (1974). Regression problems with controllable variables subject to error. *Biometrika,* **61**, 49–56.

Fernholz, L. T. (1983). *Von Mises Calculus for Statistical Functionals.* Lecture Notes in Statistics **19**. Springer-Verlag, New York.

Fieller, E. C. (1954). Some problems in interval estimation. *J. R. Statist. Soc. B*, **16**, 175–185.

Finney, D. J. (1996). A note on the history of regression. *J. Appl. Statist.,* **23**, 555–557.

Fuller, W. A. (1980). Properties of some estimators for the errors-in-variables model. *Ann. Statist.,* **8**, 407–422.

Fuller, W. A. (1987). *Measurement Error Models.* Wiley, New York.

Fuller,, W. A. (1990). Prediction of true values for measurement error model. In *Statistical Analysis of Measurement Error Models and Application* (eds P. Brown and W. Fuller). *Contemporary Mathematics*, **112**, pp. 41–57. American Mathematical Society, Providence, RI.

Geary, R. C. (1942). Inherent relations between random variables. *Proc. R. Irish Acad. Sect. A,* **47**, 1541–1546.

Geary, R. C. (1943). Relations between statistics: the general and the sampling problem when the samples are large. *Proc. R. Irish Acad. Sect. A,* **49**, 177–196.

Geary, R. C. (1953). Non-linear functional relationship between two variable when one variable is controlled. *J. Amer. Statist. Association*, **48**, 94–103.

Gibson, W. M. and Jowett, G. H. (1957). Three-group regression analysis. Part 1: Simple regression analysis. *Appl. Statist.,* **6,** 114–122.

Gleser, L. J. (1981). Estimation in a multivariate 'errors-in-variables' regression model: large sample results. *Ann. Statist.*, **9**, 24–44.

Gleser, L. J. (1983). Functional, structural and ultrastructural errors-in-variables model. In *American Statistical Association Proceedings of the Business and Economic Statistics Section*, pp. 57–66. American Statistical Assoc., Alexandria, VA.

Gleser, L. J. (1985). A note on G. R. Dolby's unreplicated ultrastructural model. *Biometrika*, **72**, 117–124.

Gleser, L. J. (1987). Confidence intervals for the slope in a linear errors-in-variables regression model. In *Advances in Multivariate Statistical Analysis* (ed. K. Gupta), pp. 85–109. D. Reidel, Dordrecht.

Gleser, L. J. (1990). Improvements of the naive approach to estimation in nonlinear errors-in-variables regression models. In *Statistical Analysis of Measurement Error Models and Application* (eds P. Brown and W. Fuller), *Contemporary Mathematics*, **112**, pp. 99–114. American Mathematical Society, Providence, RI.

Gleser, L. J. (1992). The importance of assessing measurement reliability in multivariate regression. *J. Amer. Statist. Assoc.,* **87**, 696–707.

Gleser, L. J. (1993). Estimators of slopes in linear errors-in-variables regression models when the predictors have known reliability matrix. *Statist. Probab. Lett.*, **17**, 113–121.

Gleser, L. J. and Hwang, J. T. (1987). The nonexistence of $100(1-\alpha)\%$ confidence sets of finite expected diameter in errors-in-variables and related models. *Ann. Statist.*, **15**, 1351–1362.

Goldberger, A. S. (1972). Structural equation models in the social sciences. *Econometrica,* **40**, 979–1001.

Golub, G. H. and Van Loan, C. F. (1981). An analysis of the total least squares problem. *SIAM J. Numer. Anal.*, **17**, 883–893.

Hampel, F. R. (1968). Contributions to the Theory of Robust Estimation. Ph.D. dissertation, University of California, Berkeley.

Hampel, F. R. (1971). A general qualitative definition of Robustness. *Ann. Math. Statist.,* **42**, 1887–1896.

Hampel, F. R. (1974). The influence curve and its role in robust estimation. *J. Amer. Statist. Assoc.,* **69**, 383–393.

Hampel, F. R., Ronchetti, E. M., Rousseeuw, P. J., and Stahel, W. A. (1986). *Robust Statistics.* Wiley, New York.

Hasabelnaby, N., Ware, J. H., and Fuller, W. A. (1989). Indoor air pollution and pulmonary performance: investigating errors in exposure assessment. *Statist. Med.*, **8**, 1109–1126.

Hodges, J. L. (1967). Efficiency in normal samples and tolerance of extreme values for some estimates of location. In *Proceedings of the Fifth Symposium on Mathematical Statistics and Probability* (eds L. LeCam and J. Neyman), **Vol. 1**, pp. 163–186. University of California Press, Berkeley.

Höschel, H. P. (1978a). Least squares and maximum likelihood estimation of functional relations. In *Transactions of the Eighth Prague Conference on Information Theory, Statistical Decision Functions, Random Processes*, **Vol. A**, pp. 305–317. Academia, Prague.

Höschel, H. P. (1978b). Generalized least squares estimators of linear functional relations with known error covariance. *Math. Operationsforsch. Statist. Ser. Statist.,* **9**, 9–26.

Hsiao, C. (1989). Consistent estimation for some nonlinear errors-in-variables models. *J. Econometrics*, **41**, 159–185.

Huber, P. J. (1964). Robust estimation of a location parameter. *Ann. Math. Statist.,* **35,** 73– 101.

Huber, P. J. (1967). The behavior of maximum likelihood estimates under nonstandard conditions. In *Proceedings of the Fifth Berkeley Symposium on Mathematical*

Statistics and Probability (eds L. LeCam and J. Neyman), **Vol. 1**, pp. 221–233. University of California Press, Berkeley.

Huber, P. J. (1972). Robust statistics: a review. *Ann. Math. Statist.,* **43,** 1041–1067.

Huber, P. J. (1973). Robust regression: asymptotic, conjectures, and Monte Carlo. *Ann. Statist.,* **1**, 799–821.

Huber, P. J. (1981). *Robust Statistics*. Wiley, New York.

Huber, P. J. (1983). Minimax aspects of bounded influence regression. *J. Amer. Statist. Assoc.*, **78**, 66–80.

Huwang, L. (1996). Asymptotically honest confidence sets for structural errors-in-variables models. *Ann. Statist.*, **24**, 1536–1546.

Hwang, J. T. G. (1995). Fieller's problems and resampling techniques. *Statist. Sinica*, **5**, 161–171.

Hwang, J. T. and Liu, H. K. (1992). Existence and nonexistence theorems of finite diameter sequential confidence regions for errors-in-variables model. *Statist. Probab. Lett.*, **13**, 45–55.

Johnston, J. (1972). *Econometric Methods* (2nd edn). McGraw-Hill, New York.

Jolicoeur, P. (1975). Linear regressions in fisheries research: some comments. *J. Fisheries Res. Board Canada*, **32**, 1491–1494.

Jones, T. A. (1979). Fitting straight lines when both variables are subject to error. I. Maximum likelihood and least squares estimation. *J. Int. Assoc. Math. Geol.*, **11**, 1–25.

Kelly, G. (1984). The influence function in the errors in variables problem. *Ann. Statist.*, **12**, 87– 100.

Kendall, M. G. (1951). Regression, structure, and functional relationship, I. *Biometrika*, **38**, 11– 25.

Kendall, M. G. and Stuart, A. (1979). *The Advanced Theory of Statistics*, **Vol. 2** (4th edn). Griffin, London.

Konijn, H. S. (1981). Maximum likelihood estimator and set confidence intervals for a simple errors in variables model. *Commun. Statist. A,* **10**, 983–996.

Krasker, W. S. (1980). Estimation in linear regression models with disparate data points. *Econometrica,* **48**, 1333–1346.

Kummel, C. H. (1879). Reduction of observed equations which contain more than one observed quantity. *Analyst*, **6**, 97–105.

Lakshminarayanan, M. Y. and Gunst, R. F. (1984). Estimation of parameters in linear structural relationships: sensitivity to the choice of the ratio of error variances. *Biometrika*, **71**, 569–573.

Leamer, E. E. (1978). Least-squares versus instrumental variables estimation in a simple errors in variables model. *Econometrica,* **46**, 961–968.

Lehmann, E. L. (1983). *Theory of Point Estimation*. Wiley, New York.

Lindley, D. V. (1947). Regression lines and the linear functional relationship. *J. R. Statist. Soc. Suppl.*, **9**, 218–244.

Lindley, D. V. (1953). Estimation of a functional relationship. *Biometrika*, **40**, 47–49.

Madansky, A. (1959). The fitting of straight lines when both variables are subject to error. *J. Amer. Statist. Assoc.*, **54**, 173–205.

Malinvaud, E. (1970). *Statistical Methods of Economics* (2nd rev. edn). North-Holland, Amsterdam.

Mariano, R. S. (1969). On distributions and moments of single-equation estimators in a set of simultaneous linear stochastic equations. Tech. Rep. 27. Inst. Math. Studies Social Sciences, Stanford University.

Maronna, R. A. (1976). Robust M-estimators of multivariate location and scatter. *Ann. Statist.*, **4**, 51–67.

Maronna, R. A. and Yohai, V. J. (1981). Asymptotic behaviour of general M-estimates for regression and scale with random carriers. *Z. Wahrsch. Verv. Geb.*, **58**, 7–20.

Maronna, R. A., Bustos, O. H., and Yohai, V. J. (1979). Bias- and efficiency-robustness of general M-estimators for regression with random carriers. In *Smoothing Techniques for Curve Estimation* (eds T. Gasser and M. Rosenblatt), Lecture Notes in Mathematics, **757**, pp. 91–116. Springer-Verlag, Berlin.

Martin, R. D., Yohai, V. J., and Zamar, R. H. (1989). Min-max bias robust regression. *Ann. Statist.*, **17**, 1608–1630.

Moberg, L. and Sundberg, R. (1978). Maximum likelihood estimator of a linear functional relationship when one of the departure variances is known. *Scand. J. Statist.*, **5**, 61–64.

Moon, M. S. and Gunst, R. F. (1995). Polynomial measurement error modeling. *Comput. Statist. Data Anal.*, **19**, 1–21.

Moran, P. A. P. (1971). Estimating structural and functional relationships. *J. Multivariate Anal.*, **1**, 232–255.

Morton, R. (1981). Efficiency of estimating equations and the use of pivots. *Biometrika,* **68**, 227–233.

Nair, K. R. and Banerjee, K. S. (1942). A note on fitting of straight lines if both variables are subject to error. *Sankhyā,* **6**, 331.

Nair, K. R. and Shrivastava, M. P. (1942). On a simple method of curve fitting. *Sankhyā,* **6**, 121–132.

Nakamura, T. (1990). Corrected score functions for errors-in-variable models: methodology and application to generalized linear models. *Biometrika*, **77**, 127–137.

Neyman, J. (1951). Existence of consistent estimates of the directional parameters in a linear structural relation between two variables. *Ann. Math. Statist.*, **22**, 497–512.

Neyman, J. and Scott, E. L. (1948). Consistent estimates based on partial consistent observations. *Econometrica*, **16**, 1–32.

Neyman, J. and Scott, E. L. (1951). On certain methods of estimating the linear structural relation. *Ann. Math. Statist.*, **22**, 352–361.

Nussbaum, M. (1976). Maximum likelihood and least squares estimation of linear functional relationships. *Math. Operationsforsch Statist. Ser. Statist.*, **7,** 23–49.

Nussbaum, M. (1977). Asymptotic optimality of estimators of a linear functional relation if the ratio of the error variances is known. *Math. Operationsforsch. Statist. Ser. Statist.*, **8**, 173–198.

Nussbaum, M. (1979). Asymptotic efficiency of estimators of a multivariate linear functional relation. *Math. Operationsforsch. Statist. Ser. Statist.*, **10**, 505–527.

Nussbaum, M. (1984). An asymptotic minimax risk bound for estimation of a linear functional relationship. *J. Multivariate Anal.*, **14**, 300–314.

Okamoto, M. (1983). Asymptotic theory of Brown–Fereday's method in a linear structural relationship. *J. Jap. Statist. Soc.*, **13**, 53–56.

O'Neill, M., Sinclair, L. G., and Smith, F. J. (1969). Polynomial curve fitting when abscissas and ordinates are both subject to error. *Comput. J.*, **12**, 52–56.

Pakes, A. (1982). On the asymptotic bias of the Wald-type estimators of a straight line when both variables are subject to error. *Int. Econ. Rev.*, **23**, 491–497.

Pal, M. (1980). Consistent moment estimators of regression coefficients in the presence of errors in variables. *J. Econometrics,* **14**, 349–364.

Patefield, W. M. (1978). The unreplicated ultrastructural relation: large sample properties. *Biometrika*, **65**, 535–540.

Patefield, W. M. (1981a). Multivariate linear relationships: maximum likelihood estimation and regression bound. *J. R. Statist. Soc. B*, **43**, 342–352.

Patefield, W. M. (1981b). Confidence intervals for the slope of a linear functional relationship. *Comm. Statist. Theory Meth.*, **10** (17), 1759–1764.

Pearson, K. (1901). On lines and planes of closest fit to systems of points in space. *Philos. Mag.* **2**, 559–572.

Prohorov, Y. V. (1956). Convergence of random processes and limit theorems in probability theory. *Theory Probab. Appl.*, **1**, 157–214.

Rabinowicz, E. (1970). *An Introduction to Experimentation*. Addison-Wesley, Reading, MA.

Reiersol, O. (1941). Confluence analysis by means of lag moments and other methods of confluence analysis. *Econometrica*, **9**, 1–22.

Reiersol, O. (1950). Identifiability of a linear relation between variables which are subject to errors. *Econometrica*, **18**, 375–389.

Richardson, D. H. and Wu, D. M. (1970). Least squares and grouping method estimators in the errors-in-variables model. *J. Amer. Statist. Assoc.*, **65**, 724–748.

Ricker, W. E. (1975). A note concerning Professor Jolicoeur's comments. *J. Fisheries Res. Board Canada*, **32**, 1494–1498.

Robertson, C. A. (1974). Large-sample theory for the linear structural relation. *Biometrika*, **61**, 353–359.

Rousseeuw, P. J. and Leroy A. M. (1987). *Robust Regression and Outlier Detection.* Wiley, New York.

Ronchetti, E. and Rousseeuw, P. J. (1985). Change-of-variance sensitivities in regression analysis. *Z. Wahrsch. Verv. Geb.*, **68**, 503–509.

Rothenberg, T. J. (1971). Identification in parametric models. *Econometrica*, **39**, 577–592.

Rudemo, M., Ruppert, D., and Streibig, J. (1989). Random effect models in nonlinear regression with applications to bioassay. *Biometrics*, **45**, 349–362.

Ruppert, D. (1985). On the bounded-influence regression estimator of Krasker and Welsch. *J. Amer. Statist. Assoc.*, **80**, 205–208.

Scheffé, H. (1970). Multiple testing versus multiple estimation: improper confidence sets – estimation of directions and ratios. *Ann. Math. Statist.*, **41**, 1–29.

Schneeweiss, H. (1976). Consistent estimation of a regression with errors in the variables. *Metrika*, **23**, 101–117.

Schneeweiss, H. (1982). Note on Creasy's confidence limits for the gradient in the linear functional relationship. *J. Multivariate Anal.*, **12**, 155–158.

Schneeweiss, H. (1985). Estimating linear relations with errors in the variables; the merging of two approaches. *Contributions to Econometrics and Statistics Today* (eds H. Schneeweiss and H. Strecker), pp. 207–221. Springer-Verlag, Berlin.

Schneeweiss, H. and Mittag, H. J. (1986). *Lineare Modelle mit fehlerbehafteten Daten*. Physica-Verlag, Heidelberg.

Solari, M. E. (1969). The 'maximum likelihood solution' of the problem of estimating a linear functional relationship. *J. R. Statist. Soc. B*, **31**, 372–375.

Sprent, P. (1966). A generalized least squares approach to linear functional relationships. *J. R. Statist. Soc. B*, **28**, 278–297.

Sprent, P. (1976). Modified likelihood estimation of a linear relationship. In *Studies in Probability and Statistics* (ed. E. J. Williams), pp. 109–119. North-Holland, Amsterdam.

Sprent, P. (1990). Some history of functional and structural relationships. In *Statistical Analysis of Measurement Error Models and Application* (eds P. Brown and W. Fuller), *Contemporary Mathematics*, **112**, pp. 3–15. American Mathematical Society, Providence, RI.

Stahel, W. A. (1987). Estimation of a covariance matrix with location: asymptotic formulas and optimal B-robust estimators. *J. Multivariate Anal.*, **22**, 296–312.

Staudte, R. G. and Sheather, S. J. (1990). *Robust Estimation and Testing.* Wiley-Interscience, New York.

Stefanski, L. A. (1989). Unbiased estimation of a nonlinear function of a normal mean with application to measurement error models. *Comm. Statist. A*, **18**, 4335–4358.

Stuart, A. and Ord, J. K. (1991). *Kendall's Advanced Theory of Statistics*, **Vol. 2** (5th edn). Edward Arnold, London.

Stuart, A. and Ord, J. K. (1994). *Kendall's Advanced Theory of Statistics*, **Vol. 1** (6th edn). Edward Arnold, London.

Stulajter, F. (1978). Nonlinear estimators of polynomials in mean values of a Gaussian stochastic process. *Kybernetika*, **14**, 206–220.

Theil, H. (1950). A rank-invariant method for linear and polynomial regression analysis. *Indag. Math.*, **12**, 85–91, 173–177, and 467–482.

Theil, H. and van Yzeren, J. (1956). On the efficiency of Wald's method of fitting straight lines. *Rev. Int. Statist. Inst.*, **24**, 17–26.

Tracy, D. S. and Jinadasa, K. G. (1987). On ultrastructural relationships models. In *Foundations of Statistical Inference* (eds I.B. MacNeill and G. J. Umphrey), pp. 139–147. D. Reidel, Amsterdam.

Van Huffel, S. (1997) *Recent Advances in Total Least Squares Techniques and Errors-in-Variables Modeling* (ed. S. Van Huffel). SIAM, Philadelphia.

Van Huffel, S. and Vandewalle, J. (1991). *The Total Least Squares Problem: Computational Aspects and Analysis.* SIAM, Philadelphia.

Van Montfort, K. (1988). *Estimating in Structural Models with Non-Normal Distributed Variables: Some Alternative Approaches*. M & T Series 12. DSWO Press, Leiden.

Van Montfort, K., Mooijaart, A., and de Leeuw, J. (1987). Regression with errors in variables. *Statist. Neerlandica*, **41**, 223–239.

Venables, W. N. and Ripley, B. D. (1994). *Modern Applied Statistics with S-Plus.* Springer-Verlag, Berlin.

Wald, A. (1940). The fitting of straight lines if both variables are subject to error. *Ann. Math. Statist.*, **11**, 284–300.

Wald, A. (1949). Note on the consistency of the maximum likelihood estimate. *Ann. Math. Statist.*, **20**, 595–601.

Ware, J. H. (1972). Fitting straight lines when both variables are subject to error and the ranks of the means are known. *J. Amer. Statist. Assoc.*, **67**, 891–897.

Wellman, J. M. and Gunst, R. F. (1991). Influence diagnostics for linear measurement error models. *Biometrika,* **78**, 373–380.

White, H. (1984). *Asymptotic Theory for Econometricians*. Academic Press, New York.

Willassen, J. (1979). Two clarifications on the likelihood surface in functional models. *J. Multivariate Anal.,* **9**, 138–149.

Willassen, J. (1984). Testing hypotheses on the unidentifiable structural parameters in the classical 'errors-in-variables' model with application to Friedman's permanent income model. *Econom. Lett.*, **14**, 221–228.

Williams, E. J. (1959). *Regression Analysis.* Wiley, New York.

Williams, E. J. (1973). Test of correlation in multivariate analysis. *Bull. Int. Statist. Inst. Proc. 39th Session*, **45**, Book 4, 218–232.

Wolfowitz, J. (1952). Consistent estimators of the parameters of a linear structural relation. *Skandinavisk Aktuarietidskrift,* **35**, 132–151.

Wolfowitz, J. (1953). Estimation by the minimum distance method. *Ann. Inst. Statist. Math.*, **5**, 9–23.

Wolfowitz, J. (1954). Estimation of the components of stochastic structures. *Proc. Nat. Acad. Sci.*, **40**, 602–606.

Wolfowitz, J. (1957). The minimum distance method. *Ann. Math. Statist.,* **28**, 75–88.

Wolter, K. M. and Fuller, W. A. (1982a). Estimation of the quadratic errors-in-variables model. *Biometrika*, **69**, 175–182.

Wolter, K. M. and Fuller, W. A. (1982b). Estimation of nonlinear errors-in-variables models. *Ann. Statist.*, **10**, 539–548.

Wong, M. Y. (1989). Likelihood estimation of a simple linear regression model when both variables have error. *Biometrika*, **76**, 141–148.

Yohai, V. J. and Zamar,R. H. (1990). Bounded influence estimation in the errors-in-variables model. In *Statistical Analysis of Measurement Error Models and Application* (eds P. Brown and W. Fuller), *Contemporary Mathematics*, **112**, pp. 243–248. American Mathematical Society, Providence, RI.

Zamar, R. H. (1985). Orthogonal regression M-estimates, Ph.D. Dissertation, Department of Statistics, University of Washington, Seattle.

Zamar, R. H. (1989). Robust estimation in the errors-in-variables model. *Biometrika*, **76**, 149–160.

Zamar, R. H. (1992). Bias robust estimation in orthogonal regression. *Ann. Statist.*, **20**, 1875–1888.

Author Index

Subject Index